S0-BLI-386

Calixarenes in the Nanoworld

Calixarenes in the Nanoworld

Edited by

Jacques Vicens
Laboratoire de Conception Moléculaire, ECPM-ULP-CNRS, Strasbourg, France

and

Jack Harrowfield
Laboratoire de Chimie Supramoléculaire, ISIS-ULP-CNRS, Strasbourg, France

Assistant Editor

Lassaad Baklouti
Laboratoire de Chimie des Interactions Moléculaires, FSB, Bizerte, Tunisie

A C.I.P. Catalogue record for this book is available from the Library of Congress.

ISBN-10 1-4020-5021-6 (HB)
ISBN-13 978-1-4020-5021-3 (HB)
ISBN-10 1-4020-5022-4 (e-book)
ISBN-13 978-1-4020-5022-4 (e-book)

Published by Springer,
P.O. Box 17, 3300 AA Dordrecht, The Netherlands.

www.springer.com

Printed on acid-free paper

CONTENTS

FOREWORD

The remarkable development, over approximately the past twenty years, of what has come to be referred to widely as "nanochemistry" concerns a very diverse array of physicochemical studies. The intense interest in nanochemistry is reflected in special editions of major journals (*e.g. Chemical Reviews* Vol. 105, No. 4, 2005) and the appearance of multi-volume texts (*e.g.* H. S. Nalwa (Ed.) *Encyclopaedia of Nanoscience and Nanotechnology*, 10 volumes, American Scientific Publishers, Stevenson Ranch, California, 2004 – note C. P. Rao, M. Dey in Vol. 1, pp. 475-497 review calixarene chemistry). Particles with a size between 1-100 nanometres may have properties which depend solely upon their size and which differ significantly from those of both their atomic or molecular constituents and the bulk materials which they can ultimately compose. "Quantum dots", often of very simple stoichiometry (*e.g.* CdS), provide very well-known examples of such materials. Molecules with a size in the 1-100 nm range may also have unique properties but here it is because they are a complex assembly of different functional units, ideally one where these units interact in highly specific but nonetheless modifiable ways. In the natural world, biopolymers such as proteins, sugars and nucleic acids provide obviously important examples of such "nanomaterials", while the field of supramolecular chemistry is concerned with the production of similar synthetic molecules with a wide range of properties, including those which mimic or even exceed those of biosystems. Polymer chemistry, of course, might be considered a long-established area of nanochemistry. It is another reflection of the versatility of calixarenes, a property which has long been considered one of their most distinctive, that they may be employed in all areas of nanochemistry. They are extremely valuable tools for what is usually termed the "bottom up" or modular approach towards the synthesis of nanomaterials pursued by many chemists. The chapters of this book are intended to provide significant examples of such use of calixarenes, as well as to indicate the prospects which exist for further developments.

While the identification of a field of nanochemistry and indeed of "nanoscience" generally, is a relatively recent event, its rise has occurred over a time period little briefer than that of "calixarene chemistry" and there has certainly been an important degree of mutual stimulation. For this reason we have asked David Gutsche, whose work, beginning some 40 years ago, was the stimulus to an international flowering of calixarene chemistry after what could be considered a century of nigh-dormancy, to provide in chapter one a perspective on these developments. As he recalls, one of the early stimuli for the use of calixarenes was the idea that they might provide sophisticated enzyme mimics and it is a striking feature of the present text, as distinct from earlier reviews, that the biological chemistry of calixarenes is particularly prominent. The chapters which follow that of David Gutsche are intended more to be reports of the state of the art in various areas but we hope also that there will be some lasting value in what these individual reports reveal of the motives and thinking of their authors. Not every aspect of "calixarenes in the nanoworld" is covered in detail, of course, but leading references are given as far as possible and the major issues to be confronted with the use of such chemistry are defined and explained in many ways by our various colleagues.

Lassaad Baklouti, Jack Harrowfield, Jacques Vicens

Strasbourg, February, 2006

E-MAIL CONTACTS

Chapter 1 : D.Gutsche@tcu.edu
Chapter 2 : vboehmer@mail.uni-mainz.de
Chapter 3 : ztli@mail.sioc.ac.cn
Chapter 4 : andrea.pochini@unipr.it
Chapter 5 : vicens@chimie.u-strasbg.fr
Chapter 6 : susan.matthews@uea.ac.uk
Chapter 7 : vicens@chimie.u-strasbg.fr
Chapter 8 : rudkevich@uta.edu
Chapter 9 : jfnierengarten@chimie.u-strasbg.fr
Chapter 10 : harrowfield@isis.u-strasbg.fr
Chapter 11 : j.huskens@utwente.nl
Chapter 12 : ungaro@unipr.it
Chapter 13 : Olivia.Reinaud@univ-paris5.fr
Chapter 14 : schradet@staff.uni-marburg.de
Chapter 15 : jongskim@dku.edu
Chapter 16 : neri@unisa.it

Chapter 1

CALIXARENES: A PERSONAL HISTORY

From resinous tars to nanomaterials

David Gutsche

Department of Chemistry, University of Arizona, Tucson AZ, USA. Formerly associated with: Department of Chemistry, Washington University in St. Louis 1947-1989; Texas Christian University 1989-2002 as Robert A. Welch Professor of Chemsitry

Abstract: In the belief that it is important for scientists to be cognizant of the history of their field, this introductory chapter is written in the form of an odyssey of one of its earliest practitioners.

Key words: Calixarenes, cavitands, carcerands, enzyme mimicry, polyfunctional catalysts.

1. DISCUSSION

The numerous fields of research now employing calixarenes as entities of interest have conferred on these molecules a status unimaginable in the early 1970s when they were providentially and serendipitously revealed as potentially important players in a new research program that I and my research group at Washington University in St Louis had initiated. Little did our limited imaginations foresee the degree to which they would become important players in so many other fields of research and give rise to a wide ranging endeavor now defined as "calixarene chemistry".[1] Little could we know that they would someday provide materials for the then barely nascent world of nanochemistry which now occupies a prominent place in today's arena of research.

Calixarene chemistry did not start at Washington University in 1970 but harkens all the way back to the late nineteenth century when Adolph von Baeyer carried out a reaction using a mixture of phenol, the then very rare compound formaldehyde, and a strong acid (Fig. 1 left).[2] What was produced

J. Vicens and J. Harrowfield (eds.), Calixarenes in the Nanoworld, 1–19.

was a black, resinous tar that defied characterization by the meagre analytical techniques available at that time. Baeyer's response was to abandon the experiment, throw the mess into the waste crock and go on to other pursuits for which, in due course, he was awarded the 1905 Nobel Prize. Thirty years after the aborted experiment with phenol and formaldehyde, what had started as a piece of pure academic research was transformed to a successful piece of applied research when Leo Baekland, a recent immigrant to the United States from Belgium, discovered that heating Baeyer's resinous tar turned it into a hard brittle solid. He patented his process[3] in 1908, eponymously named its product "Bakelite", and went on to establish the first plastics industry in the world.

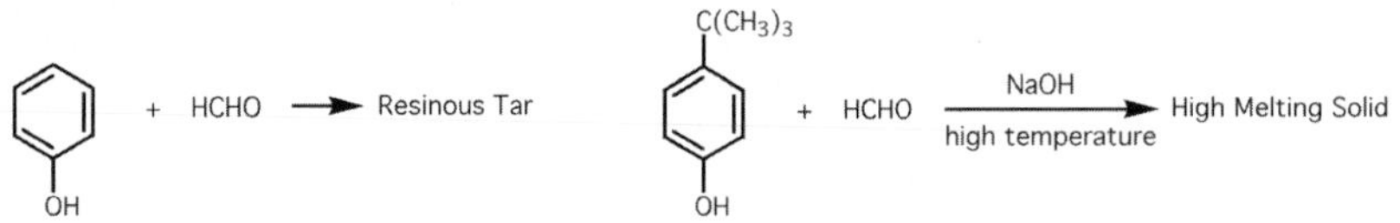

$C(CH_3)_3$

+ HCHO → Resinous Tar

+ HCHO —NaOH, high temperature→ High Melting Solid

OH

OH

Figure 1-1. Reactions of phenols with formaldehyde.

The Bakelite process inspired numerous academic studies aimed at gaining insight into its chemistry. Important among these studies were those carried out in the early 1940s by Alois Zinke in Graz, Austria who condensed formaldehyde with *p*-alkyl-substituted phenols, using sodium hydroxide as the base and linseed oil as a very high boiling solvent (Fig. 1 right).[4] What Zinke obtained from these reactions were colorless, very high melting solids, and it was these compounds that reemerged into our view thirty years later when the interests of my research group in the in St Louis shifted from "classical" problems to the area of polyfunctional catalysts, with attention focused on catalysts that mimic the action of enzymes. Having been exposed in my early days at Washington University to so many discussions involving aldolase at the Friday noon seminars, conducted by the Nobel Laureates Carl and Gerti Cori, that was the enzyme that became our choice for mimicking.

Aldolase catalyzes the condensation of dihydroxy-acetone phosphate (DHAP) with glyceradehyde phosphate (GAP) to form fructose-1,6-diphosphate (FDP) (Fig. 2). Its mechanism of action was known to involve (a) electrostatic attraction between a positively charged region of the enzyme and the anionic DHAP, (b) formation of a Schiff base with the carbonyl group of DHAP, and (c) abstraction of the α-proton of DHAP to generate the reactive intermediate.[5] Taking cognizance of these requirements, our first attempt to build a polyfunctional catalyst involved a trifunctional compound which, however, showed only a very modest enhancement in the rate of α-proton removal of

DHAP. One of the problems with this compound was that it had a linear structure, in essence a two-dimensional structure, and was very flexible. Aldolase, on the other hand, is a quite rigid three-dimensional globular protein in which the complementarity in shape between its active site and the substrate plays a key role in the overall catalytic process. Clearly, a proper enzyme mimic must include this critical feature, so coming to the fore at this point was a quest for organic compounds containing cavities sufficiently large to enfold other molecules, *i.e.* molecular baskets. But, in the early 1970s the only visible candidates were the cyclodextrins and the crown ethers, and neither seemed satisfactory for our purposes. The cyclodextrins are natural products that are not easily synthesized in the laboratory, and the crown ethers are rather flat and floppy loops rather than true baskets. Instead, what was desired was a true molecular basket that could be easily prepared from readily available starting materials. The fulfillment of this desire as mentioned above, arrived in serendipitous and unexpected fashion and provides another interesting illustration, like that of the Bakelite story, of the intermingling of academic research and industrial operations.

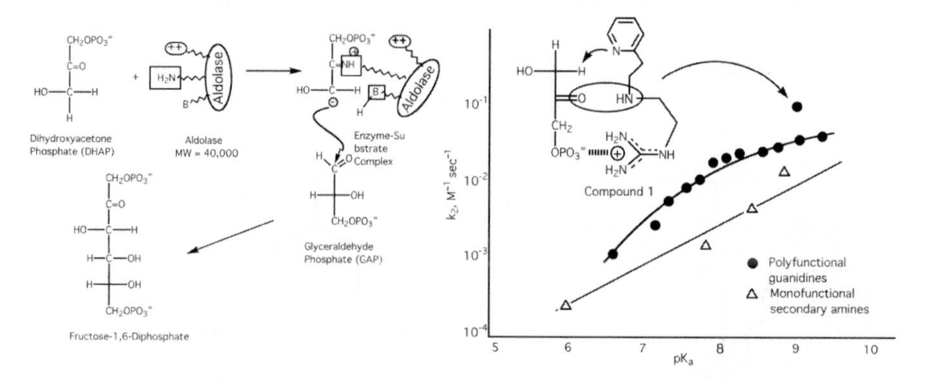

Figure 1-2. The aldolase system and a simple enzyme mimic.

For most of my 42 years at Washington University I was a consultant to the Petrolite Corporation and so was privy to their products and problems. Petrolite's business was to manufacture surfactants for use in the oil fields to break the water-oil emulsion of crude oil as it emerges from the ground, thereby allowing it to be rectified into its various components. It turns out, however, that every oil field is a bit different in the kind of demulsifier that is effective, so Petrolite marketed a large number of tailor-made surfactants for use in oil fields all over the world. One such surfactant was made by treating *p-tert*-butylphenol with formaldehyde to give what was thought to be a linear polymer which was then oxyethylated to produce the surfactant.[6] This material was dissolved/suspended in a high-boiling solvent, poured into a 55 gallon drum and shipped to the oil field for use. Shortly after its initial applications, however, it was discovered that it was extremely difficult to use because it formed intractable sludges. This problem was quickly referred to the research department whose task became that of gaining insight into the difficulty. The two chemists on whom the major burden of the investigation fell were John Munch, an excellent organic chemist, and Jack Ludwig, an equally excellent analytical chemist. What John Munch did was to simulate the method that the manufacturing plant had used to make the product, which involved heating a mixture of *p-tert*-butylphenol, formaldehyde and a small amount of sodium hydroxide in boiling xylene for a few hours. What came out was a colorless, very high melting solid which was revealed by a subsequent literature search to be very similar to the material that Zinke had obtained thirty years earlier and to which he had assigned a cyclic tetrameric structure. When molecular models of the Zinke cyclic tetamer were constructed it became immediately apparent that here were our desired molecular baskets, and glorious visions were immediately conjured up of constructing a large and variegated family of compounds by using the many readily available *p*-substituted phenols as reactants. But, such was not to be the case.

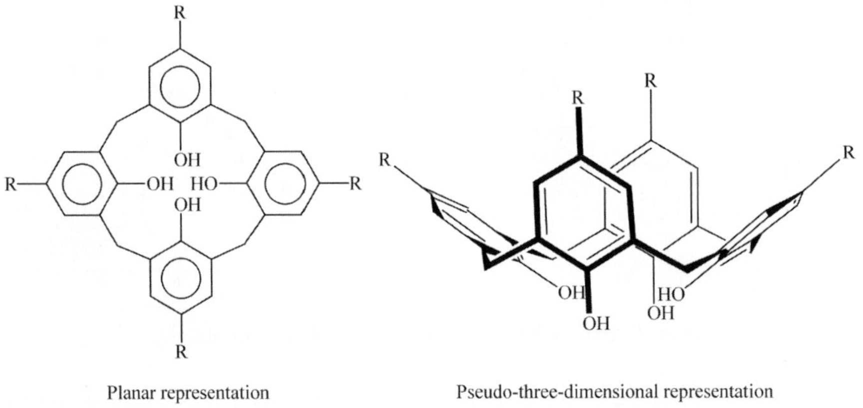

Planar representation

Pseudo-three-dimensional representation

Figure 1-3. Common representations of calix[4]arenas.

A closer inspection of the Zinke chemistry revealed disturbing problems: (a) the products proved not to be pure entities but mixtures of two or more compounds, as Cornforth had already discovered[7] in the 1950s; (b) the reaction appeared to proceed smoothly only with phenols carrying highly branched alkyl moieties in the *p*-position; (c) the reactions proved to be extremely capricious, with different results being obtained from one run to the next; and (d) the experimental evidence for the cyclic tetrameric structure was actually quite thin. It was probably this collection of difficulties that explains why these compounds languished almost unattended by other chemists for three decades. The first task of our research group, therefore, was to try to unscramble the Zinke chemistry[4]. Eventually, methods were worked out for separating the mixtures into pure components and then proving their structures via elemental analysis, spectral characterization including uv-vis, IR, NMR, and in some cases x-ray crystallography. Importantly, reliable recipes for their preparation were established by careful control of the reactants and the reaction conditions. It was also at this time that we devised a simpler name for the cyclic tetramer than pentacyclo-[19.3.1.1^{3,7}1^{9,13}1^{15,19}]-octacosa-1(25),3,5,7(28),9,11,13(27),-15,17,19(26),21,23-dodecaene 25,26,27,28-tetrol (Fig. 3). Perceiving a visual relationship between the cyclic tetramer and a Greek vase known as a "calix crater" the name "calixarene" was coined, calix serving as the prefix to denote the shape and "arene" serving as the suffix to denote the aromatic building blocks (Fig. 4). To specify the size of the calixarene a bracketed numeral was placed between "calix" and "arene"; *e.g. p-tert*-butylcalix[4]arene. This early work took place in the 1970s, and the initial announcement[8] was made at a Midwest Regional Meeting of the American Chemical Society in 1975. Subsequently, the first publication[9] appeared in the *Journal of Organic Chemistry* in 1978.

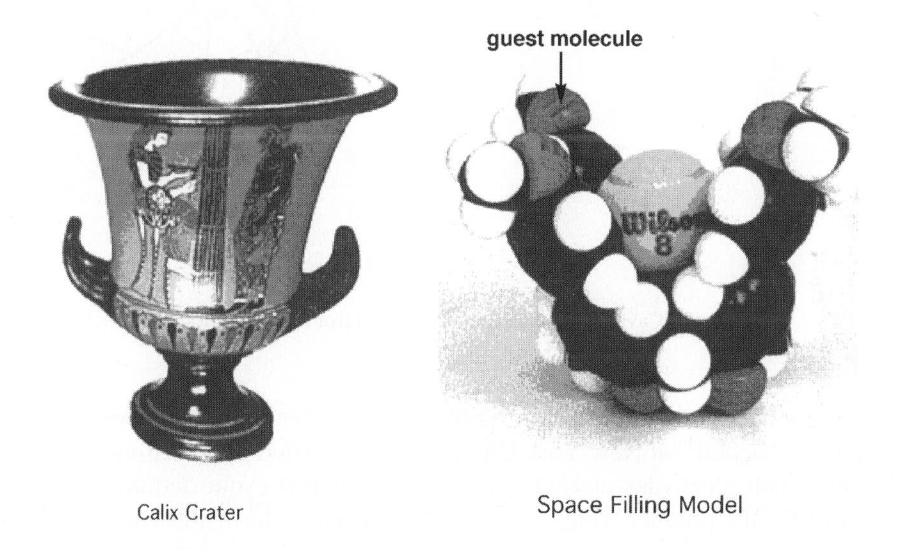

Figure 1-4. Comparison of a calix crater and a molecular model of a calix[4]arene.

Unbeknownst to us at the time, there were two other research groups also engaged in related research, one in the laboratories of Hermann Kämmerer (and, later, Volker Böhmer) at the University of Mainz in Germany and the other in the laboratories of Giovanni Andreetti, Rocco Ungaro and Andrea Pochini at the University of Parma in Italy. From the 1970s and into the early 1980s it was the work of these three pioneer groups that sustained and defined calixarene chemistry. By the early 1980s, however, other research groups, including those of Seiji Shinkai in Japan, David Reinhoudt in the Netherlands, and Javier de Mendoza in Spain, had recognized the potential of calixarenes. And, at this point the exponential phase in the growth of calixarene chemistry began. It continues to the present day with many dozens of research groups and many hundreds of scientists now working in a variety of fields, still swelling the ranks of investigators.

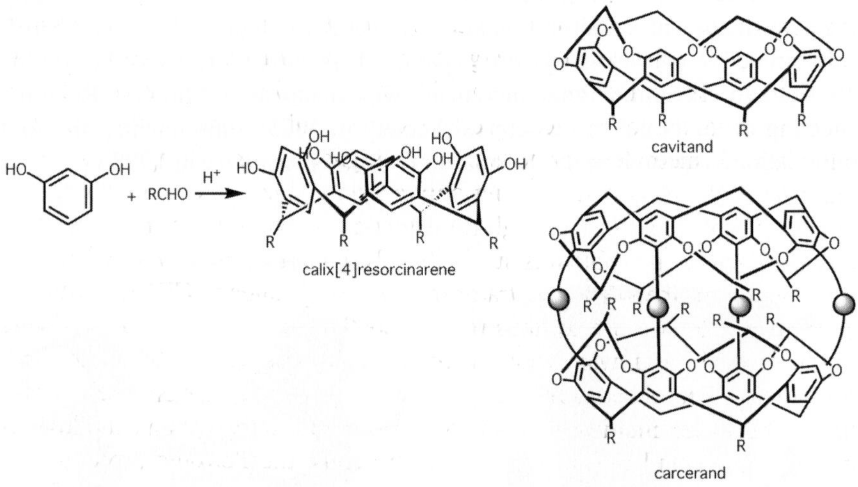

Figure 1-5. Cavitands and carcerands derived from calix[4]resorcinarenes.

Coincident with Zinke's experiments in the early 1940s were those of Joseph Niederl[10] at New York University who carried out reactions of resorcinol with aldehydes and who had also postulated cyclic tetrameric structures for the products (Fig. 5). A variety of names have subsequently been assigned to these compounds, the one that explicitly recognizes their calixarene-like structure being "calix[n]resorcinarene". However, many current authors prefer the shorter "resorc[n]arene" designation. Like the phenol-derived calixarenes, the resorcinol-derived calixarenes languished in the literature relatively unattended[11] until 1980 when Sverker Högberg at the

University of Uppsala published procedures for their facile synthesis.[12] Like the phenol-derived calixarenes, they received increasing attention thereafter, particularly by Donald Cram at the University of California at Los Angeles whose work[13] with cavitands and carcerands led to a Nobel Prize in 1987. It was my privilege and pleasure to spend part of a Guggenheim Fellowship in the Cram laboratories in 1982, although my own interests remained totally focused on the phenol-derived compounds. It was left to other chemists to study the resorcinol-derived compounds, which they did with great vigor. Together, the phenol-derived and resorcinol-derived types of cyclooligomers comprise the extended family of arenol-derived calixarenes.

Among the many problems confounding our earliest studies of calixarene chemistry was "The Petrolite Puzzle". One of the compounds from the Zinke condensation has a melting point of about 340°C, while the Petrolite product has a melting point of about 400°C, clearly an indication that they are non-identical substances. However, the room temperature NMR as well as the temperature dependent NMR spectra of the two compounds in chloroform solution are almost identical, suggesting a close structural similarity. It was not until the spectra were measured in pyridine solution that a clear difference emerged, the lower melting compound retaining a pair of doublets arising from the methylene protons and the higher-melting compounds losing this pair of doublets and showing only a singlet even at very low temperatures. This was one of those gloriously simple experiments in which, in a flash, a perplexing problem is suddenly solved, an *experimenta lucifera.* The near identity of the NMR spectra of the two compounds in $CDCl_3$ solution is a nice example of those devious roadblocks that Nature delights in throwing in the path of the unwary traveler. Spectral evidence, along with chemical, osmometric, and ultimately X-ray structural evidence[14] conclusively established the lower melting compound as the cyclic tetramer and the higher-melting compound as the cyclic octamer. Thus, the Petrolite product, first thought to be a linear polymer, then a cyclic tetramer, was ultimately found to be a cyclic octamer.

As already mentioned, a significant contribution in this early phase of calixarene research was the development of easy and reliable procedures for making the calixarenes in reproducibly good to excellent yields[15] A modified Zinke procedure, using a small amount of NaOH as the base and diphenyl ether as the high boiling solvent produces the cyclic tetramer,[16] the Petrolite procedure, involving NaOH as the base and boiling xylene as the solvent produces the cyclic octamer,[17] and a modified Zinke procedure using larger amounts of KOH produces the cyclic hexamer.[18] These three easily accessible cyclic oligomers are referred to as the "major calixarenes" (Fig. 6), but they are only three of the 17 known members of the family. The calix[5]arene[19] and calix[7]arene[20] are also relatively easily accessible, although in lower yields of 15-20% and are referred to as the "minor calixarenes" And, through the prodigious efforts of my coworker Donald Stewart, the "large calixarenes" from calix-9 to calix-20 have been isolated and characterized.[21] The calixarene family now represents one of the largest of the well characterized cyclic oligomers.[22]

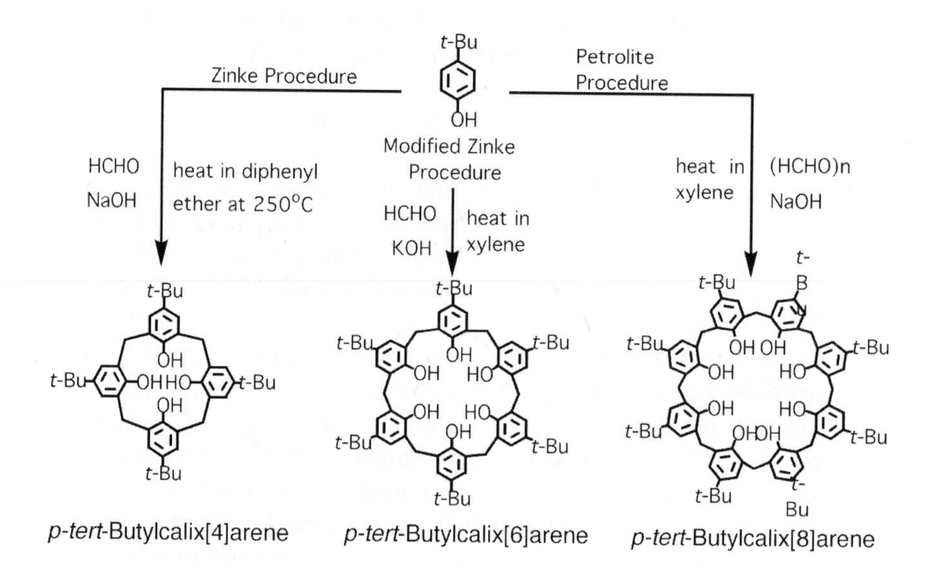

Figure 1-6. Selective pathways in calix[n]arene syntheses.

By the beginning of the 1980s, after some years of often frustrating and discouraging work, several basket-containing compounds were finally in hand, and attention could be once again focused on building an aldolase mimic. A suggested target molecule was published in an *Accounts of Chemical Research* article in 1982, but it envisaged a synthesis that was well beyond the capabilities of the day, given the paucity of knowledge about the physical and chemical behavior of the calixarenes. In fact, many years were to elapse before sufficient armamentum was available to give the synthesis of a multi-functionalized calixarene a reasonable chance for success. To reach this point it was first necessary to learn how to deal with the various shapes that the calixarenes could assume and then to devise methods for introducing functional groups in selective and controlled fashion, tasks that consumed the time and energies of an increasing number of research groups throughout the world.

The shape problem was first perceived by Sir John Cornforth[7] who was the only one to seriously investigate the Zinke compounds in the period between Zinke's work in the 1940s and our work in the 1970s. Cornforth's goal was to make tuberculostatic agents by oxyalkylating the Zinke compounds. Upon repeating Zinke's work he isolated two different compounds which he designated as the lower-melting and the higher-melting substances. Assuming both to be cyclic tetramers he postulated that the difference arose from the several possible orientations that the phenyl rings can assume around the central core, *i.e.* isomers that subsequently came to be known as conformers. The refutation of this postulate came in the early 1970s from the work of the Kämmerer Group in Mainz.[4]

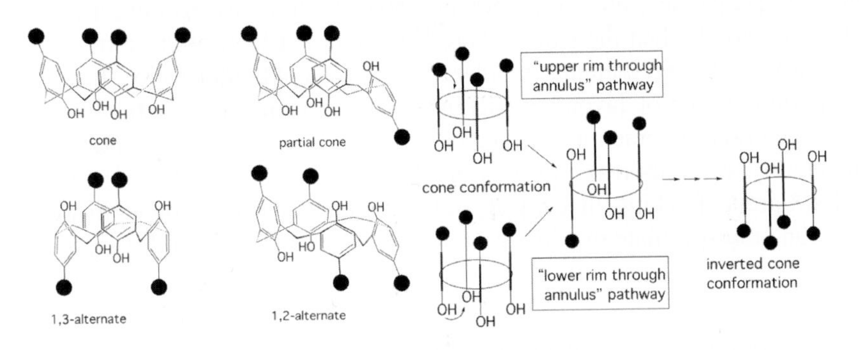

Figure 1-7. Conformations of a calix[4]arene and possible mechanisms for their interconversion.

Exploiting the multi-step synthesis of a cyclic tetramer devised by Hayes and Hunter[23] in the 1950s, Kämmerer synthesized a series of calixarenes containing 4-7 aryl units which, as mentioned above, set in motion a line of investigation in parallel with that of our group at Washington University, Kämmerer (and, later, Böhmer) pursuing the multi-step methods of synthesis and my coworkers and I pursuing the single step methods of synthesis. Research along both lines has continued in synergistic parallel to the present day and has provided a significant fraction of the basic groundwork for calixarene syntheses.

One of Kämmerer's earliest experiments was to study the variable temperature NMR spectrum of the cyclic tetramer which, as noted above, at room temperature shows a pair of doublets arising from the CH_2 groups that coalesce to a sharp singlet at elevated temperatures, interpretable as the result of a flexible cyclic tetramer undergoing transformations between what we later designated as the cone, partial cone, 1,2-alternate, and 1,3-alternate conformers. The cyclic tetramer was thus revealed to be a conformationally flexible molecule, at variance with Cornforth's assumption of a rigid conformer.

Two pathways for conformational interconversion can be envisaged, *viz* an "upper rim through the annulus pathway" and a "lower rim through the annulus pathway" (Fig. 7). In the case of the calix[4]arenes only the latter is possible, even when a simple hydrogen atom occupies the *para* position. As the number of aryl moieties around the annulus increases, however, both pathways become available. And, as the rings get larger one might reasonably expect the conformational mobility to increase in a continuous and uniform fashion. Such is not the case, however. As already mentioned, the calix[4]arene and calix[8]arene have almost identical conformational mobilities, and as one progresses through the calixarene family it is observed that the calix[12]arene, calix[16]arene and calix[20]arene are all conformationally less mobile than their immediate neighbors – a kind of "rule of 4"; *i.e.* the calix[4,8,12,16,20]arenes representing islands of conformational stability as a consequence of favorable packing modes that maximize intramolecular hydrogen bond interactions[21a]. The calix[20]arene is particularly interesting, apparently even more conformationally stable than the calix[4]arene or calix[8]arene. Lacking an x-ray crystal structure of this compound, however, one can only speculate that its structure might be that of a trefoil (Fig. 8).

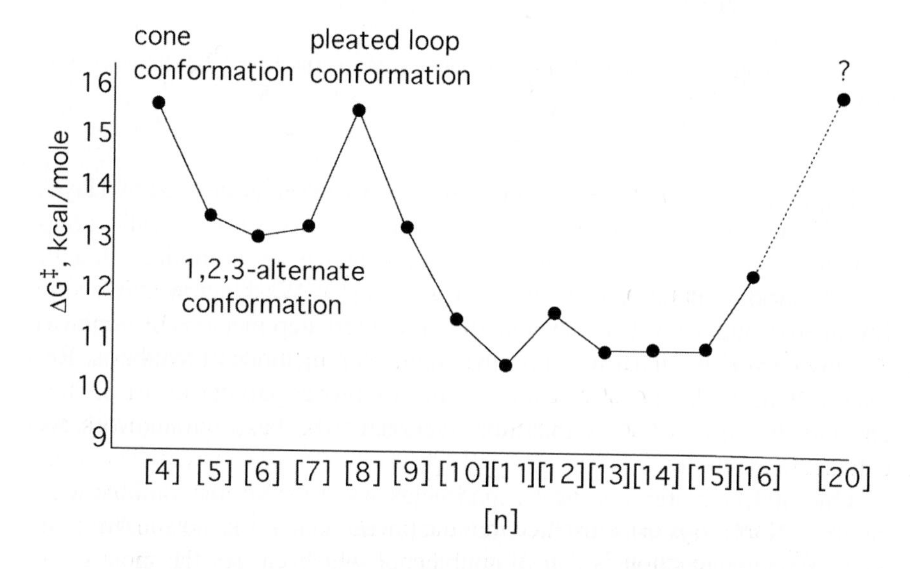

Figure 1-8. Conformational energy barriers in calix[n]arenas.

For fashioning an enzyme mimic it is presumably preferable for the calixarene to be a permanent rather than a flexible basket. A great deal of effort has been devoted to learning how to freeze the conformation, either by placing sufficiently large groups on the lower rim (and, in the case of the larger calixarenes, on the upper rim as well) or, alternatively, by building bridges on either or both of the rims.[24] For the former it has been found that by varying the derivatizing agent and the conditions for its introduction onto the lower rim, a calix[4]arene can be frozen into any one of the four possible conformations.

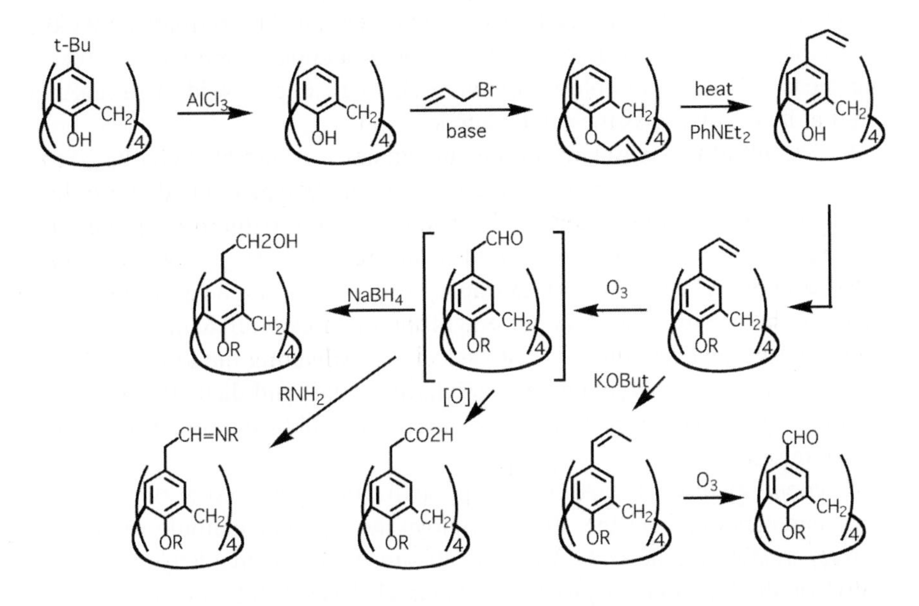

Figure 1-9. Pathways in calix[4]arene functionalisation.

The second major problem that had to be addressed was how to introduce functional groups onto the upper and/or lower rims of the calixarene. An enormous amount of effort by a large number of investigators has gone into this endeavor, and a host of functionalization procedures are now available. Two of the earliest ones, developed in our laboratories, provide representative examples (Fig. 9). The first[25] employs the Claisen rearrangement and starts with the removal of the *p-tert*-butyl groups by treatment with $AlCl_3$. It is a most fortuitous circumstance that the phenol that works best in the calixarene-forming reaction is *p-tert*-butylphenol which carries the most easily removed of the alkyl groups, thereby making the *p*-positions in the derived calixarene available for substitution. In the next step, the phenolic groups are converted to allyl ethers, and the allyl compound is heated in diethylaniline to induce the rearrangement which proceeds with migration of the allyl group to the *p*-positions, the *o*-positions being blocked. This sequence works quite nicely with the calix[4]arene, and the allyl groups provide good precursors to a variety of other functional groups. Surprisingly, however, the reaction sequence is far less successful with any of the larger calixarenes, and its application appeared to be limited to the calix[4]arene. However, this situation changed in the 1990s when my coworker Charles Gibbs devised a procedure[26] that employed a silylating agent in the rearrangement mixture, the rationale being that as the phenolic groups are generated in the course of the rearrangement they become silylated and thereby protected from any further reaction such as oxidation. The procedure works remarkably well, and by using the Gibbs modification of the Claisen rearrangement it is now possible to obtain good to excellent yields with calixarenes at least as large as the calix[8]arene.

As a result of the symmetry of the calixarenes it is a special challenge to introduce functional groups in a *selective* fashion. The methods that are described above place the same functional group on *all* of the *p*-positions, so more nuanced procedures are required to introduce fewer groups in a controlled fashion. A representative example[26] of how this can be accomplished is seen in the preparation of all but one of the possible *p*-allylcalix[5]arenes. The syntheses depends, *inter alia*, on selectively silylating the lower rim hydroxyls using appropriately hindered silylating agents and then allylating the remaining hydroxyl groups followed by use of the Gibbs modification of the Claisen rearrangement.

A second method of upper rim functionalization[27] devised in our laboratories somewhat later in the 1980s employs the "quinonemethide route". It involves treating the de-*tert*-butylated calixarene with formaldehyde and dimethyl or diallylamine to produce a Mannich base. Quaternization of the Mannich base with methyl iodide followed by treatment with two equivalents of the nucleophile of choice displaces the trialkylamine moiety with that nucleophile.

In contrast to upper ring functionalization, lower rim functionalization is generally more straightforward, and a large variety of ethers and esters have been made simply by exchanging the hydrogen of the OH group for a different moiety. A greater challenge is the replacement of the entire OH group with a different group, and only a few examples have been described in the literature. One of these, worked out by Charles Gibbs,[28] involves the treatment of a calix[4]arene with dimethyl thiocarbamyl chloride to form the dimethylthiocarbamoyl ester which is thermally rearranged via the Newman-Kwart procedure wherein the oxygen and sulfur atoms trade places. Hydrolysis or hydrogenolysis of the product yields the calix[4]arenethiol. The reaction sequence is rife with complications, however. The initial dimethylthiocarbamoylation produces a serious mixture of products that requires separation, and the thermal rearrangements require very careful control of the temperature. Nevertheless, all but one of the possible calix[4]arenethiols were ultimately obtained (Fig.10).

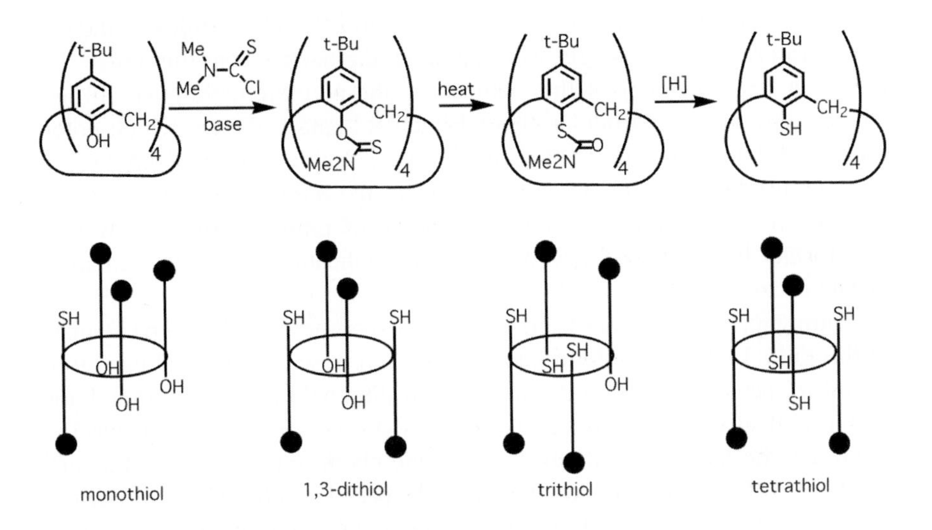

Figure 1-10. Consequences of replacing OH by SH.

As previously mentioned, the three groups of 1970s pioneers were joined in the early 1980s by a handful of other groups and then, starting in the mid 1980s, by an ever increasing number of groups. The accomplishments of these battalions of investigators have been recorded in many papers, in several books and in numerous review articles published during the last two decades. A number of major research themes have become apparent. Among these is the use of calixarenes as molecular baskets to capture other molecules, *viz.* the idea that initiated our own interest in the calixarenes. In the first of a long series of papers on solid state calixarene complexes, the Parma group reported in 1979 the x-ray crystallographic structure of *p-tert*-butylcalx[4]arene[15a] which showed that a molecule of toluene resided in its cavity. One of the earliest studies of solution state complexation was reported in 1985 by my colleague Laurenz Bauer[29] who observed a very weak interaction between toluene and calix[4]arenes in $CHCl_3$ solution. Since then numerous studies in other laboratories have dealt with solid state and solution state complexation of calixarene with molecules, but the number of such studies pales in comparison with those dealing with the complexation of calixarenes with ions, both inorganic and organic. It was the Reed Izatt group at Brigham Young University[30] that provided the opening gambit in 1985 for this area of calixarene research by showing that in strongly basic aqueous solution the three major calixarenes effectively transport Li^+, Na^+,

K^+, Rb^+, Ca^{+2}, and Ba^{+2} cations across a liquid membrane. It is this facet of calixarene chemistry that has grown to dominate the field, and for the last two decades a large fraction of the published information on the calixarenes has been concerned with ion complexation - mostly with cations in the earlier literature but increasingly with anions in the more recent literature. The ions of a large majority of the elements in the periodic table have all been studied, the ones of particular general interest being those for which potentially useful applications are envisaged. Two examples chosen from the very large number of representative examples are the removal of radioactive cesium from nuclear wastes[31] and the removal of uranium from sea water.[32] The former has already found commercial application; the latter remains a distant dream.

During the three decades from our initial experiments with Zinke chemistry in the early 1970s to my retirement from active research in 2002 calixarene chemistry remained our central focus, with the goal of building an aldolase mimic still alive but often submerged in the recesses of the subconscious mind. By the late 1990s, though, calixarene chemistry had reached a sufficient level of maturity to allow the dream expressed in the 1983 *Accounts of Research* article to be explored more effectively. Accordingly, a synthesis project was launched to assemble what was hoped would be an aldolase mimic (Fig. 11).[33] For the framework a calix[6]arene rather than a calix[4]arene was chosen in order to provide a larger cavity. Using a published procedure[34] *p-tert*-butylcalix[6]arene was converted to the 1,3,5-trimethyl ether. Taking advantage of the greater lability of a *tert*-butyl group *para* to a free phenolic group, selective de-*tert*-butylation with $AlCl_3$ was effected to yield the tris-*tert*-butyl trimethyl ether. Using the qinonemethide process cyanomethyl groups were then introduced into the three open positions on the upper rim. This was followed by the conversion of the three remaining OH groups to carbethoxymethyl ethers by treatment with ethyl bromoacetate. The three ester moieties were hydrolyzed to the corresponding carboxylic acid which was treated with thionyl chloride to afford the tris-acid chloride. Reaction of the tris-acid chloride with 1,4,7-triazacyclononane inserted the triaza ring on the lower rim of the calix[6]arene system. Treatment of this compound with a very large excess of diborane led to reduction of the three upper rim cyanomethyl groups to aminoethyl groups and the three lower rim carbamide moieties to methylene groups. Treatment with a limiting amount of *t*-BOC reagent protected two of the upper rim amino groups, leaving the third one reactive toward methyl iodide which converted it to a dimethylamino group. Finally, removal of the *t*-BOC moieties yielded the putative aldolase mimic.

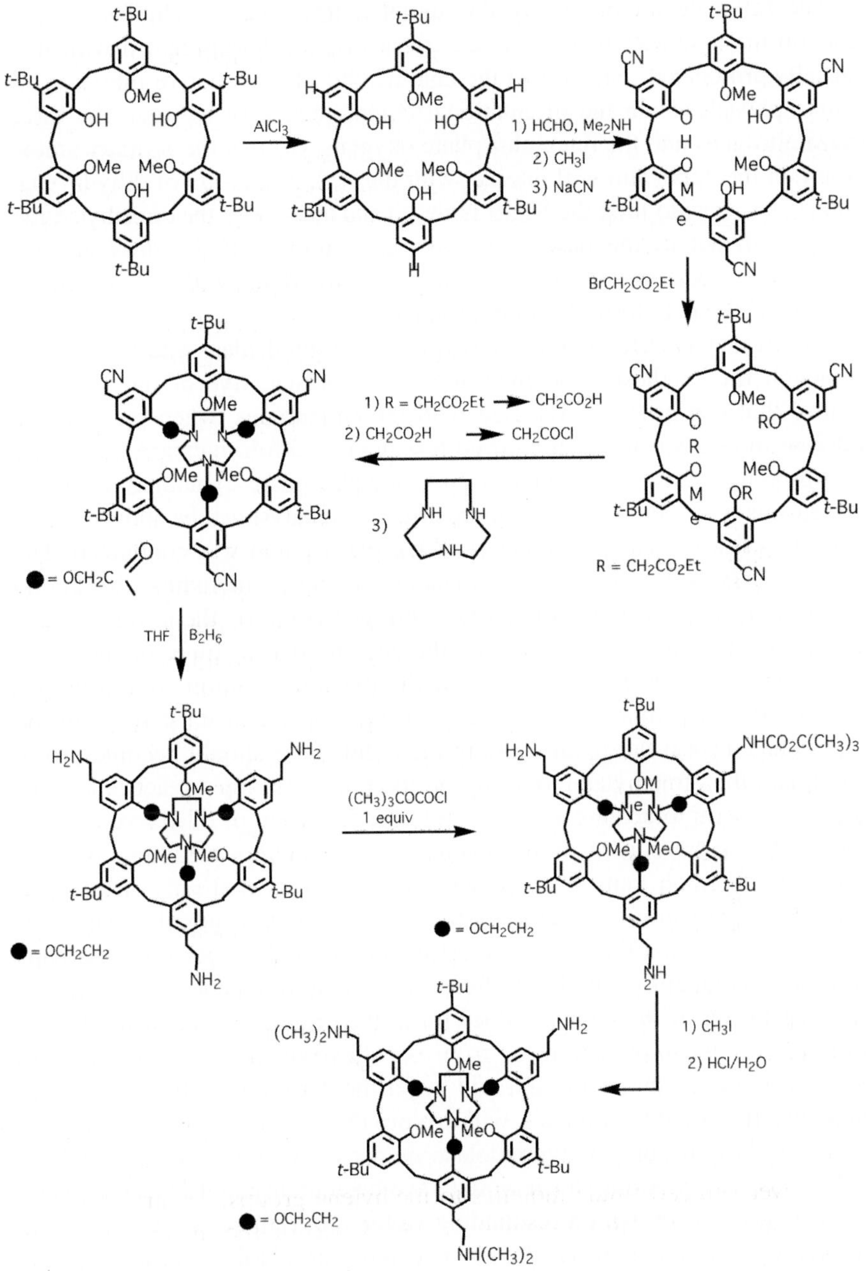

Figure 1-11. Synthesis of an aldolase mimic.

The rationale for proposing this target molecule as an aldolase mimic rests on the assumptions (a) that the triaza ring on the lower rim will, in a partially protonated state, attract the anionic phosphate group of dihydroxyacetone phosphate via the orientational complementarity between the three triaza nitrogens and the three phosphate oxygens; (b) that the primary amino group on the upper rim will interact with the carbonyl group of dihydroxyacetone phosphate to form the Schiff base; (c) and that one of the dimethylamino groups will act as the base to remove an α-proton of dihydroxyacetone phosphate to generate the reactive intermediate which will then condense with an aldehyde to form the aldol product.

Unfortunately, life does not always reward good ideas and hard work. Although the synthesis described above is representative of the synthetic control that can now be exercised with the calixarenes, when the putative aldolase mimic was tested it showed less than a 10-fold increase in the rate of α-proton removal from dihydroxyacetone phosphate. A number of possible reasons for its failure to be a more effective catalyst can be adduced. One probably critical factor is the pH at which the reaction was conducted. The system contains a variety of protonatable moieties including the amino groups at the upper rim, the three triaza nitrogen atoms on the lower rim, and the phosphate anion. To best mimic the enzyme one or more of the triaza nitrogens should be protonated while the phosphate anion should be deprotonated; the primary amino group on the upper rim must react in the unprotonated form but upon forming the Schiff base should become protonated; and the dimethylamino group on the upper rim must react in the unprotonated form to acquire a proton from the α-carbon of dihydroxyacetone phosphate in the course of the catalysis. This entails a rather complex interplay of proton transfer processes, and it is likely that only at one specific pH might significant catalysis be realized. Another possible cause for failure could be that the catalyst-substrate complex does not form, perhaps because of competitive inhibition from the tetranitromethane used as the assay reagent. Or, if the complex does form, the orientation might not be appropriate for the proton transfer from the dihydroxyacetone phosphate to occur. It is hoped that some intrepid experimenter of the future, willing to undertake the arduous synthesis, will explore this system in further detail.

The failure of our putative aldolase mimic to show significant catalytic ability was disappointing, for it would have been gratifying to be able to end this personal history with a resounding victory. But, there is comfort in the knowledge that although the dream of making an aldolase mimic remains unfulfilled, it did provide the impetus for laying a part of the groundwork for calixarene chemistry. Now, hundreds of other scientists have joined in the exploration of the calixarenes, bringing them into the mainstream of today's scientific endeavor which, *inter alia,* includes the area called nanochemistry.

From the vantage point of the chemist one of the defining characteristics of nanochemistry is size. Most of the structures that have occupied the attention of organic chemists over the years comprise molecules in which the constituent atoms are held one to another by strong intramolecular bonds. In more recent years increasing attention has been focused on structures in which such strongly intramolecularly bonded compounds are then held one to another by much weaker intermolecular forces, giving rise to assemblies much larger than their constituent molecules. This is the world of nanochemistry, a world that calixarenes have now entered through the work of Julius Rebek, David Reinhoudt, Volker Böhmer, Jerry Atwood and others who have fashioned calixarene-based molecules capable of self assembly.

When I started graduate school at the University of Wisconsin in the 1940s it was only five years after Bachman, Cole and Wilds had published a total synthesis of the steroid equilenin. This achievement is generally cited as the opening chapter in the modern era of natural product synthesis. During the more than six decades that have followed, the art of natural product synthesis has reached incredible heights and now stands as one of the major hallmarks of twentieth century chemistry. Today, even the most convoluted of structures with countless stereogenic centers yield to total synthesis in the hands of some remarkably talented chemists, and undoubtedly this field will continue to attract many of the best and brightest. It seems likely, though, that in the years ahead more of these exceptional talents will be drawn to the rapidly expanding area of nanochemistry, which is destined to stand as one of the major hallmarks of twenty first century chemistry. The part that calixarenes can play in this adventure is the subject of the remaining chapters in this book.

2. REFERENCES

1. Several books on calixarene chemistry have been published, including: (a) C. D. Gutsche, "Calixarenes", Monographs in Supramolecular Chemistry No 1; J. F. Stoddart, Ed.; Royal Society of Chemistry, Cambridge, U. K. (1989); (b) J. Vicens and V. Böhmer, eds. "Calixarenes: A Versatile Class of Macrocyclic Compounds", Kluwer Academic Publishers (1991); (c) C. D. Gutsche, "Calixarenes Revisited", Royal Society of Chemistry (1998); (d) L. Mandolino and R. Ungaro eds., "Calixarenes in Action", Imperial College Press (2000); (e) G. J. Lumetta, R. D. Rogers and A. S. Gopalan, eds."Calixarenes for Separations", American Chemical Society (2000); (f) Z. Asfari, V. Böhmer, J. Harrowfield and J. Vicens, eds. "Calixarenes 2001", Kluwer Academic Publishers (2001). For a complete list of all the review articles and books published prior to 1998 cf. page 4 of ref 1c.
2. A. Baeyer, *Ber.* **5**, 25, 280, 1094 (1872).
3. L. H. Baekland, U. S. Patent 942,699 (October 1908).

4. A. Zinke and E. Ziegler, *Ber.* **B74**, 1829 (1941); *ibid.* **77**, 264 (1944); A. Zinke, G. Zigauer, K. Hössinger and G. Hoffmann, *Monatsh.* **79**, 438 (1948); A. Zinke, R. Kretz, F. Leggewie and K. Hössinger, *ibid.* **83**, 1213 (1952).
5. For a recent x-ray crystallographic structure of aldolase and a discussion of the mechanism of proton transfer *cf.* K. H. Choi, J. Shi, C. E. Hopkins, D. R. Tolan, and K. N. Allen, *Biochemistry*, **40**, 12868 (2001).
6. R. S. Buriks, A. R. Fauke, and J. H. Munch, U S. Patent 4, 259, 464 (Filed 1976; Issued 1981).
7. J. W. Cornforth, P. D'Arcy Hart, G. A. Nicholls, R. J. W. Rees, and J. A. Stock, *Br. J. Pharmacol*, 73 (1956); J. W. Cornforth, E. D. Morgan, K. T. Potts, and R. J. W. Rees, *Tetrahedron*, **29**, 159 (1973).
8. C. D. Gutsche, T. C. Kung, and M-I. Hsu, Abstracts of 11th Midwest Regional Meeting of the American Chemical Society, Carbondale, IL, No. 517 (1975).
9. C. D. Gutsche and R. Muthukrishnan, *J. Org. Chem.* **43**, 4905 (1978).
10. J. H. Niederl and H. J. Vogel, *J. Am. Chem. Soc.* **62**, 2512 (1940).
11. X-Ray crystallogaphic structures were determined by Erdtmann and coworkers at the University of Stockholm; *e.g.* H. Erdtman, S. Högberg, S. Abrahamsson and B. Nillsen, *Tetrahedron Lett.* 1679 (1968).
12. A. G. S. Högberg, *J. Org. Chem.* **45**, 4498 (1980); *idem. J. Am. Chem. Soc.* **102**, 6046 (1980).
13. D. J. Cram and J. Cram, "Container Molecules and Their Guests", Monographs in Supramolecular Chemistry No 4; J. F. Stoddart, Ed.; Royal Society of Chemistry, Cambridge, U. K. (1994).
14. C. D. Gutsche and L. J. Bauer, *Tetrahedron Lett.* **22**, 4763 (1981).
15. (a) The x-ray structure of the cyclic tetramer was established by D. Andreetti, R. Ungaro, and A. Pochini, *J. Chem. Soc. Chem. Commun.* 1005 (1978); (b) The x-ray structure of the cyclic octamer was established with crysals painstakingly prepared by my wife Alice (C. D. Gutsche, A. E. Gutsche and A. I. Karaulov, *J. Inclusion Phenom.* **3**, 447 (1985).
16. C. D. Gutsche and M. Iqbal, *Org. Syn.* **68**, 234 (1990).
17. J. H. Munch and C. D. Gutsche, *ibid.* **68**, 243 (1990).
18. C. D. Gutsche, B. Dhawan, M. Leonis and D. Stewart, *ibid.* **68**, 238 (1990).
19. D. R. Stewart and C. D. Gutsche, *Org. Prep. Proced. Int.* 25, 137 (1993); K. Iwamoto, K. Araki and S. Shinkai, *Bull. Chem. Soc. Jpn*, **67**, 1499 (1994); I. Dumazet, N. Ehlinger, F. Vocanson, S. Lecocq, R. Lamartine and M. J. Perrin, *J. Incl. Phenom.* 29, 175 (1997).
20. F. Vocanson, R. Lamartine, P. Lanteri, R. Longeray and J. Y. Gauvrit, *New J. Chem.* **19**, 825 (1995).
21. (a) D. R. Stewart and C. D. Gutsche, Third International Calixarene Conference, Fort Worth, TX, Abstract P-42 (1995); *idem. J. Am. Chem. Soc.* **121**, 4136 (1999). (b) I. Dumazet, J. B. Regnouf-de-Vains and R. Lamartine, *Synth. Commun.* **27**, 2547 (1997).
22. The mechanism of calixarene formation remains an unsolved puzzle. Mechanisms for formation of linear oligomers are more obvious than those for cyclisation. Kinetic analyses indicate that the cyclic tetramer is not formed directly from the linear tetramer but, rather, comes by way of the cyclic octamer. The cyclic octamer appears to be the *kinetic product*, and the cyclic tetramer the *thermodynamic product*. The cyclic hexamer, may be viewed as the *template controlled product.* It is postulated that the linear tetramer, instead of cyclizing, forms an intermolecularly hydrogen-bonded dimer, a "hemicalixarene" and that the hemicalixarene subsequently splits out water and formaldehyde at its extremities to form the cyclic octamer. Then, under more strenuous conditions the cyclic octamer transforms to the cyclic tetramer. For this process a "molecular mitosis pathway" was

envisaged in which a pinched conformation splits into a pair of cyclic tetramers. The following experiment suggests, however, that this is, at most, a very minor pathway:

A sample of fully deuterated cyclic octamer was mixed with an equal amount of ordinary protiated cyclic octamer, and the mixture was subjected to the reversion conditions which transform the cyclic octamer to the cyclic tetramer. If molecular mitosis is the sole pathway the product should consist only of fully deuterated cyclic tetramer and fully proteated cyclic tetramer. If a fragmentation-recombination pathway is followed the label will become scrambled, and in the ultimate case the product will be a 1:4:6:4:1 mixture of variously deuterated and protiated cyclic tetramers. In the actual experiment the product was found to contain all of these labeled compounds, and the observed ratio – obtained by mass spectral analysis – corresponded to 25% or lower molecular mitosis pathway. Clearly, fragmentation-recombination pathways are dominant.

23. B. T. Hayes and R. F. Hunter, *Chem. Ind.* 193 (1954); *idem. J. Appl. Chem.* **8**, 743 (1958).
24. Bridges do not always impede conformational inversion, however, as demonstrated by a calix[6]arene bridged between the 1 and 4 positions. Whereas the parent compound exists in the cone conformation, the tetramethyl ether assumes a 1,2,3-alternate conformation with the bridge threaded through the annulus from the upper rim to the lower rim to form what was designated as a "self-anchored rotaxane" [S. Kanamathareddy and C. D. Gutsche, *J. Am. Chem. Soc.* **115**, 6572 (1993)].
25. C. D. Gutsche and J. A. Levine, *J. Am. Chem. Soc.* **104**, 2652 (1982); C. D. Gutsche, J. A. Levine and P. K. Sujeeth, *J. Org. Chem.* **50**, 5802 (1985).
26. C. G. Gibbs, J-S. Wang, and C. D. Gutsche, "Calixarenes for Separations", ACS Symposium Series 757; G. J. Lunetta, R. D. Rogers, and A. S. Gopalan eds., 313 (2000).
27. C. D. Gutsche and K. C. Nam, *J. Am. Chem. Soc.* **110**, 6153 (1988).
28. C. G. Gibbs, P. K. Sujeeth, J. S. Rogers, G. C. Stanley, M. Krawiec, W. H. Watson and C. D. Gutsche, *J. Org. Chem.* **60**, 8394 (1995).
29. L. J. Bauer and C. D. Gutsche, *J. Am. Chem. Soc.* **107**, 4314 (1985).
30. R. M. Izatt, J. D. Lamb, R. T. Hawkins, P. R. Brown, S. R. Izatt and J. J. Christensen, *J. Am. Chem. Soc.* **107**, 63 (1983).
31. (a) *Cf.* Ref 1e: J. F. Dozol, V. Lamare. N. Simons, R. Ungaro and A. Casnati, p. 12; P V. B. Bonnesen, T. J. Haverlock, N. I. Engle, R. A. Sachleben and B. A. Moyer, p. 26; (b) J. Vicens *Journal of Inclusion Phenomena* DOI 10.1007/s10847-005-9021-x (2005).
32. S. Shinkai, Y. Shiramama, H. Satoh, O. Manabe, T. Arimura, K. Fujimoto and K. Matsuda, *J. Chem. Soc. Perkin Trans.* **2**, 1167 (1989).
33. C. D. Gutsche and G. Liu, unpublished work.
34. R. G. Janssen, W. Verboom, D. N. Reinhoudt, Al. Casnati, M. Freriks, A. Pochini, F. Ugozolli, R. Ungaro, P. N. Nieto, M. Marramolino, F. Cuevas, P. Prados, and J. deMendoza, *Synthesis*, 380 (1993).

Chapter 2

TETRAUREA CALIX[4]ARENES

From dimeric capsules to topologically novel molecules

Yuliya Rudzevich, Anca Bogdan, Myroslav O. Vysotsky, and Volker Böhmer
Johannes Gutenberg-Universität, Fachbereich Chemie, Pharmazie und Geowissenschaften, Abteilung Lehramt Chemie, Duesbergweg 10-14, D-55099 Mainz

Abstract: Hydrogen-bonded, dimeric capsules are formed via self-assembly of calix[4]-arenes, substituted at their wide rim by four urea functions. The thermodynamic and the kinetic stability of such capsules and the internal mobility of the hydrogen bonded belt and of the included guest are briefly reviewed. Half-lives for the guest exchange range from seconds to months. The selective formation of hetero- or homo-dimers in mixtures of tetraurea derivatives can be used for the construction of self-assembled polymers or structurally uniform dendrimers. Alkenyl groups attached to the urea residues may be oriented by the dimerisation and connected via metathesis reaction in a controlled way. The two-fold use of a calix[4]arene as template leads, via multimacrocyclic tetra-loop derivatives as intermediates, to huge macrocycles containing up to 100 atoms in the rings. Bis- and tetra-loop compounds are also the building blocks for hitherto unknown bis[2]-, bis[3]- and cyclic [8]-catenanes, of tetra[2]rotaxanes and of further topologically novel compounds.

Key words: Calixarenes, capsules, hydrogen bonds, multiple catenanes, multiple rotaxanes, olefin metathesis.

1. INTRODUCTION

Ten years ago, it was shown by Rebek and Shimizu that tetra alkoxy calix[4]arene derivatives, substituted at their wide rim by four urea functions (general formula **1**, Fig. 1) are self-complementary and form dimeric capsules.[1] These capsules are held together by a seam of hydrogen bonds between the –NH and C=O groups of urea functions attached to the two calixarenes. The volume of the cavity thus created is in the range of 190-200 $Å^3$ and the inclusion of a suitable guest (molecule) which ideally occupies about 55% of this space[2] is a necessary condition of the dimerisation.

J. Vicens and J. Harrowfield (eds.), Calixarenes in the Nanoworld, 21–46.

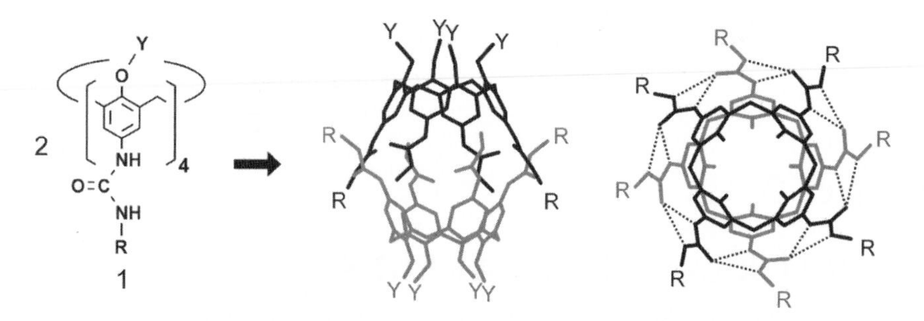

Figure 2-1. The dimeric capsule formed by tetraurea derivatives **1**; the included guest is not shown.

NMR-spectroscopy provided the first evidence for the formation of these dimeric capsules. The ^{1}H NMR spectrum of **1**, showing time averaged C_{4v}-symmetry in hydrogen bond breaking solvents, shows two m-coupled doublets for the aryl protons of the calixarene and a strongly separated (up and down field shifted) pair of signals for the urea NH-protons. High-field-shifted signals for an included guest ($\Delta\delta$ = 3-4 ppm) completed the picture of an S_8-symmetrical capsule composed of two calix[4]arenes with C_4-symmetry. Immediate and unambiguous confirmation of the occurrence of dimerisation was provided by the observation of an additional set of NMR resonances indicating heterodimer formation in a mixture of two different calixarenes of type **1**.[3] The first single crystal X-ray analysis of such a dimer[4] finally confirmed in all details, and in agreement with molecular mechanics calculations, the structure initially proposed by Rebek *et al.*

While this dimerisation of tetraurea calix[4]arenes is well-established, several basic aspects of the chemistry warrant further study. Larger hydrogen-bonded capsules have been described, including examples based on calixarenes[5] or resorcarenes,[6,7] and the tetraurea system has been used for other purposes, *e.g.* to build up larger structures via self-assembly[8] or to preorganize functional groups for further reactions. The present chapter will focus on three key topics:

- the thermodynamic and kinetic stability of calixurea capsules.
- the application of (selective) dimerisation of calixureas to build up special structures via self-assembly.
- the synthesis of topologically novel molecules using the special geometry of calixurea dimers.

Since these properties are interdependent, this listing does not (necessarily) represent a mutually exclusive categorisation. We also do not aim to give an exhaustive survey, but rather to highlight typical properties and to indicate future possibilities offered by these sophisticated systems.

1.1 Symmetry Properties

It is useful to start with a short discussion of the (intriguing) symmetry properties of dimeric capsules which are connected with or caused by the mutual orientation of the two calix[4]arenes.[9] For this purpose, we represent each calix[4]arene by a square where the corners are the phenolic units, indicate different substituents on the phenolic units by capital letters A, B[10] and take a view along the fourfold axis (Fig. 2). The directionality of the hydrogen-bonded urea belt is indicated by arrows, but it is important to discuss also the symmetry of the dimeric (= supramolecular) assembly alone, without this directionality.

It is apparent that a heterodimer (C_{4v}) becomes chiral (C_4) because of this directionality, while a homodimer (D_{4d}) remains achiral (S_8). To isolate enantiomers of these C_4-symmetrical heterodimers, the direction of the hydrogen bonds should be stable on a time scale which allows their separation.

All homodimers of tetraurea calix[4]arenes containing two different phenolic units A and B are chiral simply because their assembly limits any symmetry elements to pure rotations and thus they may be said to exhibit "***supramolecular chirality***". Especially interesting are tetraureas of the alternating ABAB-type which form only ***one*** homodimer, while in the other cases two regioisomers are possible. The symmetry is reduced also here by the directionality of the hydrogen bonds but it should be emphasized that this directionality does not create further stereoisomers. The even more complicated situation of calixarenes consisting of three or four different phenolic units will not be discussed here.

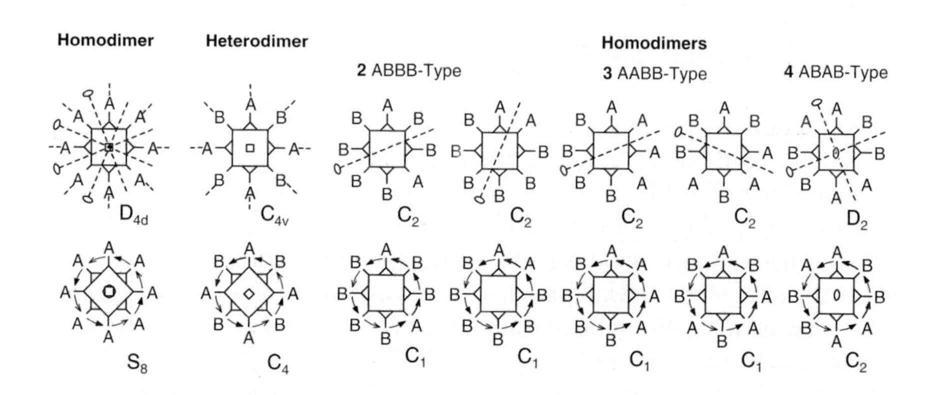

Figure 2-2. Schematic representation of different dimers and their symmetry properties. The directionality of the hydrogen-bonded belt cancels all two-fold axes lying in the drawing plane, and also sigma planes perpendicular to it.

2. STABILITY AND SELECTIVITY

Various factors may determine the thermodynamic and the kinetic stability of the dimeric capsules:

a) Most obviously, *the solvent*, since only monomeric species exist under hydrogen-bond-breaking conditions.

a) *The structure of the tetraurea molecule*, primarily the residues R attached to the urea but also the ether residues Y.

c) *The included guest*, which may be different from the solvent.

Assuming a simple equilibrium, the dimerisation may be described by the equation $M + M + G \rightleftarrows D(G)$ or more simply by $M + M \rightleftarrows D(S)$ if the solvent S is included as guest. Equilibrium constants for these reactions are high, and cannot be directly determined by NMR spectroscopy at the concentrations normally used. In typical solvents like benzene-d_6 or chloroform-d_1, only the dimer can be observed by ^{1}H NMR, even at elevated temperature. Association constants therefore were estimated from rate constants for the dissociation and association processes, as determined by FRET measurements[11] (see below). Values of 10^6 M^{-1} to 10^8 M^{-1} for homodimers of tetraaryl ureas explain why monomeric compounds are not observed under the conditions of usual NMR experiments.

To compare the relative "stability" of different dimers in a given solvent, increasing amounts of a hydrogen bond breaking solvent like DMSO-d_6 can be added to determine how much is needed to destroy the dimer. In mixtures where signals for both species appear, integration can be used to estimate the concentrations.[12] The formal equilibrium constants thus calculated vary with the total concentration, however, indicating that a description in terms of a single equilibrium is not appropriate.

2.1 Thermodynamics

For capsular assemblies, it is perhaps automatic to associate the term "selectivity" with the inclusion of guests.[13] However, the relative stability of different dimers may lead also to useful selectivities in the combination of tetraureas. Assuming the same thermodynamic stability for both homodimers and their heterodimer, the three species should be present in the ratio 1:2:1, and, with only slight deviations, this has been found for various combinations. The mixture of equimolar amounts of an aryl- and an arylsulfonyl-tetraurea, however, leads to the "quantitative" formation of the heterodimer, although both tetraurea derivatives form homodimers alone.[14] The reason for this pronounced selectivity (which is not found for alkylureas) is not exactly known. A favourable combination of the increased acidity of $-SO_2$-NH- with the basicity of the urea carbonyl group has been tentatively suggested, but the different geometry of the two urea groups seems to be also important.[15]

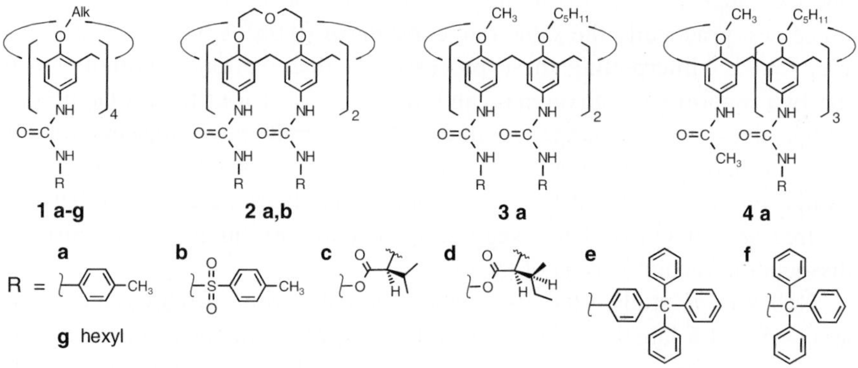

Tetraurea derivatives **2** derived from a calix[4]arene rigidified in a very regular cone conformation by two short crown-3 bridges[16] show enhanced (thermodynamic) stability towards DMSO-d_6 in mixtures with $CDCl_3$.[12] (For kinetic stabilities see below.) Surprisingly at first sight, they **do not** form heterodimers with "usual" tetraureas **1**, even in an 8-fold excess. This can be understood, however, on an entropic basis, since dimers **2•2** show a much lower "internal flexibility" than **1•1**; while heterodimers **1•2** would be of similar rigidity to **2•2**, and thus entropically less favored.

For "cross combinations" of **1a** with **2b** or **1b** with **2a** the "enthalpic effect", expressed by the formation of tolyl/tosyl heterodimers, dominates in CD_2Cl_2, while in $CDCl_3$ heterodimers are formed only for the combination of the flexible tolyl derivative **1a** with the rigid tosyl derivative **2b**.[17]

The combination of tetraureas may be also steered by chiral residues R attached to the urea functions. In a series substituted by α-aminoacid methyl esters the β-branched derivatives of L-valine and L-isoleucine (**1c,d**) form homodimers and *exclusively* heterodimers with tetratolylurea **1a** in benzene-d_6 (the selectivity is not so pronounced in $CDCl_3$), while the situation is less clear for other amino acids.[18] Surprisingly, *no* heterodimers are formed with tetratosyl ureas.

2.2 Kinetics

The first detailed kinetic data were derived from NOESY-NMR-studies, using the C_{2v}-symmetrical calixarene **3a** consisting of two phenolic units in an ABAB arrangement. The respective homodimer has C_2-symmetry (compare Fig. 2) and four positions (A, A', a, a' and B, B', b, b') exist for the aromatic protons of the units A and B, respectively. This means that three distinct exchange processes can be observed by ^{1}H NMR spectroscopy. (The S_8-symmetrical homodimer of a C_{4v}-symmetrical tetraurea shows only one exchange process involving the two possible positions.) For all three processes, practically the same rate constant of $(0.065 \pm 0.015)\ s^{-1}$ at 298 K was found.[19] Interpreting these processes as dissociation/recombination of the two monomeric calixarenes, and allowing for a fourth, similar process which does not exchange positions, the overall rate constant for the dissociation/recombination can be calculated as $k = (0.26 \pm 0.06)\ s^{-1}$. A similar rate $(k = 0.46 \pm 0.1)\ s^{-1}$ was obtained for benzene exchange between its free and included states, suggesting that this exchange occurs also via dissociation/recombination of the dimer.

Rebek *et al.* studied the dissociation/recombination of tetraurea calix[4]-arenes **5** and **6** labelled by coumarin dyes (***A, D***) covalently attached to the narrow rim. The two dyes were chosen so that the emission spectrum of ***D*** overlaps with the absorption spectrum of ***A***. Dimerisation **5•6** brings them within a distance of ~20 Å, enabling observation of fluorescence resonance energy transfer (FRET) within a dimer. Such FRET experiments, carried out at much lower concentrations (0.05-0.5 μM) than our NMR-studies (3-5 mM), led to much lower reaction rates for the dissociation of the dimer.[11] One of the explanations offered was that the reaction rate must increase with smaller ether residues, especially when alkoxy groups are (partly) replaced by methoxy, as in **3a**. This was confirmed in preliminary kinetic experiments, similar to those described below.[20]

Since the FRET studies also allowed the determination of rate constants for association, stability constants could be derived for the dimers (Table 1).

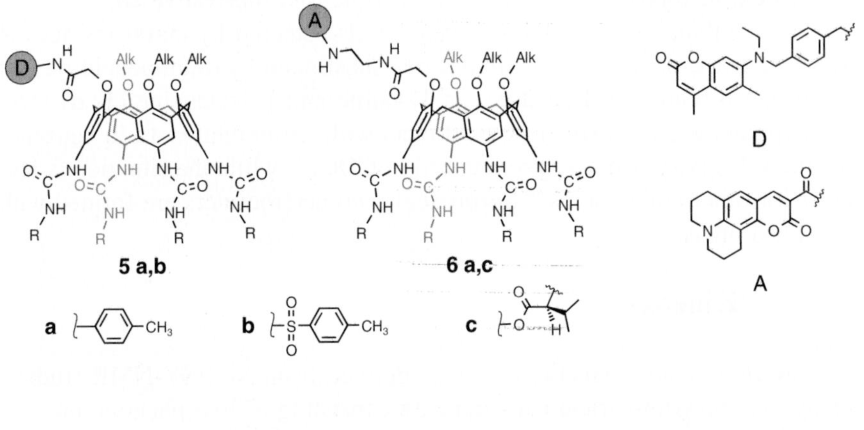

5 a,b **6 a,c**

D A

a **b** **c**

Table 2-1. Rate constants for the formation and dissociation, as measured by FRET at concentrations of 0.05-0.5 μM, and stability constants of tetraurea calixarene dimers.[11]

	k_{ass} ($M^{-1}s^{-1}$)	k_{diss} (s^{-1})	K_A (M^{-1})
5b·6b	< 50	$5.4 \cdot 10^{-5}$	$<10^6$
5b·6a	$3 \cdot 10^3$	$5 \cdot 10^{-6}$	$6 \cdot 10^8$
5a·6a	$2.4 \cdot 10^4$	$6 \cdot 10^{-4}$	$4 \cdot 10^7$

Using cyclohexane-d_{12} (a potential guest) as solvent, the escape of a guest (or better its exchange against cyclohexane-d_{12}) is slow enough to be easily followed by recording conventional ^{1}H NMR spectra as a function of time, most conveniently by monitoring the decreasing signal of an included "non-deuterated" guest. For various guests, half-lives between several months (tetrachloromethane, (methyl)cyclohexane, 1,4-difluorobenzene), several days (benzene, fluoro-, chloro-, bromobenzene) and several hours (chloroform, toluene) have been observed in clear first order reactions (Fig. 3).[21]

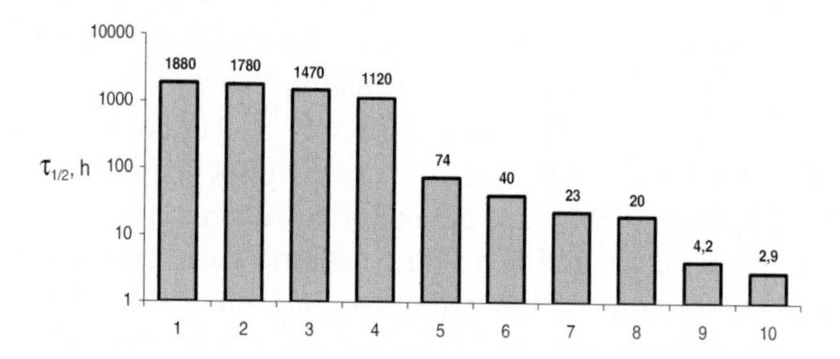

Figure 2-3. Half-lives for the exchange of an included guest against the solvent C_6D_{12} for cyclohexane (1), methylcyclohexane (2), 1,4-difluorobenzene (3), tetrachloromethane (4), fluorobenzene (5), chlorobenzene (6), bromobenzene (7), benzene (8), toluene (9), chloroform (10).[21]

Bulkier urea residues R make such measurements possible in benzene-d_6 also.[22] For heterodimers of tetra-trityl urea **1f** with tetra-tritylphenyl urea **1e** and tetra-tolyl urea **1a**, the signal for included C_6H_6 (4.03 ppm) disappears with $\tau_{1/2}$ = 60 h and $\tau_{1/2}$ = 130 min, respectively. Other spectral changes are not observed.[23] However, for toluene-d_8 or p-xylene-d_{10} as solvent, the signals of the heterodimer **1e•1f** disappear at the same rate, being replaced by signals of the homodimer **1e•1e**. This is clear additional evidence that the guest exchange occurs (at least mainly) via dissociation/recombination, and that the cavity offered by heterodimers with **1f** is too small to include a larger solvent molecule (*e.g.* toluene) under these conditions. Since the homodimers **1e•1e** are obviously not too small, the size of the cavity must be influenced by the size of the residues R.

This observation led to the idea that the homodimerisation of **1f**, which is not observed in benzene (where benzene would have to be the guest) could be initiated by the addition of a smaller, but more favourable guest, *e.g.* an organic cation (see also below). In fact, dimeric capsules were obtained by refluxing **1f** in dichloromethane with tetramethylammonium chloride or iodide.[24] Very similar ^{1}H NMR spectra are observed for the dimeric assembly in $CDCl_3$ and in DMSO-d_6, where the high-field shifted signal of the methyl protons of the included cation appears at –1.1 ppm. Over longer times in DMSO-d_6, signals for the dimer and its guest disappear (while the signals of monomeric **1f** appear) with first order kinetics. Half-lives of 40 h and 93 h for the chloride and iodide are unprecedented for a hydrogen-bonded dimer in a typical hydrogen-bond breaking solvent like DMSO.[25]

2.3 Cations as Guest, Internal Mobility

As shown by Rebek *et al.* organic cations of appropriate size usually are preferred over neutral molecules as guest. This was the basis of the first reliable and detailed study of the encapsulation by mass spectroscopy.[26] The additional stabilisation of the whole assembly by cation-π interactions can compensate even for the steric strain caused by inclusion of larger cations such as tetraethylammonium (see below) or cobaltocenium.[27]

At room temperature, a dimer with a cationic guest shows only a singlet for the aromatic protons of the calixarene, and thus one of the important criteria of dimerisation is missing. However, the observation of heterodimers (Fig. 4) is an unambigous proof for the dimerisation, and at the same time it shows that the observed singlet cannot be the result of rapid **intermolecular** exchange by dissociation/recombination and that it must be due to an **intramolecular** process.[28] The only possible explanation is that the direction of the hydrogen-bonded urea belt changes rapidly on the NMR time scale. At low temperature, the signal of the calixarene aryl-protons splits into two signals (m-coupled doublets) and thus, the energy barrier for this process can be determined by VT-NMR.[29,30]

A second process can be observed with tetraethylammonium cation as guest. The signal for the methyl protons, appearing as a high-field-shifted signal (~ – 1.7 ppm) at slightly elevated temperature, splits into two broad signals (~ –0.4 ppm and ~ –3.2 ppm) at lower temperature. This observation can be explained by a reorientation of the included cation. Methyl groups facing the aromatic walls of the capsule are much more strongly shielded than those in the "equatorial" plane (the hydrogen bonded belt) of the capsule

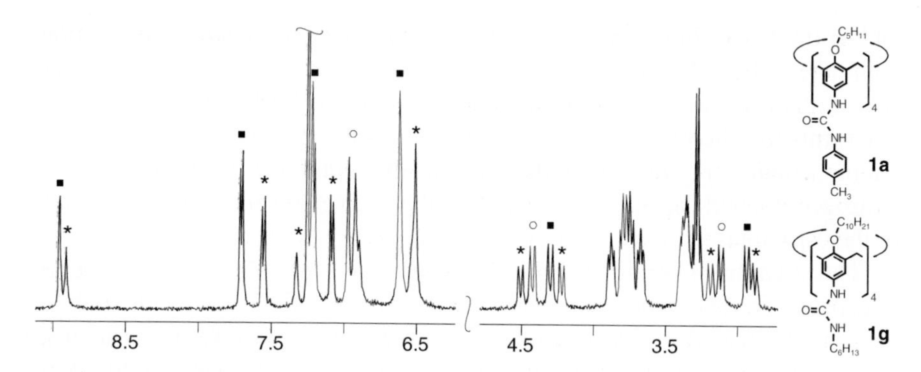

Figure 2-4. Homodimers **1a•1a** (■), **1g•1g** (○), and heterodimer **1a•1g** (*) with the tetraethyl ammonium cation as guest: Representative section(s) of the ^{1}H NMR spectra.

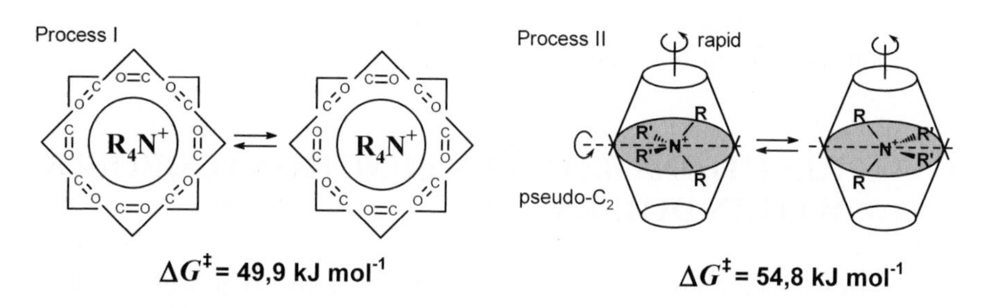

Figure 2-5. Schematic representation of the two (main) rearrangements observed within a dimeric capsule of **1a•1a** containing tetraethylammonium as guest. The energy barriers $\Delta G^{\neq}$ at the coalescence temperature T_c are also indicated.

Both processes are schematically illustrated in Fig. 5, and the energy barriers $\Delta G^{\neq}$ at the respective coalescence temperature T_c are given for tetratolylurea **1a•1a** and Et_4N^+ PF_6^- in $CDCl_3$.

Examined more closely, the two signals of the methyl protons of Et_4N^+ show a further splitting at low temperature which may be due to the fact that the two equatorial and the two polar methyl groups are now "distinctly diastereotopic", due to the directionality of the hydrogen bonded belt. A single-crystal X-ray structure confirms the postulated orientation of the included guest.[31] It shows, also, that the tetraethylammonium cation fills the dimeric capsule so tightly that (according to standard criteria) only 8 of 16 possible hydrogen bonds are present, a result which is in agreement with MD-simulations.

Changing the direction of the hydrogen bonded belt requires a rotation around the σ-bond connecting the urea groups with the calixarene skeleton. Nothing is known about the mechanism in detail (*e.g.* concerted or "step-by step") but it is reasonable to assume that the carbonyl oxygen (the negative end of the C=O dipole) will pass through the cavity occupied by the cation while the N-H hydrogens pass the exterior. A decrease of the energy barrier $\Delta G^{\neq}$ with decreasing radius of the (mono)anion (increasing charge density) is not surprising therefore.[30,32] Small anions, such as chloride (and, depending on the conditions, also bromide) may even inhibit the formation of the hydrogen bonded capsules by competing themselves for the H-bonding sites of the calixurea.

The strength of organic cations as guests may even induce the formation of capsules in cases where other assemblies are preferred. In solvents such as benzene or chloroform, triurea monoacetamide derivatives **4a**, for instance, form hydrogen bonded tetrameric assemblies which do not include any guest. Their complex structure was derived from ^{1}H NMR spectra in combination with an X-ray structure.[33] Addition of tetraethylammonium bromide or hexafluorophosphate to a solution of **4a** in $CDCl_3$, however, leads to the quantitative formation of one of the two possible hydrogen bonder dimers with Et_4N^+ as guest.

3. REACTIONS BETWEEN FUNCTIONAL GROUPS WITHIN DIMERS

The dimerisation of tetraurea calix[4]arenes leads to a well defined arrangement of the eight (2 x 4) urea residues in space. This mutual orientation may be used for (selective) reactions between functional groups attached to the urea groups. Possible modes for connections between these functional groups are illustrated in Fig. 6. Intra-dimer connections could involve adjacent urea residues of the same calix[4]arene (α-connection) or residues belonging to different calix[4]arenes (β-connection). In the latter case, it might be possible also to connect residues which are not immediately neighbouring (β'-connection).

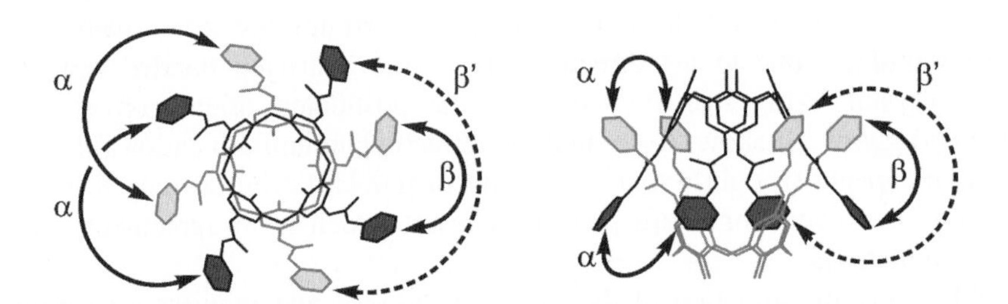

Figure 2-6. Possible connections between functional groups attached to the urea residues within a tetraurea dimer.

3.1 First Attempts

A reaction often used to stabilize complex molecular architectures (*e.g.* dendrimers[34]) or to connect the single parts of a self-assembled aggregate[35] is the metathesis between ω-alkenyl groups.[36] Naively we tried this reaction for homodimers formed by tetraurea **1h•1h**(n=5) in order to produce novel catenanes by α-connection. However, after hydrogenation of the crude mixture to avoid complications due to the presence of cis/trans isomers, we isolated three products (shown by MS to be isomers) in an overall yield of nearly 60%.

They were identified as formed in the ratio (roughly) of 1:2:1 and their NMR spectra in aprotic solvents are consistent with C_1, C_2, and C_4-symmetry.[37]

As shown in Fig. 7, the bis[2]catenane (**I**) formed by four (2 x 2) α-connections should have C_2-symmetry under conditions where the directionality of the hydrogen bonded belt is kept. Four β-connections would result in the tetra-bridged capsule (**III**) with C_4-symmetry, while two α- and two β-connections should lead to the doubly bridged mono[2]catenane (**II**).

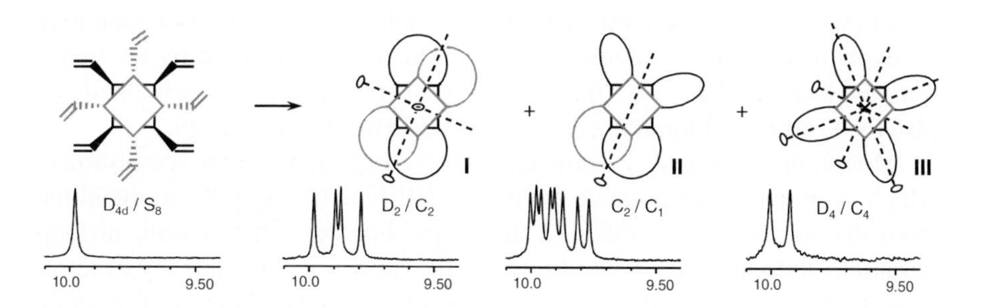

Figure 2-7. Reaction products obtained by metathesis and hydrogenation of dimers **1h•1h**(n=5). The low field section of the ^{1}H NMR-spectra in benzene-d_6 reflects the symmetry properties expected for directional hydrogen bonds.

Table 2-2. Energy barriers of the hindered rotation of the included Et_4N^+ $\Delta G^{\neq}$ (kJ mol^{-1}) within corresponding complexes, rate constants k (h^{-1}) and half-lives $\tau_{1/2}$ [h] (25°C).

Compound	$\Delta G^{\neq}$ (T_c[K])	k [h^{-1}]	$\tau_{1/2}$ [h]
1h•1h	47 (265)	0.090	7.7
I	55 (306)	0.045	15
II	52 (292)	0.048	14
III	50 (278)	0.059	12

In spite of the mechanical and/or covalent connection of the two calix[4]arenes, compounds **I-III** are still able to exchange the included guest. Thus, it was possible to study the kinetics of the guest exchange (*e.g.* cyclohexane against benzene-d_6) and the internal mobility of an included tetraethylammonium cation as described above. These values are compared in Table 2 with those obtained for the starting dimer **1h•1h.**

3.2 Improved Syntheses of Bis[2]catenanes

The yield of bis[2]catenanes can be improved by starting with tetraurea derivatives in which one or two pairs of urea residues are already connected to a macrocyclic loop. Such mono- or bisloop derivatives **7** and **8** are available by conventional macrocyclisation under high dilution conditions as outlined in Fig. 8.

Starting with tetraamine **10**[38] in which two adjacent amino groups are Boc-protected, an initial cyclisation with an activated bis-urethane leads to the monoloop compound **11**. Deprotection and acylation gives the monoloop tetraurea **7** bearing two alkenyl groups. The bisloop urea **8** is obtained by direct cyclisation of the tetraamine. Lower yields in the cyclisation step may be explained by "trans cavity" bridging. This can be avoided by an analogous cyclisation of **12** (after deprotection), a synthetic pathway which allows also the synthesis of tetraurea derivatives with two different loops.

Monoloop tetraureas **7** form exclusively one of the two regioisomers which in principle are possible for the AABB type (see Fig. 8), to avoid the sterically unfavorable overlap of the loops. For the same reason, bisloop tetraureas **8** do not form homodimers at all. However, the tendency of tetraureas to dimerize in aprotic, apolar solvents leads to the exclusive formation

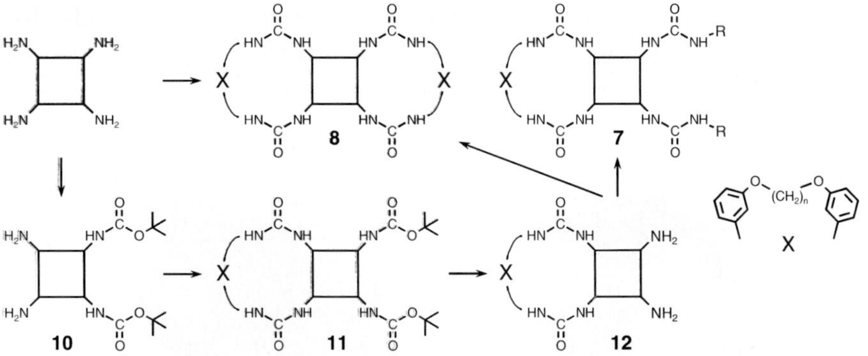

Figure 2-8. Synthesis of mono- or bisloop tetraurea calix[4]arenes (schematic).

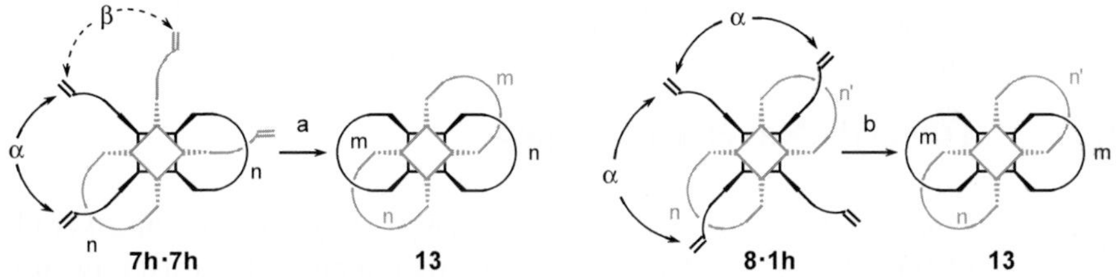

Figure 2-9. Improved strategies for the selective preparation of bis[2]catenanes from mono- and bisloop tetraureas.

of heterodimers if the stoichiometric amount of an "open chain" tetraurea is added, since this is the only way to "saturate" all urea functions by hydrogen bond.

As indicated schematically in Fig. 9 only one (wrong) β-connection is possible in homodimers **7h•7h**, while only "correct" α-connections can occur within the heterodimers **8•1h**.[39] Consequently the yield of bis[2]catenanes was increased to about 50% starting with the monoloop derivative (pathway **a**) and up to 80% starting with the bisloop derivative (pathway **b**). It must be emphasized that these pathways are complementary if isomeric bis[2]catenanes with different sizes of the loops are envisaged. Pathway **a** leads to catenanes consisting of two identical calix[4]arenes (m/n) bearing two loops of different size, while pathway **b** leads to a catenane formed by different calix[4]arenes (n/n + m/m) bearing two identical loops. (Three different ring sizes are also available in a particular way (n/n'+ m/m) see Fig. 9.)

While disorder within the aliphatic chains complicated the analysis of the first X-ray structure determination[30] for such a bis[2]catenane, a second example (Fig. 10) provided unambiguous definition of its form.[40] We also tried the chiral resolution of these C_2-symmetrical compounds by chromatography on Chiralpak AD-H but a complete separation was not always obtained. Separation factors ranged from 1 to 2.4.[41]

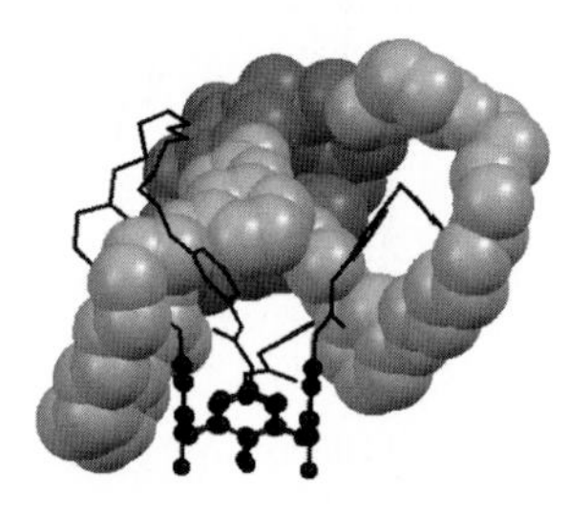
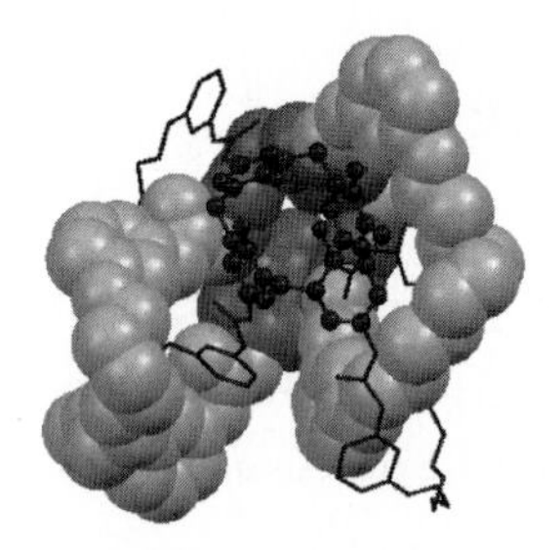

Figure 2-10. Single crystal X-ray structure of **13** (n = 10). Crystals obtained from chloroform/methanol show the bis[2]catenane with an "open capsule".

3.3 Towards Novel Topologies

If a β-connection is formed in homodimers of monoloop bisalkenyl tetraureas of type **7h**, two isolated double bonds remain, which usually (up to octenyl residues) cannot be connected via metathesis. While metathesis does occur with longer undecenyl residues, it is accompanied by formation of an isomeric product which made the purification of the desired bis[2]catenane difficult or (in some cases) even impossible.

To determine conditions under which a β'-connection might be favoured, we synthesized monoloop tetraurea **14**, bearing one alkenyl-substituted urea and one urea group with a substituent too large to pass the loop. Thus, only one of the two regioisomeric dimers, in which the alkenyl groups are in β'-position, could be formed. This was easily confirmed by recording the ^{1}H NMR spectrum of this homodimer. (It is interesting to note that **14** is chiral, and that only the same/identical enantiomers can combine to a dimer as a consequence of the C_2-symmetry of the bis[2]catenane skeleton).

In fact, the metathesis reaction (followed by hydrogenation) produces a product, which, according to the MS spectrum, is the expected compound with a β'-connection.[42] The ^{1}H NMR spectrum is consistent with the presence of a single C_1-symmetrical compound, although a single-crystal X-ray structure has not yet been solved.

As shown in Fig. 11, the topology of the reaction product contains elements of catenanes and rotaxanes and even of knots.

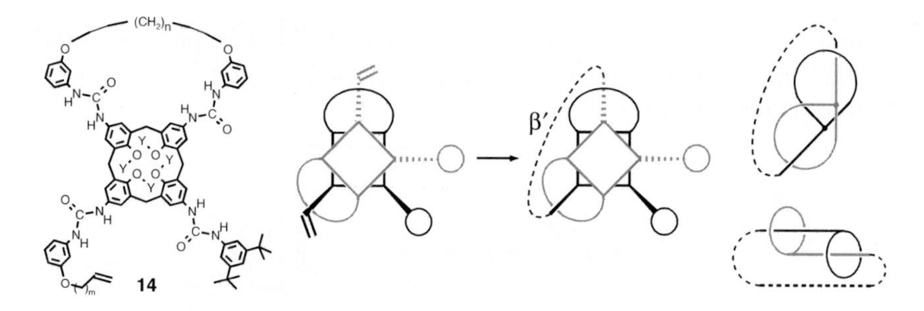

Figure 2-11. β'-Connection via metathesis. To illustrate the topology of the product, the bulky urea group is omitted and the non-interlocking (!) calixarene macrocycles are "graphically reduced" to a branching point.

3.4 Multimacrocycles

Bisloop tetraureas **8** were obtained from tetraamino calix[4]arenes by reaction with an active bisurethane. The use of high dilution conditions was successful in suppressing intermolecular connections, but not undesired intramolecular connections (Fig. 12), such as the bridging of two opposite amino groups across the cavity. Other complications arise if the connection of a given phenolic group to both its neighbours is desired and of course it must be remembered that with small ether substituents, such as methyl groups, the calix[4]arene becomes conformationally labile and does not necessarily adopt the *cone* conformation needed for the desired intramolecular coupling.[43]

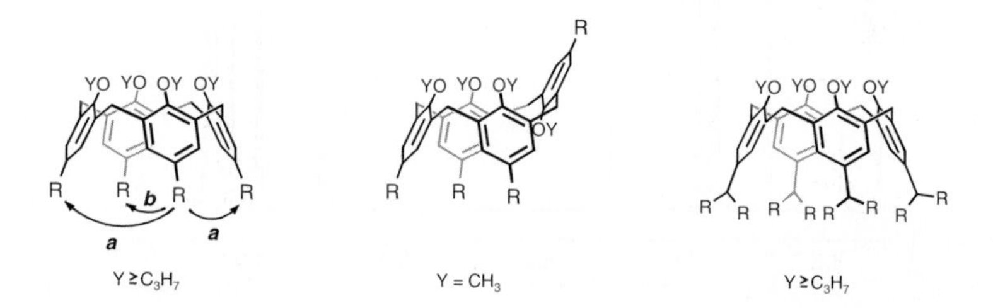

Figure 2-12. Potential difficulties for the controlled intramolecular connection of functional groups R attached to the wide rim of a calix[4]arene.

For reactions between functional groups attached to the urea residues of a tetraurea calix[4]arene, the prearrangement within a heterodimer, *e.g.* with a tetratosyl urea, may be used to overcome these difficulties. A reaction sequence leading to bis- and tetraloop derivatives is outlined in Fig. 13. Again we used the metathesis reaction (with subsequent hydrogenation) for the intramolecular connection of ω-alkenyl groups,[44] but in principle other reactions may be used, provided they can be carried out in conditions under which the heterodimer is stable.

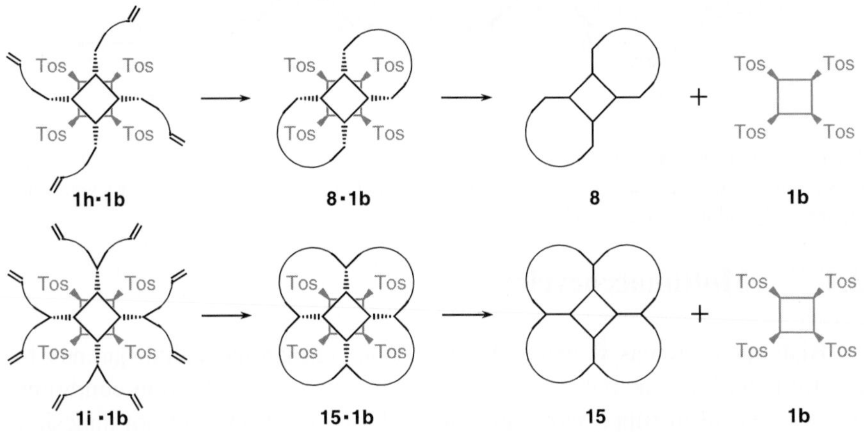

Figure 2-13. Template synthesis of bis- and tetraloop tetraurea calix[4]arenes.

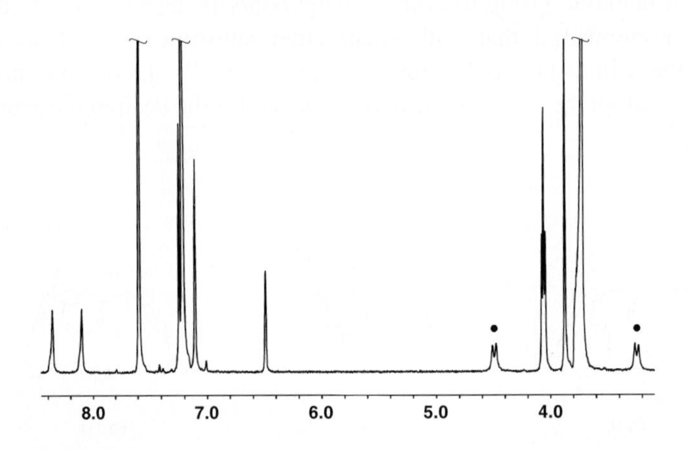

Figure 2-14. High temperature ^{1}H NMR spectrum of **15** (n = 14). The pair of doublets for the methylene protons (•) indicates a kinetically stable cone conformation.

The yields for the multicyclic compounds thus prepared (and isolated easily after splitting the heterodimer by a hydrogen bond breaking solvent) are in the range of 60-92% for bisloop compounds (n = 8, 10, 14, 20) and 55-80% for tetraloop derivatives (n = 10, 14, 20) respectively. The second set, especially, is remarkable for the formation of four intramolecular links in a well defined way.

The data presented in Figs. 14 and 15 show that the unusual cone conformation of a tetramethylether is fixed by the loops, which are too short to allow the molecule to assume the partial cone conformation usually preferred by tetramethyl ethers of calix[4]arenes.

The potential of the method is best demonstrated by the selective synthesis of tetraloop compounds with different length and/or structure of the linkers between the phenolic units.[45]

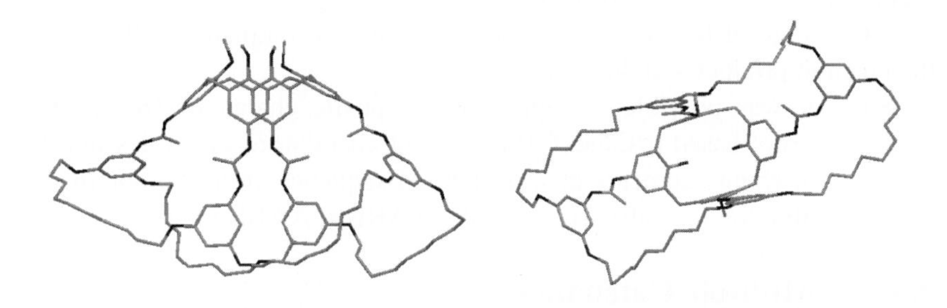

Figure 2-15. Single crystal X-ray structure of **15** (n = 10) showing the molecule in a "pinched" cone conformation.

3.5 Huge Macrocycles

The template synthesis described above for tetraloop compounds can be extended if, in a further step, the urea links to the calix[4]arene can be cleaved. After various attempts, it turned out that a clean cleavage is possible simply by refluxing the tetraloop tetraurea with acetic acid. Thus, macrocyclic molecules **16** with (presently) up to 100 atoms in the ring could be prepared with yields of up to 75% in the last step (Fig. 16).

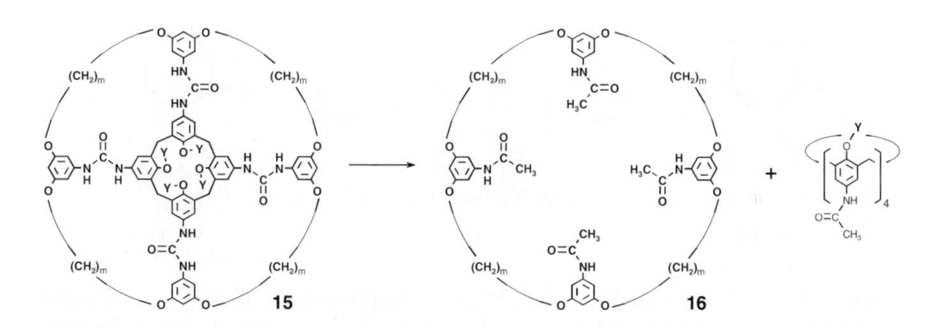

Figure 2-16. Detachment of the macrocyclic molecule **16** from the calix[4]arene used as template.

The whole reaction sequence, twice using a calix[4]arene as a template, may be summarized and generalized as follows:[46]

- Bifunctional building blocks (here bisalkenylethers) are covalently attached (via urea links) to a (first) template, which thus determines the number of building blocks connected in the final product.
- A second template (the tetratosylurea) is used to arrange the functional groups in the correct position within a hydrogen bonded complex.
- After the connecting reaction (here metathesis/hydrogenation) the second template is removed by breaking the hydrogen bonds.
- Cleavage of the covalent linkages to the first template finally releases the reaction product and the template.

Since it was possible to replace the aliphatic chains in **16** by oligo-ethylene glycol based residues[46] it is evident also that macrocycles build up by different chain segments in well defined sequence could be obtained by starting with tetraurea calix[4]arenes of the ABBB, AABB or ABAB type.

3.6 Multiple Catenanes

Tetraloop compounds **15**, easily available as shown before, can be used as starting materials for the preparation of even more elaborate molecular structures. Since they cannot form homodimers (see above) their stoichio-metric combination with tetra-O-alkenyl and octa-O-alkenyl compounds

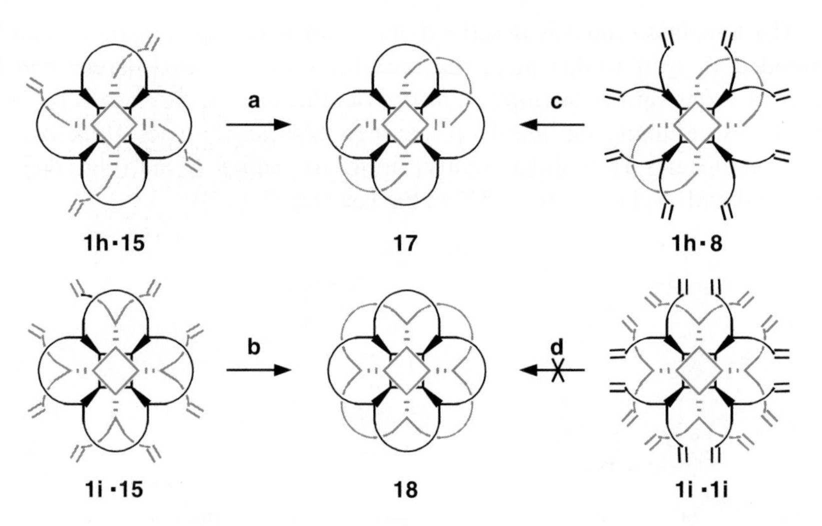

Figure 2-17. Synthesis of bis[3]catenanes **17** and cyclic [8]catenanes **18**. Pathways a) and b) have been realized, c) seems to be a reasonable alternative for a), while the selective formation of a defined product seems unlikely for pathway d).

leads to the exclusive formation of heterodimers, for reasons discussed above. Metathesis reactions of these heterodimers (followed as usual by hydrogenation) lead to catenanes[47] in which two times three rings are interlocked (**17**), or even all eight annulated rings (**18**), as shown in Fig. 17.

We therefore suggest the names bis[3]catenanes and cyclic [8]catenanes for these compounds, in analogy to bis[2]catenanes used already above. The number in brackets refers here to the number of interlocking rings. In all cases only two molecules (the two multicyclic calix[4]arenes) are mechanically interlocked.

The structure of these multiple catenanes can be unambiguously deduced from their ^{1}H NMR spectra, which in the case of the cyclic [8]catenane shows the same S_8-symmetrical pattern as a usual homodimer of a tetraurea with four identical phenolic units (Fig. 18, left). For **18**, single crystals were obtained from CH_2Cl_2/hexane. An X-ray structure thus gave the definite proof for the eightfold interlocking ring system (Fig. 18, right).

In spite of this multicatenated cage structure, the included guest can still be exchanged. This follows from ESI mass-spectra where doubly charged peaks for $(M + 2Na + C_6H_4F_2)$ were found for **17** and **18**. 1,4-Difluorobenzene, known as a good guest, was not present during synthesis and work up. It was added for the mass spectrum and must have been included (exchanged against the originally present guest) during the ionisation process.

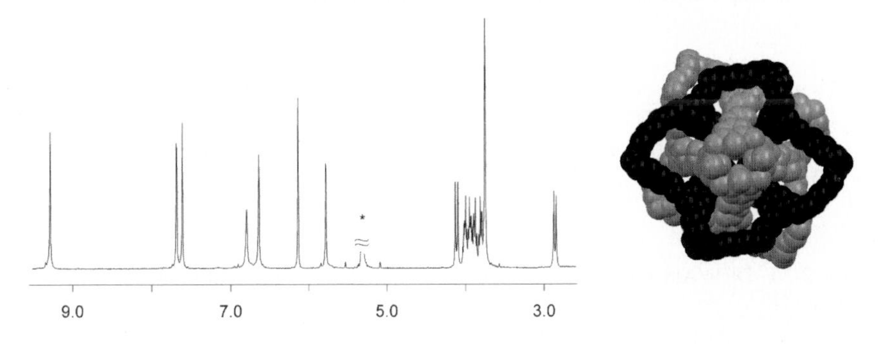

Figure 2-18. Section of the ^{1}H NMR spectrum and molecular shape in the crystalline state of a cyclic [8]catenane **18** (n = 14).

3.7 Rotaxanes

A heterodimer between a tetraloop and an open chain tetraurea may be regarded as a “pseudo rotaxane”. Introduction of stoppers at the open ends of the urea residues would convert it into a rotaxane (Fig. 19), which should be stable also under conditions where hydrogen bonds are broken. As a stoppering reaction we have chosen the Diels-Alder cyclo-addition between maleic imide groups and a 1,4,5,8-tetraalkoxy anthracene (Fig. 20).

As expected "open chain" tetraureas **1j**, bearing maleic imide groups at the end of the urea residues R form clean heterodimers with the tetraloop **15** in benzene-d_6 or toluene-d_8, and this dimerisation is not disturbed by the anthracene derivative (Fig. 21). When such a solution is heated, the singlet for the –CH=CH– protons disappears, and two new signals appear in the ^{1}H NMR spectrum due to the newly formed aliphatic protons H_a and H_b. Thus, the progress of the reaction can be easily monitored.[48] Although an apparently quantitative conversion was reached after 72 h, the isolated yield of the pure product is still only in the range of 50%.

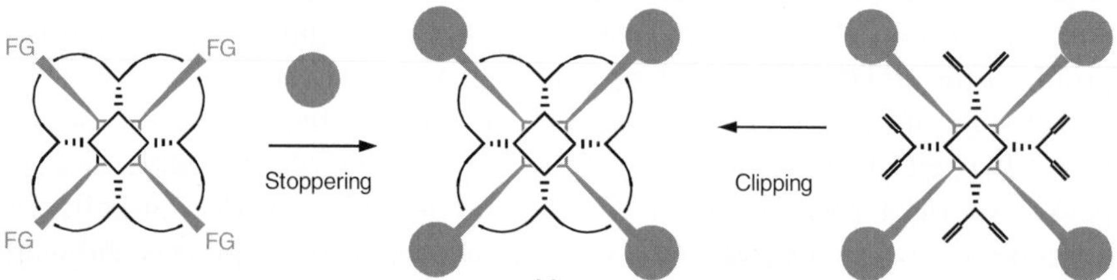

FG FG FG FG Stoppering Clipping 19

Figure 2-19. Synthesis of tetra[2]rotaxanes by stoppering (a) or clipping (b); bis[2]rotaxanes based on bis-loop compounds should be available analogously.

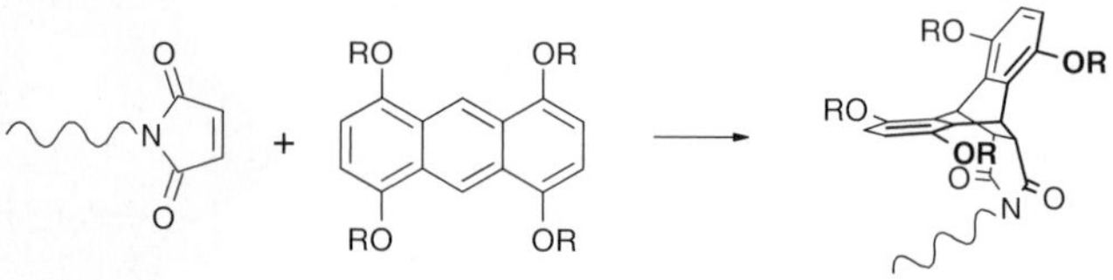

O N O + RO OR RO OR RO OR RO OR O N O

Figure 2-20. Diels-Alder cyclo-addition used to introduce stoppers during the rotaxane synthesis.

ESI- and MALDI-TOF-MS prove that the two calix[4]arenes in **19** are mechanically interlocked. Addition of increasing amounts of THF-d_8 to a solution of **19** in benzene-d_6 leads to an upfield shift for the signals of the most strongly hydrogen bonded NH-groups but, even in pure THF-d_8, the values for the free calix[4]arenes are not reached. The observation of two signals for the aromatic protons of each calixarene skeleton which nearly coalesce at 50°C indicates that the rotation of the urea functions and the change of the directionality of the hydrogen bonded belt are hindered in rotaxanes **19**.

As indicated in Fig. 18, similar tetra[2]rotaxanes should be available also by clipping procedures. This requires the selective (or at least predominant) formation of heterodimers between an octaalkenyl tetraurea and a tetraurea with bulky residues. Attempts with the tetraurea pair **1e**(X = *t*-Bu)/**1h** led to

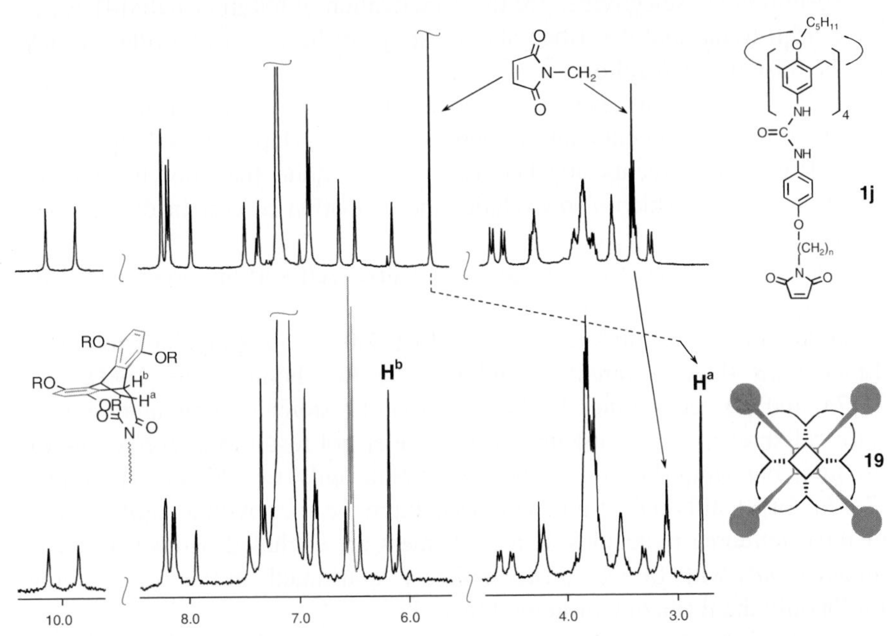

Figure 2-21. Main spectral changes during the synthesis of a tetra[2]rotaxane via "stoppering" by Diels-Alder cycloaddition; ^{1}H NMR spectra in benzene-d_8.

only 5% of the desired rotaxane, after difficult work up.[49] The obvious explanation is that the homodimerisation of the bulky **1e** is not sufficiently disfavored to shift the dimerisation equilibria sufficiently towards the heterodimer.

As discussed above, tetratosylureas **1b** form heterodimers exclusively with tetraaryl ureas, and this can be used to form tetraloop compounds. For n ≥ 10 these tetraloop compounds were easily detached from the template by hydrogen bond breaking solvents, *e.g.* THF. For n = 8 we faced for the first time difficulties for this splitting and for shorter chains the two calixarenes could not be separated,[50] although an exchange of the included guest molecule (ethylacetate against THF-d_8) is still possible.

4. OUTLOOK

4.1 Self-Sorting

Several novel selectivities for the dimerization of tetraurea calix[4]arenes have been found and described above. They are based on two rules, which can be formulated as follows:

a) If adjacent urea functions are covalently connected/linked by a longer chain, dimerisation occurs only if there is not an overlap of these loops.

b) Bulky substituents attached to a urea residue may not be able to penetrate a loop, which also excludes the formation of certain dimers (see above).

These restrictions can be used to realize self-sorting processes,[51] as explained by a simple example in Fig. 22.

A stoichiometric mixture of a monoloop **7a** and an open chain tetraurea **1a** contains three different assemblies, the two homodimers **1a•1a** and **7a•7a**, and the heterodimer **1a•7a**. Addition of a stoichiometric amount of a third calixarene, the bisloop derivative **8**, does not result in a more complex mixture, but simplifies its composition. Now only two dimers are found (**7a•7a**, **1a•8**), although in a general case there are six possible combinations of three tetraurea molecules, if regioisomers are excluded. However, **8** can dimerise only with **1a**, which is completely "consumed" in this way, leaving for **7a** only the formation of homodimers.[42]

The tendency to have all urea functions "saturated" by hydrogen bonds thus leads to the formation of only two of the four possible dimers **1a•1a**, **7a•7a**, **1a•7a**, and **1a•8**.

4.2 Dendrimers

In principle, such self-sorting processes should enable the construction of well defined dendritic assemblies from dendrons consisting of covalently connected multiple tetraurea molecules. The first example realised was based also on the dimerisation of triphenylmethane based triureas, which is not disturbed by tetraureas and vice versa (see Fig. 23).[52]

Covalent connection of triurea **21** with three tetraureas **22** via the narrow rim leads to the building block **A**. This molecule can form dimeric structures via the tri- and the tetraurea parts, which could lead to more or less branched/cross linked assemblies. Addition of the required amount of a tetratosyl urea **23 (=B)** prevents this "polymerisation" by the formation of heterodimers.

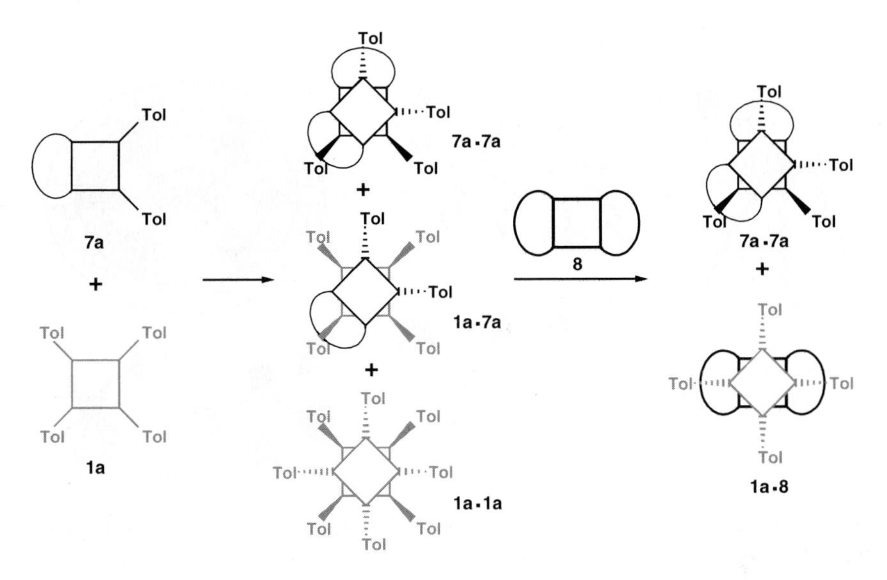

Figure 2-22. Self-sorting of open chain (**1a**), monoloop (**7a**) and bisloop (**8**) tetraureas into two dimers **7a•7a** and **1a•8**.

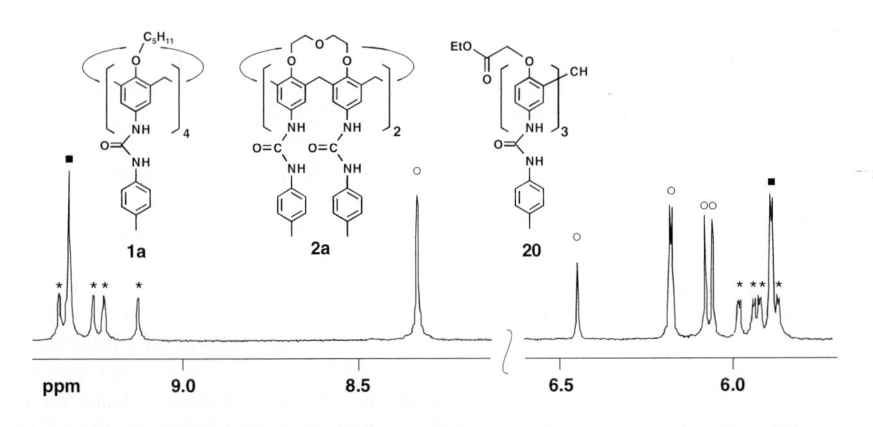

Figure 2-23. Independent dimerisation of tri- and tetraureas. A mixture of **1a**, **2a** and **20** shows ***only*** the signals for the three homodimers **1a•1a** (■), **2a•2a** (∗) and **20•20** (○).

Thus, as illustrated in Fig. 24, 2 + 6 building blocks (+ 6 guest molecules) form a structurally uniform dendrimer with a molar mass of >25 000 g/mol, bearing 24 functional groups (here the phthalimido groups) in its outer sphere. The correctness of the assembly was confirmed by ^{1}H NMR, including DOSY spectra, and by dynamic light scattering.[53]

We are presently trying to extend these studies and to build up larger structurally uniform dendrimers using various combinations of the selectively formed dimeric assemblies described in this essay.

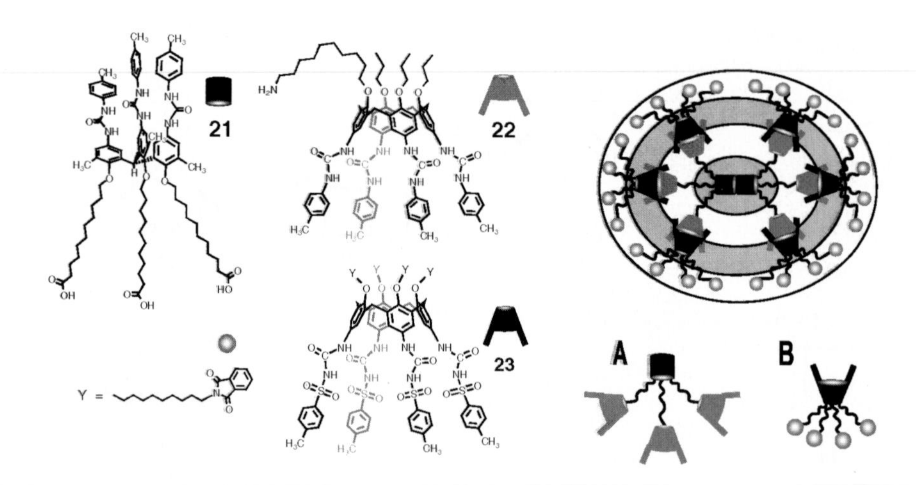

Figure 2-24. Structurally uniform dendrimers via self-sorting self-assembly of 2 + 6 building blocks **A** and **B**.

5. REFERENCES

1. K. D. Shimizu and J. Rebek, Jr., *Proc. Natl. Acad. Sci. USA* **92**, 12403-12407 (1995); B. C. Hamann, K. D. Shimizu and J. Rebek, Jr., *Angew. Chem. Int. Ed.* **35**, 1326-1329 (1996).
2. S. Mecozzi and J. Rebek, Jr., *Chem. Eur. J.* **4**, 1016-1022 (1998).
3. O. Mogck, V. Böhmer and W. Vogt, *Tetrahedron* **52**, 8489-8496 (1996).
4. O. Mogck, E. F. Paulus, V. Böhmer, I. Thondorf and W. Vogt, *Chem. Commun.*, 2533-2534 (1996).
5. J. M. C. A. Kerckhoffs, F. W. B. van Leeuwen, A. L. Spek, H. Kooijman, M. Crego-Calama and D. N. Reinhoudt, *Angew. Chem. Int. Ed.* **42**, 5717-5722 (2003).
6. For some examples of cylindrical capsules based on cavitands and their complexes, see: T. Heinz, D. M. Rudkevich and J. Rebek, Jr, *Angew. Chem., Int. Ed.* 1999, **38**, 1136-1139 (1999); A. Scarso, A. Shivanyuk, J. Rebek, Jr., *J. Am. Chem. Soc.* **125**, 13981-13983 (2003).
7. For hexameric capsules, see: L. R. MacGillivray and J. L. Atwood, *Nature* **389**, 469-472 (1997); T. Gerkensmeier, W. Iwanek, C. Agena, R. Fröhlich, S. Kotila, C. Näther and L. Mattay, *Eur. J. Org. Chem.* 2257-2262 (1999); A. Shivanyuk and J. Rebek, Jr., *Proc. Natl. Acad. Sci. USA* **98**, 7662-7665 (2001).
8. R. K. Castellano, D. M. Rudkevich and J. Rebek Jr., *Proc. Natl. Acad. Sci. USA* **94**, 7132-7137 (1997); R. K. Castellano and J. Rebek, Jr., *J. Am. Chem. Soc.* **120**, 3657-3663 (1998).
9. A. Pop, M. O. Vysotsky, M. Saadioui and V. Böhmer, *Chem. Commun.*, 1124-1125 (2003).
10. This difference may be due to different ether residues Y, or different urea residues R, or both.

11. R. K. Castellano, S. L. Craig, C. Nuckolls and J. Rebek, Jr., *J. Am. Chem. Soc.* **122**, 7876-7882 (2000).
12. M. O. Vysotsky, O. Mogck, Y. Rudzevich, A. Shivanyuk, V. Böhmer, M. S. Brody, Y. L. Cho, D. M. Rudkevich and J. Rebek, Jr., *J. Org. Chem.* **69**, 6115-6120 (2004).
13. This topic will not be treated in detail in this article.
14. R. K. Castellano, B. H. Kim and J. Rebek, Jr., *J. Am. Chem. Soc.* **119**, 12671-12672 (1997).
15. The exclusive formation of heterodimers is also observed for aryl phosphoryl instead of aryl sulfonyl ureas. V. Brusko and V. Böhmer, unpublished.
16. A. Arduini, M. Fabbi, M. Mantovani, L. Mirone, A. Pochini, A. Secchi and R. Ungaro, *J. Org. Chem.* **60**, 1454-1457 (1995).
17. Y. Rudzevich, M. O. Vysotsky, V. Böhmer, M. S. Brody, J. Rebek, Jr., F. Broda and I. Thondorf, *Org. Biomol. Chem.* **2**, 3080-3084 (2004).
18. R. K. Castellano, C. Nuckolls and J. Rebek, Jr., *J. Am. Chem. Soc.* **121**, 11156-11163 (1999).
19. O. Mogck, M. Pons, V. Böhmer, and W. Vogt, *J. Am. Chem. Soc.* **119**, 5706-5712 (1997).
20. M. O. Vysotsky and V. Böhmer, unpublished.
21. M. O. Vysotsky and V. Böhmer, *Org. Lett.* **2**, 3571-3574 (2000).
22. Rigidification of the calix[4]arene skeleton by two short crownether bridges also increases the kinetic stability of the capsules by a factor 30-38 for the exchange of benzene or cyclohexane against benzene-d6.[12]
23. M. O. Vysotsky, I. Thondorf and V. Böhmer, *Angew. Chem., Int. Ed.* **39**, 1264-1267 (2000).
24. M. O. Vysotsky, I. Thondorf and V. Böhmer, *Chem. Commun.*, 1890-1891 (2001).
25. The half-lifetime is much lower in THF, showing that the complete explanation is not quite as straightforward; M. O. Vysotsky and V. Böhmer, unpublished.
26. C. A. Schalley, R. K. Castellano, M. S. Brody, D. M. Rudkevich, G. Siuzdak and J. Rebek, Jr., *J. Am. Chem. Soc.* **121**, 4568-4579 (1999).
27. L. Frish, M. O. Vysotsky, V. Böhmer and Y. Cohen, *Org. Biomol. Chem.* **1**, 2011-2014 (2003).
28. The expressions intra- and intermolecular are used here for the whole dimeric capsule.
29. M. O. Vysotsky, A. Pop, F. Broda, I. Thondorf and V. Böhmer, *Chem. Eur. J.* **7**, 4403-4410 (2001).
30. L. Frish, M. O. Vysotsky, S. E. Matthews, V. Böhmer and Y. Cohen, *J. Chem. Soc., Perkin Trans.* **2**, 88-93 (2002).
31. I. Thondorf, F. Broda, K. Rissanen, M. O. Vysotsky and V. Böhmer, *J. Chem. Soc., Perkin Trans.* **2**, 1796-1800 (2002).
32. M. O. Vysotsky, J. Lacour and V. Böhmer, unpublished.
33. A. Shivanyuk, M. Saadioui, F. Broda, I. Thondorf, M. O. Vysotsky, K. Rissanen, E. Kolehmainen and V. Böhmer, *Chem. Eur. J.* **10**, 2138-2148 (2004).
34. M. S. Wendland, S. C. Zimmerman, *J. Am. Chem. Soc.* **121**, 1389-1390 (1999); S. C. Zimmerman, M. S. Wendland, N. A. Rakow, I. Zharov, K. S. Suslick, *Nature* **418**, 399-403 (2002).
35. L. J. Prins, K. A. Jolliffe, R. Hulst, P. Timmerman, D. N. Reinhoudt, *J. Am. Chem. Soc.* **122**, 3617-3627 (2000); J. M. C. A. Kerckhoffs, M. G. J. ten Cate, M. A. Mateos-Timoneda, F. W. B. van Leeuwen, B. Snellink-Ruel, A. L. Spek, H. Kooijman, M. Crego-Calama, D. N. Reinhoudt, *J. Am. Chem. Soc.* **127**, 12697-12708 (2005).

36. R. H. Grubbs, *Tetrahedron* **60**, 7117-7140 (2004); T. M. Trnka and R. H. Grubbs, *Acc. Chem. Res.* **34**, 18-29 (2001).
37. M. O. Vysotsky, M. Bolte, I. Thondorf and V. Böhmer, *Chem. Eur. J.* **9**, 3375-3382 (2003).
38. M. Saadioui, A. Shivaniuk, V. Böhmer and W. Vogt, *J. Org. Chem.* **64**, 3774-3777 (1999).
39. A. Bogdan, M. O. Vysotsky, T. Ikai, Y. Okamoto and V. Böhmer, *Chem. Eur. J.* **10**, 3324-3330 (2004).
40. M. Bolte, A. Bogdan, M. O. Vysotsky and V. Böhmer, unpublished.
41. T. Ikai, Y. Okamoto, O. Molokanova, A. Bogdan, M. O. Vysotsky and V. Böhmer, unpublished.
42. A. Bogdan and V. Böhmer, unpublished.
43. J. Blixt and C. Detellier, *J. Am. Chem. Soc.* **116**, 11957-11960 (1994).
44. M. O. Vysotsky, A. Bogdan, L. Wang and V. Böhmer, *Chem. Commun.*, 1268-1269 (2004).
45. Y. Cao, M. O. Vysotsky and V. Böhmer, unpublished.
46. Y. Cao, L. Wang, M. Bolte, M. O. Vysotsky and V. Böhmer, *Chem. Commun.*, 3132-3134 (2005).
47. L. Wang, M. O. Vysotsky, A. Bogdan, M. Bolte and V. Böhmer, *Science* **304**, 1312-1314 (2004).
48. C. Gaeta, M. O. Vysotsky, A. Bogdan and V. Böhmer, *J. Am. Chem. Soc.* **127**, 13136-13137 (2005).
49. M. O. Vysotsky and V. Böhmer, unpublished.
50. O. Molokanova, M. O. Vysotsky, Y. Cao, I. Thondorf and V. Böhmer, unpublished.
51. A. X. Wu, L. Isaacs, *J. Am. Chem. Soc.* **125**, 4831-4835 (2003); P. Mukhopadhyay, A. X. Wu, L. Isaacs, *J. Org. Chem.* **69**, 6157-6164 (2005).
52. Y. Rudzevich, V. Rudzevich, D. Schollmeyer, I. Thondorf and V. Böhmer, *Chem. Lett.* **7**, 613-616 (2005).
53. Y. Rudzevich, V. Rudzevich, C. Moon, I. Schnell, K. Fischer and V. Böhmer, *J. Am. Chem. Soc.* **127**, 14168-14169 (2005).

Chapter 3

CALIX-ROTAXANES AND -CATENANES

Interlocking with cavities

Zhan-Ting Li, Xin Zhao, and Xue-Bin Shao
State Key Laboratory of Bioorganic and Natural Products Chemistry, Shanghai Institute of Organic Chemistry, Chinese Academy of Sciences, 354 Fenglin Lu, Shanghai 200032, China.
E-mail: ztli@mail.sioc.ac.cn

Abstract: The self-assembly of rotaxanes and catenanes incorporating calix[4]arene, calix[6]arene and structurally related homooxacalix[3]arene units is reviewed. The function of the calixarene moiety in new supramolecular architectures is briefly described.

Key words: Calixarenes, catenanes, rotaxanes, self-assembly.

1. INTRODUCTION

Rotaxanes and catenanes are dynamic, interlocked supramolecular systems (Fig. 1).[1,2] In early statistical and multistep syntheses of such types of molecular architectures, they had been considered as laboratory curiosities[3,4] but the introduction of template syntheses of catenanes and subsequently also of rotaxanes by Sauvage *et al.*[5] in 1984 and by Stoddart *et al.*[6] in 1989 opened a new era. In the past two decades, the construction of these two classes of interlocked molecular architectures has received increasing attention not only because of their topological beauty but also because of their potential applications in molecular electronics as molecular switches, motors, machines and related devices.[7,8] In recent years, the chemistry of rotaxanes and catenanes has extended to the field of calixarene chemistry.[9-12] In these new interlocked assemblies, calixarene moieties not only are versatile rigid skeletons but also provide a unique kind of three-dimensional platform for stimulating the imagination of supramolecular chemists.

J. Vicens and J. Harrowfield (eds.), Calixarenes in the Nanoworld, 47–62.

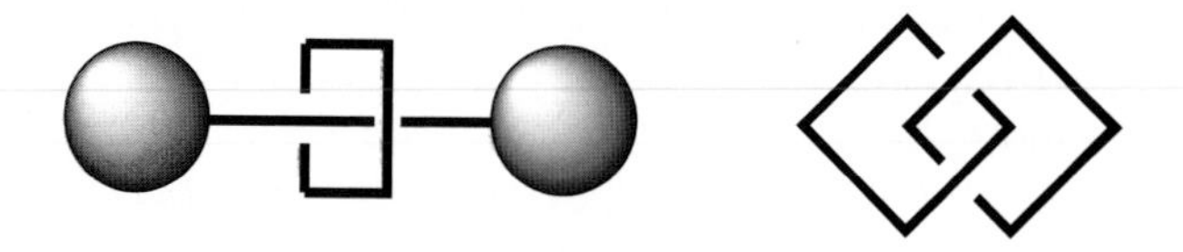

Figure 3-1. General structures of [2]rotaxanes and [2]catenanes.

2. [2]ROTAXANES WITH CALIX[4]ARENE STOPPERS

A [2]rotaxane is composed of a cyclic molecule threaded by a linear molecule. To retain this threading, the latter must bear "stoppers" on its ends which are too large to pass through the cavity of the cyclic component. Various rigid porphyrin units, for example, have been extensively used as stoppers of rotaxanes.[13] In 1998, Vögtle et al. described the [2]rotaxanes **4**, the stoppers of which are two calix[4]arene units (Scheme 1).[14]

The driving force for the formation of the [2]rotaxane is intermolecular hydrogen bonding between the amide oxygen and hydrogen atoms of the two components.[15] In addition to [2]rotaxane **4**, the reaction also afforded dumbbell **5** in 55% yield. No [2]rotaxanes could be produced when compound **1** was replaced with **6a** or **6b**, obviously because shorter spatial separation between the bulky calix[4]arene unit and the first amide unit formed prevented the intermolecular hydrogen bonding and consequently the formation of the pseudo[2]rotaxane intermediate. Surprisingly, the reaction of **2** with **7**, which is smaller in size than **1**, in the presence of **3** under identical conditions resulted in the formation of the corresponding dumbbell molecule **5** in 68% yield with only trace of [2]rotaxane similar to **4**. This result has been attributed to the flexibility of **7** due to the introduction of the methylene group. 3-Chlorosulfonylbenzoyl chloride **8** has been found to be an excellent precursor to a number of non-calixarene [2]rotaxanes[16] but no [2]rotaxane product was detected from its reaction with **1** in the presence of **3**, perhaps because of steric effects.

More recently, Smithrud *et al.* reported the construction of [2]rotaxanes **11** and **12** (Scheme 2).[17] The self-assembly of the unsymmetrical [2]rotaxane was achieved by making use of the complexation between dibenzo-24-crown-8 and dialkylammonium, previously extensively studied by Busch[18] and Stoddart *et al.*[19]. The calix[4]arene unit in [2]rotaxane **12** acted not only as a stopper but also as a hydrophobic recognition domain, which, together with the two ionic arginine units attached to the dibenzo-24-crown-8 moiety, was designed to bind a variety of small guest species. ^{1}H NMR and fluorescence measurements revealed that both the calix[4]arene hydrophobic pocket and the ring of [2]rotaxane **12** contributed to complexation for a number of guests, including *N*-Ac-Trp, indole, *N*-Ac-Gly, fluorescein, pyrene and 1-(dimethylamino)-5-naphthalenesulfonate in DMSO or DMSO-H_2O mixtures. In contrast, covalently bonded host **13** exhibited substantially weaker binding affinity, presumably due to poor matching of the three binding sites.

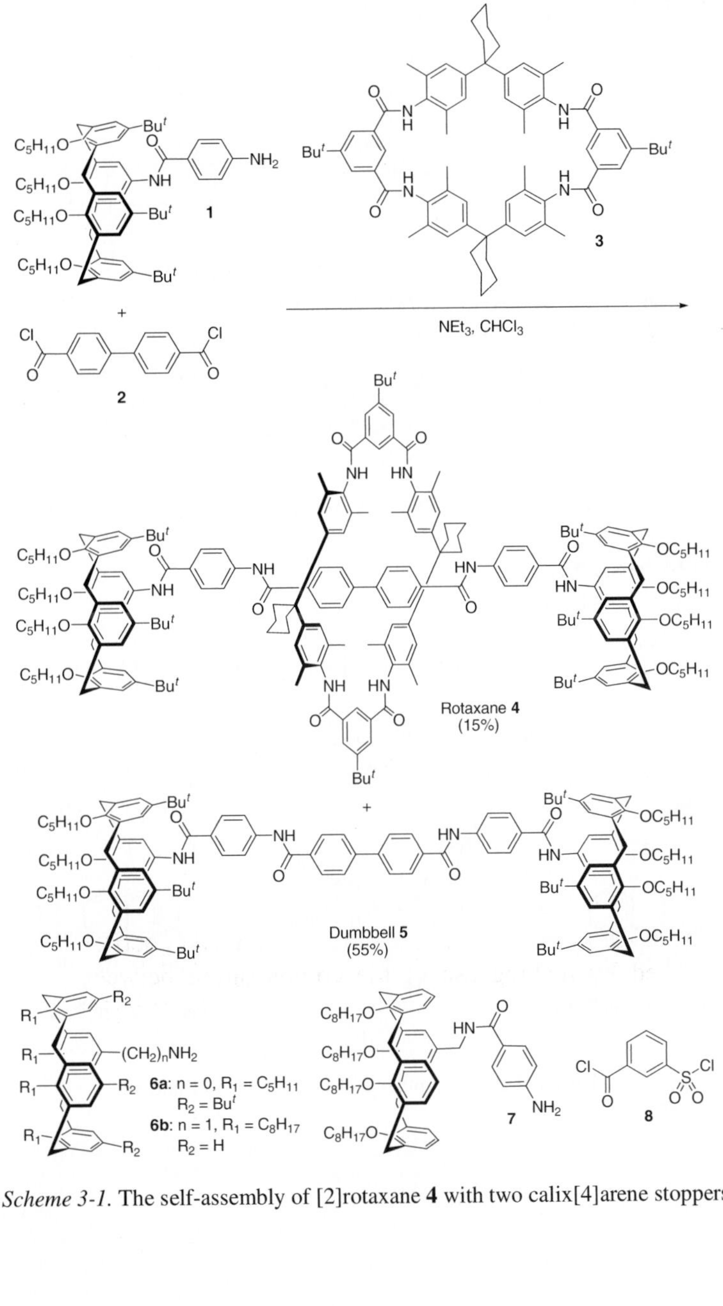

Scheme 3-1. The self-assembly of [2]rotaxane **4** with two calix[4]arene stoppers.

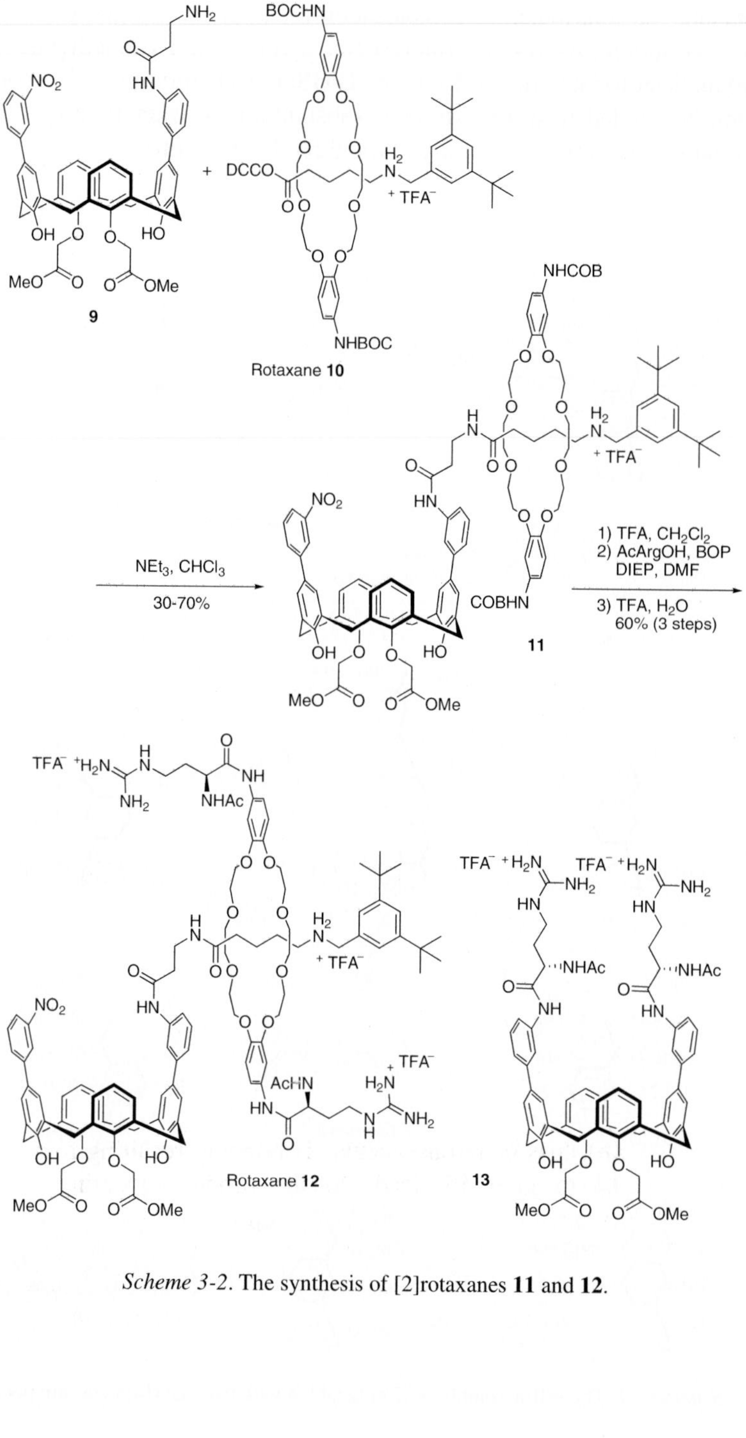

Scheme 3-2. The synthesis of [2]rotaxanes **11** and **12**.

3. SELF-ANCHORED [2]ROTAXANES

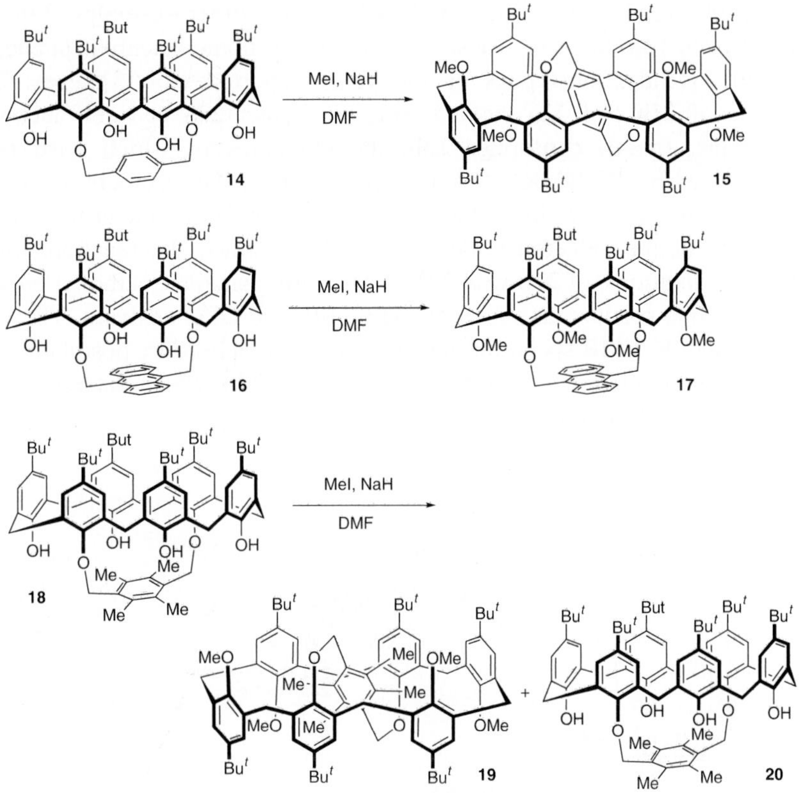

Scheme 3-3. The methylation of 1,4-bridged calix[6]arenes affords products with different conformations, depending on the size of the bridge. Compounds **15** and **19** represent examples of self-anchored [2]rotaxanes.

For most rotaxanes, the stoppers are covalently connected to the linear component only. However, some calix[6]arene derivatives can adopt a special conformation in which a bridging moiety threads the cavity of the calix[6]arene skeleton. Thus, Gutsche *et al.*[20,21] found that although all three 1,4-bridged calix[6]arene derivatives **14**, **16** and **18** adopt a cone conformation (u,u,u,u,u,u), loss of intramolecular H-bonding resulting from tetramethylation of **14** to give **15**, leads to the product adopting the the (u,u,d,d,d,u) conformation (Scheme 3). The location of the phenylene bridge in the cavity of the calix[4]arene and the resulting (u,u,d,d,d,u) conformation were established by ^1H NMR studies.[22] In contrast, tetramethylation of **16** only led to the formation of the cone product **17** due to the increased size of the anthrylene moiety, whereas a mixture of conformational isomers **19** and **20** were obtained from the reaction of **18**, containing a durylene bridge, the size of which is between phenylene and anthrylene. Compounds **15** and **19** may be regarded as self-anchored [2]rotaxanes.[23,24]

In 1999, Shinkai *et al.*[25] reported (Scheme 4) further examples, **23b** and **24b,** of such [2]rotaxanes. These were based on homooxacalix[3]arenes, a class of extended analogues of the yet unreported calix[3]arenes. The coupling of **21a** and **22a** yielded cone conformer **23a** and self-anchored [2]rotaxane, (u,u,d) conformer **23b**, the structures of which have been characterized by ^{1}H NMR and mass spectroscopy. The isomers are stable and cannot interconvert due to the presence of the bulky amide groups on the lower rim of the calixarene moieties.[26,27] In contrast, the reaction, under identical conditions, of **21b** with **22b**, both bearing smaller methoxyl groups, gave only one product, as shown by TLC, MS and GPC analyses. The very complicated ^{1}H NMR spectrum in $CDCl_3$ was rationalised by postulating the presence of a mixture of conformers **24a** and **24b** which interconverted slowly on the NMR time scale.

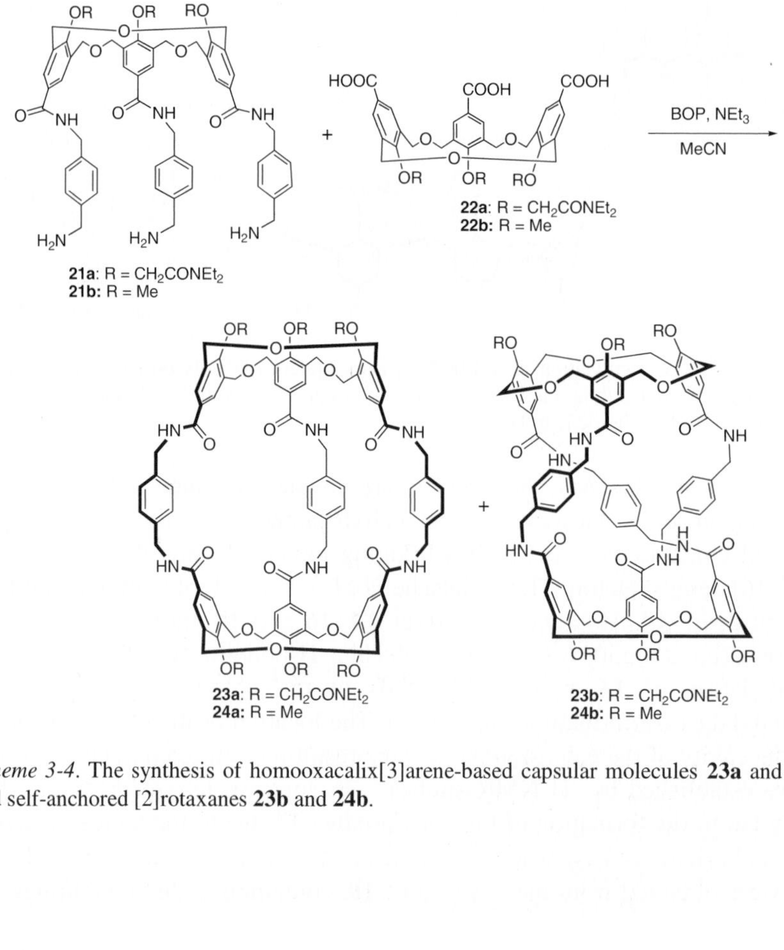

Scheme 3-4. The synthesis of homooxacalix[3]arene-based capsular molecules **23a** and **24a** and self-anchored [2]rotaxanes **23b** and **24b**.

4. [2]ROTAXANES WITH CALIX[6]ARENE AS A WHEEL

The calix[6]arene platform possesses an annulus large enough to allow a linear guest molecule to thread and therefore can be potentially used as a molecular wheel for the construction of [2]rotaxanes. In 2000, Pochini *et al.* reported that calix[6]arene derivative **25**, which was preorganized by 1,3,5-trimethoxy-2,4,6-trioctyloxy groups on the lower rim and three phenylureido groups on the upper rim, was able to strongly complex dioctylviologen ditosylate **26a** or diiodide **26b** in chloroform and acetonitrile, as evidenced by ^{1}H NMR spectroscopy and the X-ray structure of **27b** (Scheme 5).[28] The formation of complexes **27a** or **27b** had been attributed to the intermolecular hydrogen bonding between the ureido protons of **25** and the anions of the linear molecules. However, as suggested later by the authors,[29] the donor-acceptor interaction between the dipyridinium unit and the calix[6]arene unit might also play an important role because the solution of the 1:1 complexes was deep purple, indicating the formation of a charge transfer complex.[30]

Pseudo[2]rotaxanes, similar to **27**, could also be formed from **25** and **28** under identical conditions. Acylation of the two hydroxyl groups of **28** of the pseudo[2]rotaxane with diphenylacetyl chloride **29** in chloroform resulted in the formation of [2]rotaxane **30** in 25% yield (Scheme 5).[28] The long aliphatic chains in **28** provided it with a good solubility in less polar solvents and also reduced the possible steric hindrance during the formation of the [2]rotaxane.

Calix[6]arenes in a cone conformation resemble cyclodextrins[31] in that they may be threaded by an unsymmetrical linear molecule in two senses, giving rise to two constitutional isomers of pseudo[2]rotaxanes. In 2005, Arduini *et al.* reported that calix[6]arene-based constitutional pseudo[2] rotaxanes of such kind could be obtained by tuning the structure of the linear component (Scheme 6).[32] This was based on the observation that, when a mixture of **31** and **32a** or **32b** was heated in toluene, stable pseudo[2] rotaxanes **33a** and **33b** could be produced exclusively.[29] Treatment of **33a** or **33b** with an excess of acyl chloride afforded the corresponding [2]rotaxane isomers **34a** or **34b**, exclusively. The orientation of the linear component within the pseudo[2]rotaxanes and [2]rotaxanes was established by ^{1}H NMR spectroscopy. It was proposed that **32a** and **32b** were present as contact ion pairs during the initial formation of the pseudo[2]rotaxanes and therefore could not fit into the calix[6]arene cavity of **31**. The hydrogen bonding of the ureido NH groups on the upper rim of the calix[6]arene towards the countered anion of the linear molecules pivoted the linear portion of the axle to thread through the cavity of **31** exclusively from the upper rim. The pseudo[2]rotaxane isomers are kinetically stable, as a result, the OH group of the linear components protruded from the lower rim of the wheel, leading to the formation of [2]rotaxanes **34a** and **34b** exclusively. Because the X-ray analysis revealed that the viologen unit of the linear component was actually located within the three-benzene units and the calix[6]arene cavity of **31**,[29] steric repellency between the bulky ester stopper and the calix[6]arene and the longer chains on the lower rim might also play a role for the selective formation of the pseudo[2]rotaxane.

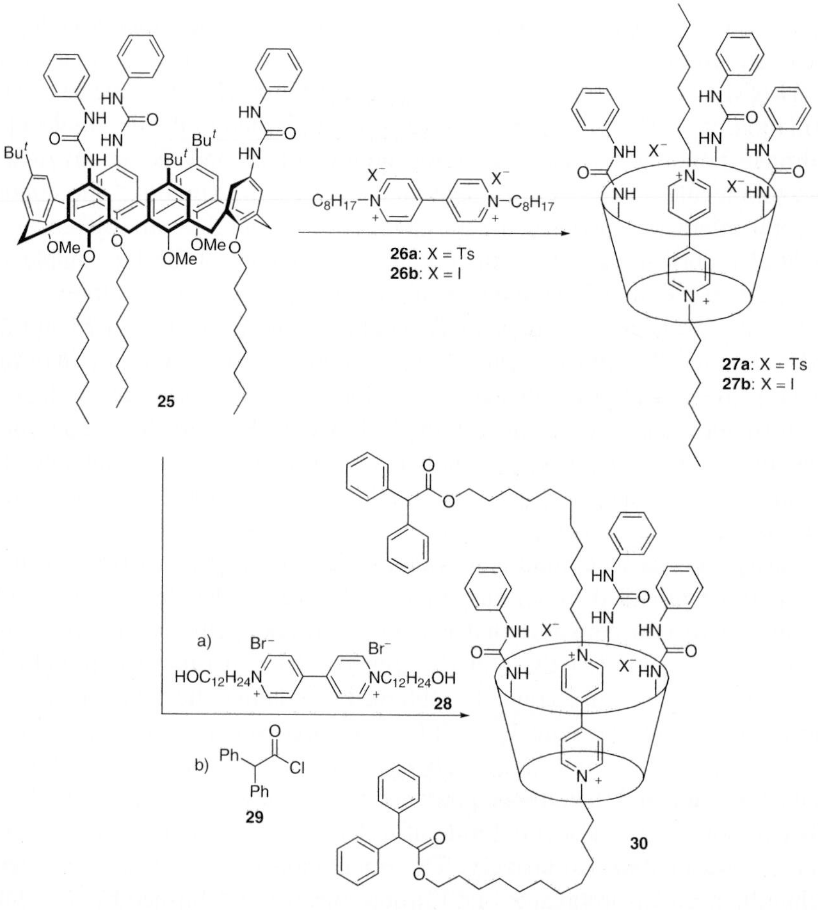

Scheme 3-5. The formation of pseudo[2]rotaxanes **27** and [2]rotaxane **30** with calix[6]arene as a cyclic component.

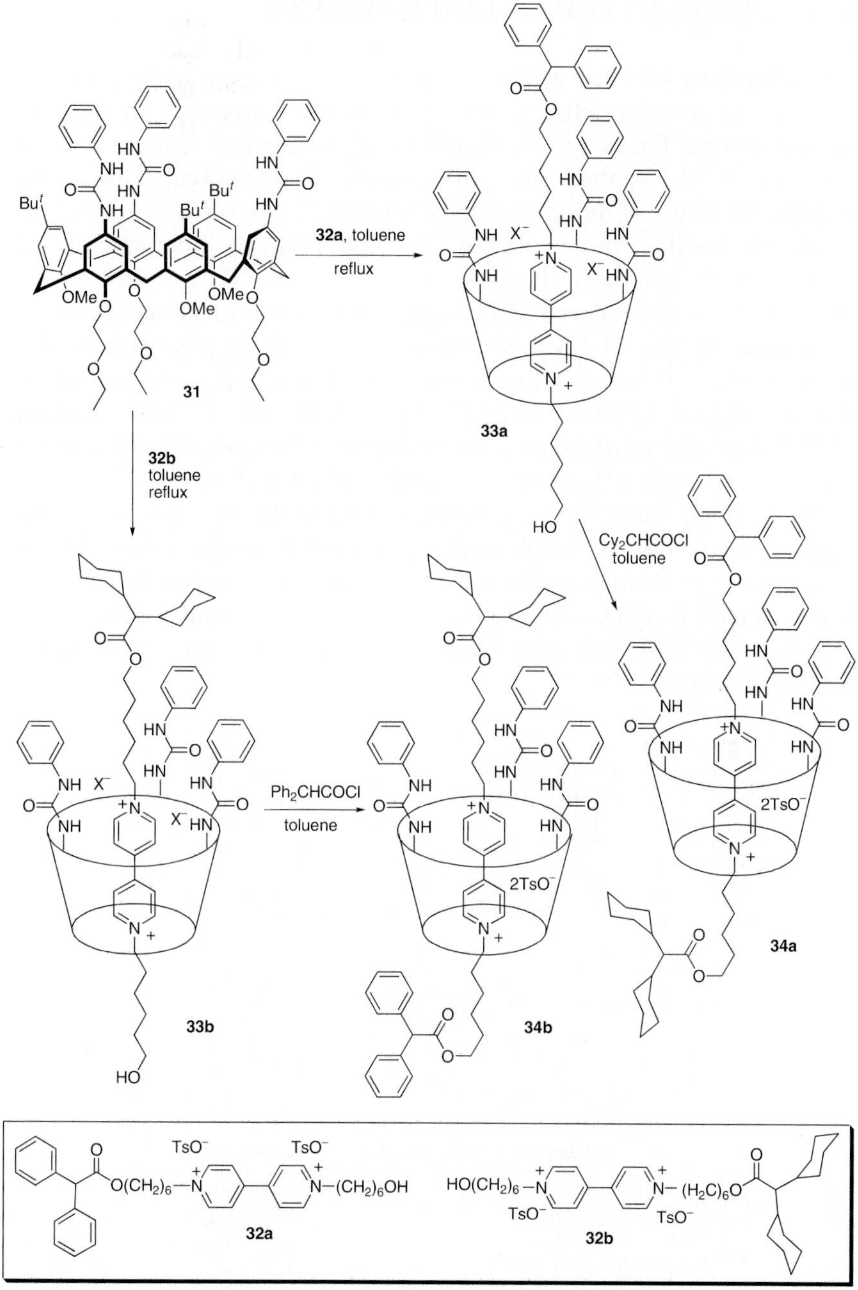

Scheme 3-6. Self-assembly of constitutionally isomeric calix[6]arene [2]rotaxanes **34a** and **34b**.

5. CATENATED CALIX[4]ARENES

π-π Stacking between electron rich dioxybenzene or dioxynaphthalene units and an electron deficient 4,4'-dipyridinium unit is one of the most efficient driving forces for the formation of interlocked structures.[33,34] In 1998, Li *et al.* reported the self-assembly of calix[4]arene-based [2] catenanes by virtue of this approach (Scheme 7).[35] Two bipyridyl units were introduced to the lower rim of the calix[4]arene platform of **35**. Intramolecular hydrogen bonding of the hydroxyl groups of the calix[4]arene moiety induced **35a** to adopt a cone conformation exclusively, as indicated by the ^{1}H NMR spectrum. As a result, a preorganized cleft precursor was generated, which greatly facilitated the formation of the corresponding calix[4]arene-based [2]catenanes **38**. The ^{1}H NMR spectrum in CD_3CN indicated that the cone conformation of the calix[4]arene unit was retained in all three [2]catenanes, a result rationalized by considering the flexibility of the calix[4]arene cyclophane due to the presence of the two aliphatic chains. The ^{1}H NMR spectrum in CD_3CN revealed that **35b** was conformationally flexible. Under the same conditions as for the formation of **38**, no catenane products could be obtained from the reaction of **35b** with **36** in the presence of **37**, indicating that the intramolecular hydrogen bonding in **35** was essential.

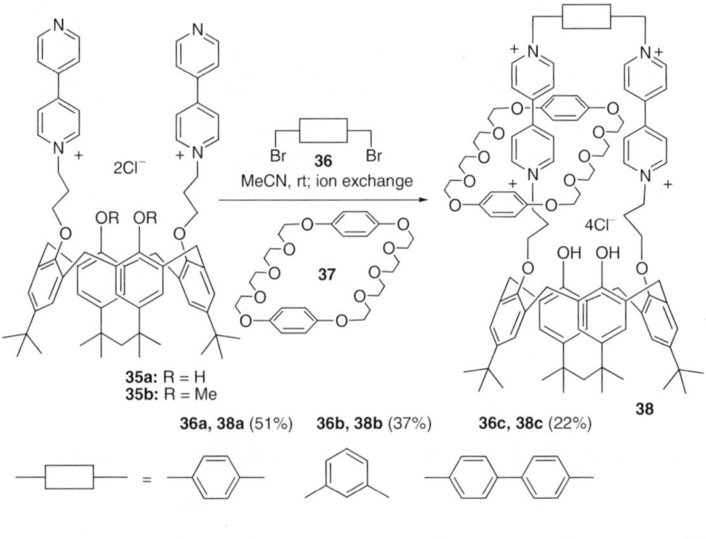

Scheme 3-7. The self-assembly of calix[4]arene-incorporated [2]catenanes **38**.

Changing the length of the aliphatic chains connecting the calix[4]arene unit and the dipyridyl units has an important effect on the conformation of the calix[4]arene unit in the [2]catenanes (Scheme 8).[36] For example, treatment of **39**, bearing two ethylene chains, with **36a** and **37** in acetonitrile afforded the 1,3-alternate [2]catenane **40**, but not the cone [2]catenane similar to **38**, whereas the reaction of **41**, bearing two butylene chains, with **36a** in the presence of **37** yielded both cone [2]catenane **42a** and partial cone isomer **42b**. The conformation of the [2]catenanes could be determined conveniently by their ^{1}H and ^{13}C NMR spectra.[37] The dynamic processes associated with the calix[4]arene [2]catenanes were investigated by ^{1}H NMR spectroscopy, which revealed that introduction of the calix[4]arene unit to the [2]catenane skeleton generally reduced the donor-acceptor interaction between the hydroquinone units and the viologen units due to the increased flexibility of the acceptor cyclophane. The fact that [2]catenanes incorporating calix[4]arene units of different conformation could be generated from the reactions of different precursors suggests that catenation represents a useful method to tune the conformation of the calix[4]arene platform, which may find further applications in supramolecular self-assembly in the future.

Macrocyclic crown ethers **43a** and **43b**, in which a calix[4]arene platform is incorporated on the lower rim, could also be used to template the formation of [2]catenanes,[38] as shown in Scheme 9. The ^{1}H NMR spectrum in CD_3CN showed that both **43** and **45** are in the cone conformation, suggesting the existence of the intramolecular hydrogen bonding on the lower rim in the [2]catenanes. In order to explore the effect of the intramolecular hydrogen bonding on the catenane assembly, **46a** and **46b** were also prepared. As expected, the calix[4]arene moiety in compound **46a** was a mixture of conformational isomers at rapid equilibrium, while the cone conformation of the calix[4]arene moiety in **46b** was kept unchanged.[39] Catenation of both **46a** and **46b** did not cause important change for the conformation of their calix[4]arene moiety. Nevertheless, the yield of the [2]catenanes was pronouncedly reduced compared to that of **45b**. This result shows that the intramolecular hydrogen bonding within the calix[4]arene is crucial for efficient self-assembly of the [2]catenanes.

From the reaction of **35a** and **36a** in acetonitrile in the presence of an excess of **43b**, [2]catenane **48**, incorporating two calix[4]arene moieties, could also be obtained (Scheme 10).[39] Due to the bulk of the calix[4]arene moiety, the circumrotation of the two cyclic components could be prevented. However, the introduction of the two calix[4]arene moieties into both components also led to the decrease of the donor-acceptor interaction due to the increased flexibility of both components.

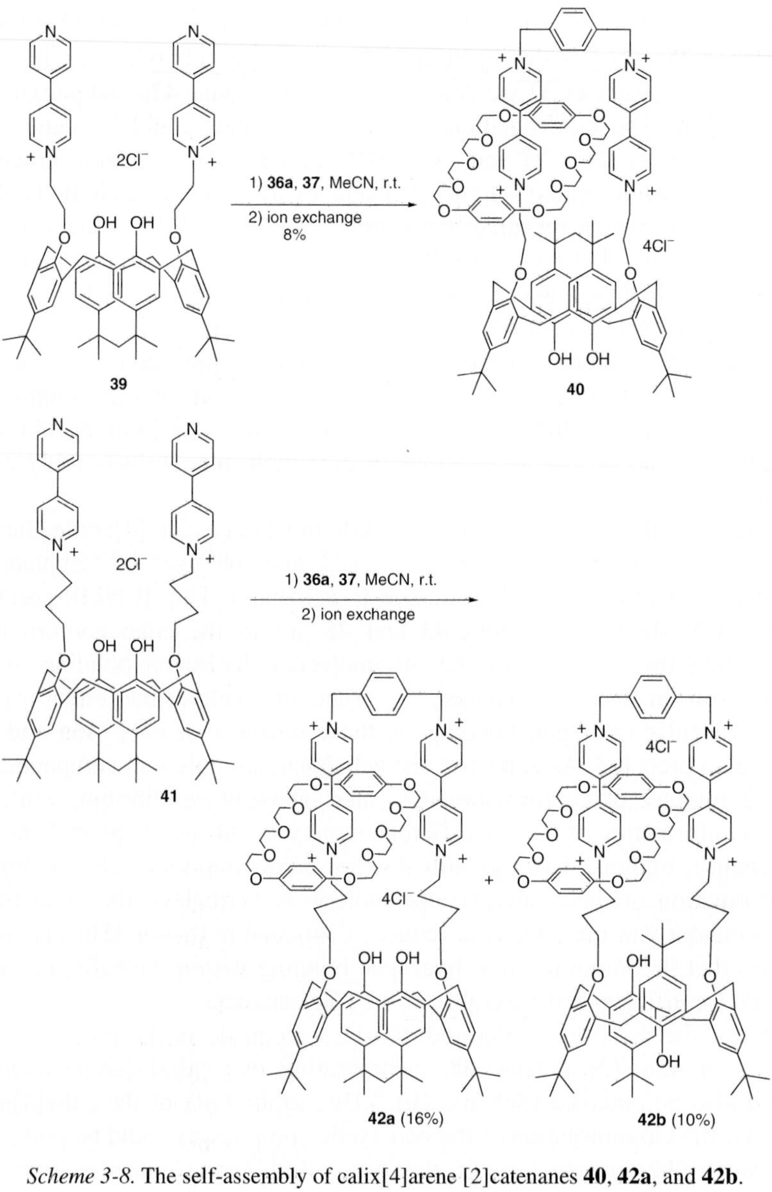

Scheme 3-8. The self-assembly of calix[4]arene [2]catenanes **40**, **42a**, and **42b**.

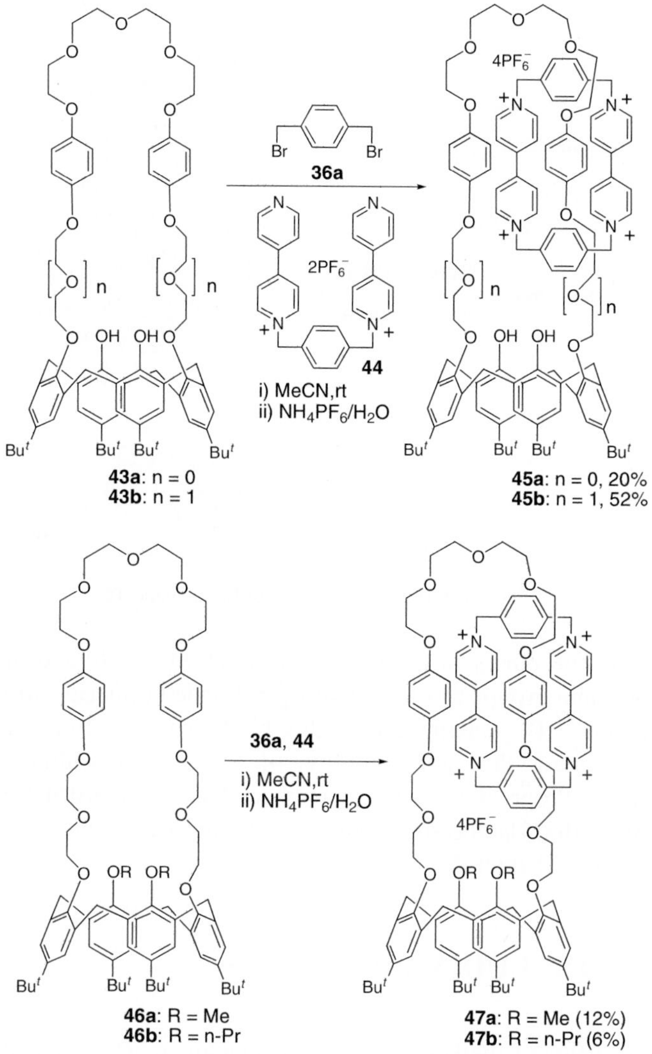

36a

44

$2PF_6^-$

i) MeCN,rt
ii) NH_4PF_6/H_2O

$4PF_6^-$

43a: n = 0
43b: n = 1

45a: n = 0, 20%
45b: n = 1, 52%

36a, **44**

i) MeCN,rt
ii) NH_4PF_6/H_2O

46a: R = Me
46b: R = n-Pr

47a: R = Me (12%)
47b: R = n-Pr (6%)

Scheme 3-9. The formation of [2]catenanes **45** and **47**.

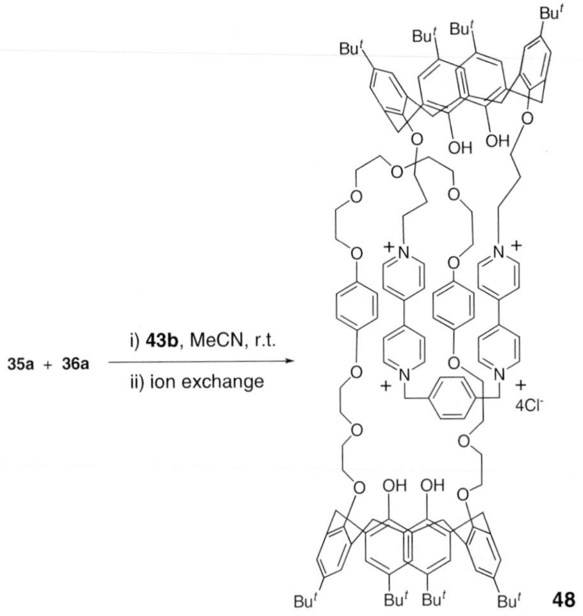

35a + 36a → i) 43b, MeCN, r.t. ii) ion exchange → 48

Scheme 3-10. The formation of [2]catenane **48**.

Calix[4]arene ethers in the cone conformation substituted at the upper rim by four urea groups can form hydrogen bonded dimeric capsules in less polar solvents.[40] By making use of this binding mode, Böhmer *et al.* has recently reported the self-assembly of several structurally elaborate catenands.[41,42] Progress in the line may lead to the construction of more complicated calix[4]arene supramolecular architectures with new molecular topologies (See Chapter 2).

6. CONCLUDING REMARKS

Several efficient methods have been developed which enable the self-assembly of a variety of structurally interesting [2]rotaxanes and [2] catenanes incorporating one or two calix[4]arene or calix[6]arene moieties. The function of the calixarene platform in the interlocked supramolecular architectures obviously needs further exploration in the future. Considering the unique structural features and well-established methods of modification of both lower and upper rims, a variety of functional units can be relatively easily introduced to the calixarene moiety, which may be used to regulate the shuttling behavior of the calixarene-incorporated interlocked systems and lead to the construction of new controllable supramolecular devices. The three-dimensional structural feature of the calix[4]arene or calix[6]arene moieties also make them ideal precursors for constructing complicated, topologically interesting architectures.

ACKNOWLEDGEMENT

We thank the National Natural Science Foundation and the Ministry of Science and Technology of China for financial support of our work.

7. REFERENCES

1. G. Schill, *Catenanes, Rotaxanes, and Knots* Academic Press. (1971).
2. J.-P. Sauvage, C. Dietrich-Buchecker (Eds.) *Molecular Catenanes, Rotaxanes and Knots: a Journey Through the World of Molecular Topology* Wiley-VCH, Weinheim, (1999).
3. I. T. Harrison, S. Harrison, *J. Am. Chem. Soc.* **89**, 5723 (1967).
4. G. Schill, H. Zollenkopf, *Nachr. Chem. Tech.* **15**, 149 (1967).
5. C. O. Dietrich-Buchecker, J. P. Sauvage, J. M. Kern, *J. Am. Chem. Soc.* **106**, 3043 (1984).
6. P. R. Ashton, T. T. Goodnow, A. E. Kaifer, M. V. Reddington, A. M. Z. Slawin, N. Spencer, J. F. Stoddart, C. Vicent, D. J. Williams, *Angew. Chem. Int. Ed. Engl.* **28**, 1396 (1989).
7. J. Liu, M. Gomez-Kaifer, A. E. Kaifer, *Struct. & Bond.* **99**, 141 (2001).
8. S. J. Rowan, S. J. Cantrill, G. R. L. Cousins, J. K. M. Sanders, J. F. Stoddart, *Angew. Chem. Int. Ed.* **41**, 899 (2002).
9. C. D. Gutsche, *Calixarenes* RSC, Cambridge, 1989.
10. J. Vicens, V. Böhmer, *Calixarenes: a Versatile Class of Macrocyclic Compounds* Kluwer, Dordrecht, 1991.
11. C. D. Gutsche, *Calixarenes Revisited* RSC, Cambridge, 1998.
12. L. Mandolini, R. Ungaro, *Calixarenes in Action* Imperial College Press, London, (2000).
13. M. J. Gunter, *Eur. J. Org. Chem.* 1655 (2004).
14. C. Fischer, M. Nieger, O. Mogck, V. Böhmer, R. Ungaro, F. Vögtle, *Eur. J. Org. Chem.* 155 (1998).
15. F. Vögtle, R. Hoss, *Angew. Chem. Int. Ed. Engl.* **33**, 375 (1994).
16. F. Vögtle, R. Jäger, M. Händel, S. Ottens-Hildebrandt, W. Schmidt, *Synthesis* 353 (1996).
17. I. Smukste, B. E. House, D. B. Smithrud, *J. Org. Chem.* **68**, 2559 (2003).
18. A. G. Kolchinski, D. H. Busch, N. W. Alcock, *J. Chem. Soc. Chem. Commun.* 1289 (1995).
19. P. R. Ashton, P. J. Campbell, E. J. T. Chrystal, P. T. Glink, S. Menzer, D. Philp, N. Spencer, J. F. Stoddart, P. A. Tasker, D. J. Williams, *Angew. Chem. Int. Ed. Engl.* **34**, 1865 (1995).
20. A. Casnati, P. Minari, A. Pochini, R. Ungaro, *J. Chem. Soc. Chem. Commun.* 1413 (1991).
21. P. Neri, S. Pappalardo, *J. Org. Chem.* **58**, 1048 (1993).

22. S. Kanamathareddy, C. D. Gutsche, *J. Am. Chem. Soc.* **115**, 6572 (1993).
23. A. Siepen, A. Zett, F. Vögtle, *Liebigs. Ann.* 757 (1996).
24. A. Casnati, P. Jacopozzi, A. Pachini, F. Ugozzoli, R. Cacciapaglia, L. Mandolini, R. Ungaro, *Tetrahedron* **51**, 591 (1995).
25. Z. Zhong, A. Ikeda, S. Shinkai, *J. Am. Chem. Soc.* **121**, 11906 (1999).
26. K. Araki, K. Inada, H. Otsuka, S. Shinkai, *Tetrahedron* **49**, 9465 (1993).
27. K. Araki, N. Hashimoto, H. Otsuka, S. Shinkai, *J. Org. Chem.* **58**, 5958 (1993).
28. A. Arduini, R. Ferdani, A. Pochini, A. Secchi, F. Ugozzoli, *Angew. Chem. Int. Ed.* **39**, 3453 (2000).
29. A. Arduini, F. Calzavacca, A. Pochini, A. Secchi, *Chem. Eur. J.* **9**, 793 (2003).
30. D. B. Amabilino, F. M. Raymo, J. F. Stoddart, in *Comprehensive Supramolecular Chemistry* Ed.: M. W. Hosseini, J.-P. Sauvage, Vol. 9, 85-130 (1996).
31. B. C. Gibb, *J. Supramol. Chem.* **2**, 123 (2002).
32. A. Arduini, F. Ciesa, M. Fragassi, A. Pochini, A. Secchi, *Angew. Chem. Int. Ed.* **44**, 278 (2005).
33. M. C. T. Fyfe, J. F. Stoddart, *Acc. Chem. Res.* **30**, 393 (1997).
34. F. M. Raymo, J. F. Stoddart, *Chem. Rev.* **99**, 1643 (1999).
35. Z.-T. Li, G.-Z. Ji, S.-D. Yuan, A.-L. Du, H. Ding, M. Wei, *Tetrahedron Lett.* **39**, 6517 (1998).
36. Z.-T. Li, G.-Z. Ji, C.-X. Zhao, S.-D. Yuan, H. Ding, C. Huang, A.-L. Du, M. Wei, *J. Org. Chem.* **64**, 3572 (1999).
37. E. Ghidini, F. Ugozzoli, R. Ungaro, S. Harkema, A. A. El-Fadi, D. N. Reinhoudt, *J. Am. Chem. Soc.* **112**, 6979 (1990).
38. Z.-T. Li, X.-L. Zhang, X.-D. Lian, Y.-H. Yu, Y. Xia, C.-X. Zhao, Z. Chen, Z.-P. Lin, H. Chen, *J. Org. Chem.* **65**, 5136 (2000).
39. K. Iwamoto, K. Araki, S. Shinkai, *J. Org. Chem.* **56**, 4955 (1991).
40. M. M. Conn, J. Rebek, Jr., *Chem. Rev.* **97**, 1647 (1997).
41. M. O. Vysotsky, M. Bolte, I. Thondorf, V. Böhmer, *Chem. Eur. J.* **9**, 3375 (2003).
42. L. Wang, M. O. Vysotsky, A. Bogdan, M. Bolte, V. Böhmer, *Science* **304**, 1312 (2004).

Chapter 4

MOLECULAR MACHINES AND NANODEVICES

An ion-pair recognition approach

Arturo Arduini, Andrea Secchi, and Andrea Pochini

Dipartimento di Chimica Organica e Industriale, Università di Parma, Parco Area delle Scienze 17/A, I-43100 Parma, Italy

Abstract: The parameters that affect the binding efficiency of calix[4]arene derivatives toward quaternary ammonium salts have been investigated in apolar media. The achievements of these studies have been transferred to the construction of prototype of molecular machines belonging to the class of pseudorotaxanes and rotaxanes and to the synthesis of new multivalent calixarene-coated gold nanoparticles that are able to recognize quaternary ammonium cations in apolar and aqueous media.

Key words: Calixarenes, QUATS recognition, ion pairs, oriented rotaxanes, gold nano-particles, multivalency.

1. INTRODUCTION

Molecular self-assembly is a process in which molecules, through noncovalent interactions, spontaneously form ordered aggregates whose structure, in equilibrium conditions, is determined by the chemical information present in the components of the assembly.[1] The focus on self-assembly as a strategy for synthesis has been confined largely to molecules since chemists have been generally concerned with manipulating the structure of matter at the molecular scale. The expanding contact of chemistry with biology and materials science and the direction of technology toward nanometer and micrometer scale structures, however, have begun to broaden this focus to include matter at scales larger than the molecular one. On this basis, self-assembly is scientifically interesting and technologically important in a range of fields: chemistry, physics, biology, materials science, and nanoscience.

J. Vicens and J. Harrowfield (eds.), Calixarenes in the Nanoworld, 63–88.

Intense activity has been devoted to the explicit application of molecular recognition processes to control the formation of organized supramolecular aggregates employing hydrogen bonding, donor acceptor, and metal coordination interactions for controlling these processes and holding the entities together. The clever exploitation of templating and self-organization has given access to a range of molecular and supramolecular entities showing specific physical and chemical properties of truly impressive structural complexity, which otherwise would have been considered impossible to construct, such as interlocked molecules, whose components are mechanically held together.[2]

In this perspective the present chapter will mainly describe three key topics: i) the study of the parameters that affect the recognition properties of the aromatic cavity of calix[4]arene derivatives toward quaternary ammonium ions in apolar media; ii) the application of these studies to the preparation and the study of pseudorotaxanes and rotaxanes, using calix[6]arene derivatives as wheels and the exploitation of their chemical information as control elements in the threading processes; iii) the preparation and utilization of monolayer protected gold nanoclusters, characterized by the presence of calix[4]arene receptors.

In this chapter we will not present an exhaustive survey of organic ion-pair recognition using the calixarenes as receptors, but rather highlight the control elements present in these processes, indicating the future possibilities offered by these supramolecular systems.

2. RECOGNITION OF QUATERNARY AMMONIUM CATIONS THROUGH THE AROMATIC CAVITY OF CALIX[4]ARENES

Calix[4]arenes[3] in the *cone* conformation present an internal cavity potentially able to host guest molecules of complementary size. Whereas in the solid state the inclusion of guests has been extensively studied, in apolar media any attempt to exploit their π donor cavity as binding site for organic guests has given very poor results. This has been attributed to solvation phenomena and to the weak preorganization of these hosts. Starting from the observation that in solution, a residual conformational flexibility of tetraalkoxy calix[4]arenes *cone* isomer still exists (see Fig. 1),[4] approaches to fix these macrocycles in a rigid *cone* structure have been pursued.

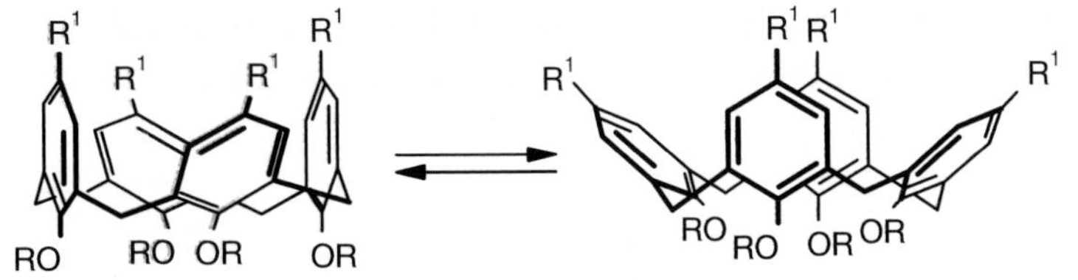

Figure 4-1. Residual conformation mobility of cone tetralkoxycalix[4]arene derivatives.

Very rigid and undistorted calix[4]arene-biscrown-3 *cone* derivatives were obtained in our laboratories by the linkage of two proximal phenol rings with short diethylene glycol bridges (see Scheme 1). These hosts were found to recognize both small neutral organic guests bearing acid CH_3 or CH_2 groups and alkylammonium cations in apolar organic solvents.[5] These findings demonstrated for the first time the role of the host rigidity as a control element in these recognition processes. They also indicated that the design of a more complex receptor acting through weak host-guest interactions where the binding site is the apolar cavity of a calix[4]arene, should contain *rigidity* as structural information stored within the architecture of the host.

Different strategies, based on non-covalent interactions, are also effective in rigidifying calix[4]arene hosts. Stibor and coworkers, for instance, showed that 1,3-dialkoxy-calix[4]arenes are efficient hosts for neutral guests bearing acidic C-H bonds.[6] The explanation of these results can be found in theoretical studies on the conformational distribution and interconversion of lower rim partially alkylated calix[4]arenes performed by Reinhoudt and coworkers[7] that have been experimentally confirmed by Böhmer and co-workers.[8] From these studies it emerged that the conformations of partially methylated derivatives are far less mobile than either the tetrahydroxy- or the tetramethoxycalix[4]arenes and that therefore, for these hosts, the rigidity derives from a combination of hydrogen bonding interactions and steric repulsions. More recently, partially alkylated calix[4]arenes were also employed by Abraham and co-workers to obtain hosts able to efficiently recognize organic cations.[9]

With the aim to enhance the recognition efficiency toward organic ammonium cations through the aromatic cavity of calixarenes, we started a systematic investigation of the parameters that affect these processes by looking at the structure of the host and the nature of the guest, keeping the apolar medium as a constant. In fact, while polar aprotic solvents, by solvating the interacting species, strongly depress binding, apolar media favour these recognition processes so that the highest binding constants have been observed in these latter solvents.[10] In the literature, a plethora of not directly comparable binding data, obtained in several more polar, or in mixtures of apolar and polar solvents have been collected.

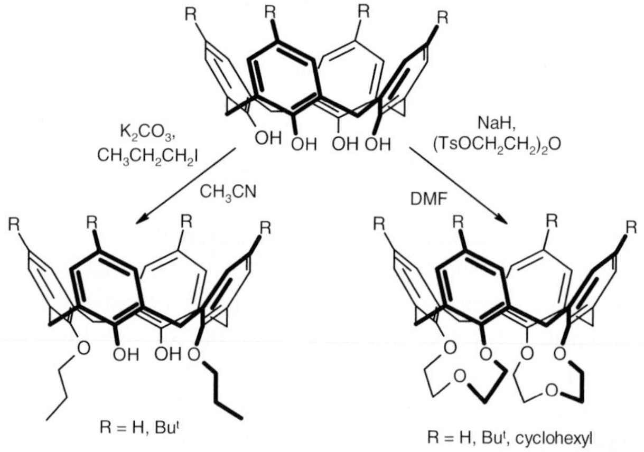

Scheme 4-1.

In apolar media however, the binding process between the host and cations or anions as guests is often limited by solubility of the salts that, in addition, in these media, are present as tight ion pairs. Ion pairs or their aggregates are thus the actual guest species present in these media.[11] In simple instances, the Coulombic interaction between the two ions in the ion pair modifies the charge density present on the ion that should be recognized by the host so that the counterion can strongly, and usually adversely, affect the extent of binding of the ion under investigation. Quite often, to enhance binding of the cation or the anion under investigation, weakly coordinating counterions, e.g., picrate and tetrabutylammonium, respectively, have been employed. Less studied and less straightforward is the situation where the coordinating power of the counterion cannot be neglected. Only recently have studies aimed at disclosing the role of counterions on these recognition processes been reported.[5,12]

In particular, by studying the interaction of a lipophilic cyclophane with several tetramethylammonium (TMA) salts in deuterated chloroform to ascertain the influence of the counterion on the cation-π interaction, it was verified that changing the anion for a more charge-dispersed species improves cation binding substantially, indicating that charge dispersion is a major factor determining the influence of the anion on the cation-π interaction.[12e]

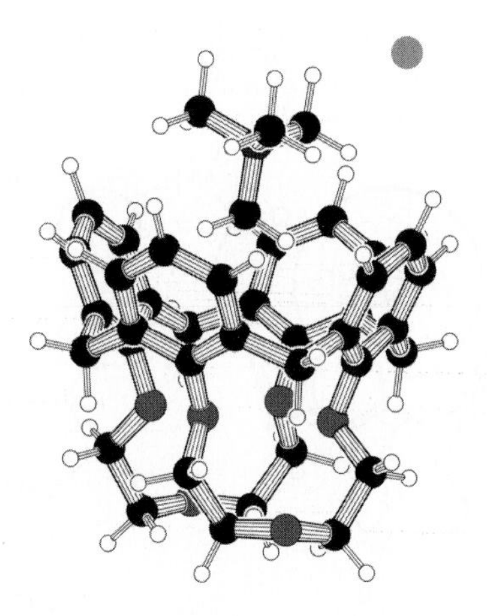

Figure 4-2. Energy minimized structure (PM3 semi empirical method) of the complex formed by calix[4]arene-biscrown-3 (**1**) with tetramethylammonium chloride.

In addition, computational studies indicated that the variation of the binding free energy of the TMA cation with its counterion is closely accounted for by the electrostatic potential (EP) of the ion pair. Guest binding appears to respond to the cation's charge density exposed to the receptor, which is determined by the anion's charge density through a polarization mechanism. Therefore host-guest association is governed by Coulombic attraction, as long as factors (steric, entropic, solvation, etc.) other than pure electrostatics, are not prevalent.[12e]

The crucial point to rationalize these data on the exclusive base of electrostatic interactions is the assumption that the region of the ion pair exposed to the receptor for binding is the van der Waals surface of the cation remote from the anion contact. This assumption was confirmed by our studies on the recognition of tetramethylammonium salts using calix[4]-arene-biscrown-3 (**1**) as host (see Figs. 2 and 3).[12d]

The binding ability toward a series of TMA salts having different counterions [tosylate (TsO^-), chloride (Cl^-), acetate (Ac^-), trifluoroacetate (TFA^-), picrate (Pic^-)] was evaluated by ^1H NMR titrations in $CDCl_3$. To disclose the role of the anion on these recognition processes, the titrations of TMACl were also performed in the presence of the tri-n-butylthioureido derivative of tren **3** (see Fig. 3)[13] that forms a stable complex with the chloride anion (*dual-host* approach).

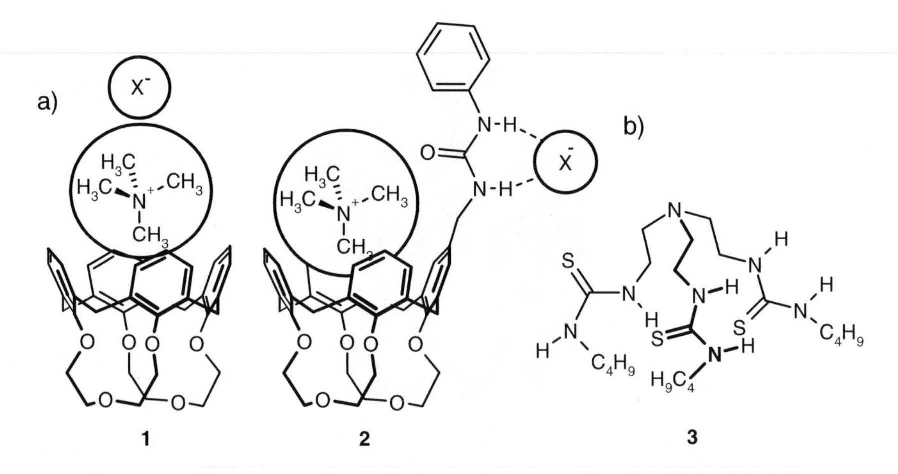

Figure 4-3. a) Schematic representation of tetramethylammonium ion-pairs complexation by rigid calix[4]arene-biscrown-3 derivatives **1** and **2** in apolar solvents; b) ligand **3** specific for the chloride anion.

From the data reported in Table 1, it emerges that host **1**, which possesses a rigid *cone* structure, recognizes the tetramethylammonium cation as a tight ion pair, thus experiencing a lower binding ability toward salts having the highly interacting anions. This is reflected by the values of the K_a measured for the different salts, which increase when the anion interacts less extensively with the cation and follow the order $Cl^- \subset$ **3** > TFA^- = Pic^- > Ac^- > Cl^- > TsO^-. This order is comparable with that observed, with a similar anion series, for example by Roelens and co-workers[12e] with cyclophanes, and Mandolini and co-workers[12c] with crown derivative of calix[5]arene, excluding the data obtained with picrate. All these achievements suggest that the effect of the anion in these recognition processes is mainly due to electrostatic interactions between the two ions.

Table 4-1. Association constants K_a (M^{-1}) of the 1:1 complexes of tetramethylammonium salts (TMAX) with hosts **1** and **2** in $CDCl_3$.

Host/X^-	TsO^-	Cl^-	Ac^-	TFA^-	$Cl^- \subset$ **3**
1	$0.3 \pm 0.1 \times 10^2$	$0.8 \pm 0.2 \times 10^2$	$2.5 \pm 0.3 \times 10^2$	$3.7 \pm 0.8 \times 10^2$	$4.3 \pm 0.9 \times 10^2$
2	$0.7 \pm 0.2 \times 10^3$	$8.8 \pm 0.8 \times 10^3$	5.0 ± 1.0 x	$1.3 \pm 0.4 \times 10^4$	$3.8 \pm 0.6 \times 10^3$
K_a(**2**)/K_a(**1**)^a	23	110	20	35	9

^{a} K_a(**2**)/K_a(**1**) = cooperative heteroditopic effect

A further step was the study of the recognition properties of an heteroditopic host,[14-17] characterized by the rigid aromatic calix[4]arene cavity as binding site for organic cations and bearing an additional H-bond donor binding site able to interact cooperatively with the anion. In fact it appeared to us that, although the problem of ion-pair recognition is well documented, very few efforts have been made so far to study in a systematic way, the heteroditopic cooperative effects in apolar solvents. This precludes a clear and direct understanding of the role played by the different parameters that affect ion-pair recognition.

The synthesis of the receptor **2** was based on preliminary molecular modeling studies which showed that the introduction of a methylene-phenylureido moiety onto the upper rim of the rigidified calix[4]arene-biscrown-3 (**1**) would result in a receptor having the appropriate arrangement of binding sites able to interact with TMA ion pairs as guests without the requirement of appreciable rearrangement of the binding sites upon ion-pair recognition.[12d] Host **2**, which was previously synthesized and utilized for the recognition of amides, can be easily obtained from the calix[4]arene-biscrown-3 (**1**) by a Tscherniac-Einhorn amidomethylation reaction using *N*-hydroxymethyl-*N*'-phenylurea in a trifluoroacetic acid dichloromethane mixture.[18]

The binding efficiency of **2** toward TMA salts evidenced a strong increase of the binding constants, up to 1.3×10^4, as consequence of the cooperative action of the two binding sites present on the host. A particular aspect is that, with the exclusion of the tosylate anion, strong binding with no discernible trend is observed with all the anions tested.

Very interesting is however the comparison of the association constants (K_a) experienced by **1** and **2** toward the same series of TMA salts (see Table 1) that evidences the heteroditopic effect experienced by **2**. The increase of two orders of magnitude of the K_a experienced by **2** toward TMACl points out the importance of the ion-pair structure in determining the heteroditopic cooperative effect, which reaches its maximum with the spherical Cl^-.[16]

A perusal of the data collected in these studies indicates that to enhance binding, the use of heteroditopic receptors represents a better approach than that making use of the dual host. The exploitation of synthesis directed towards highly efficient hosts with simple and easily accessible receptors, bearing binding sites able to cooperate in the recognition of both the cation and the anion, seems to be the better tool for the efficient recognition of ion pairs.

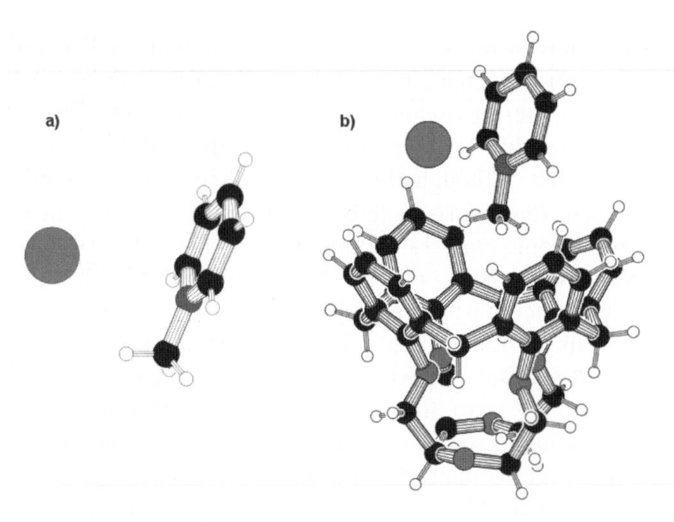

Figure 4-4. a) X-ray structure[19] of *N*-methylpyridinium iodide and b) energy minimized structure (PM3 semi empirical method) of the complex formed by calix[4]arene-biscrown-3 (**1**) and the same guest.

To study the effect of the cation structure on the cooperative heteroditopic effects, *N*-methylpyridinium (NMP) salts were selected as guests employing **2** as host. These organic salts are characterized, as evidenced by their X-ray crystal structure[19] and by computational studies,[20] by stronger stereoelectronic requirements for the interacting ions present in the ion pairs. In particular, the anion finds the most energetically favoured location over the plane of the pyridinium ion and in particular over the nitrogen atom, which delocalize its positive charge among the C2, C4, C6 of the pyridinium ring and the methyl group (see Fig. 4a). The structure of this cation and the delocalization of its positive charge could favour also the formation of π-π interactions, in addition to charge-π and CH-π that are those responsible for TMA interaction with the aromatic cavity of calix[4]arenes.

Table 4-2. Association constants K_a (M^{-1}, 300 K, $CDCl_3$) for hosts **1** and **2** with tetramethylammonium (TMA) and *N*-methylpyridinium (NMP) tosylates.

Guest/Host	**1**	**2**	$K_a(\mathbf{2})/K_a(\mathbf{1})^a$
TMA	$0.3 \pm 0.1 \times 10^2$	$0.7 \pm 0.2 \times 10^3$	23
NMP	$1.1 \pm 0.1 \times 10^2$	$3.2 \pm 1.0 \times 10^5$	2900

[a] $K_a(\mathbf{2})/K_a(\mathbf{1})$ = cooperative heteroditopic effect.

From molecular modelling, NMP enters the aromatic cavity of receptor **1** with the methyl group as head, although this process is strongly hindered by the presence of the anion that interacts repulsively with the upper rim of the calix[4]arene (see Fig. 4b). To reduce these repulsive interactions the pyridinium ring must assume a bent orientation with respect to the plane of the methylene bridges of the calixarene. From the data reported in Table 2 it emerges that **2** experiences, toward NMP tosylate, a heteroditopic cooperative effect of about 3000. Preliminary studies on the stoichiometry of these recognition processes evidenced the possible formation of 2 : 1 Hosts/Guest complexes at higher concentrations of the analytes, as suggested by the solid state structure of the complex formed by **2** and NMP chloride (see Fig. 5).[21]

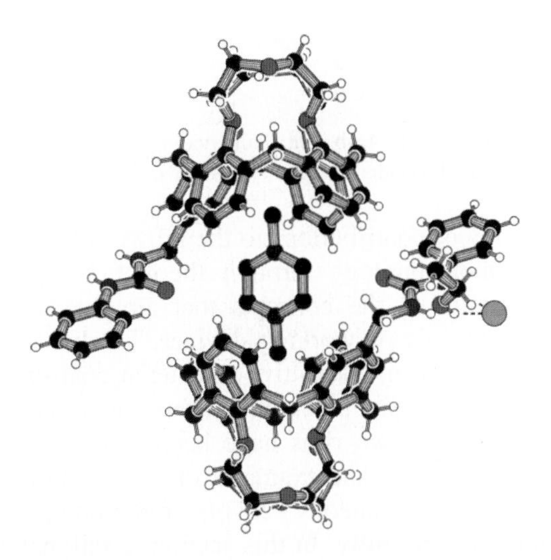

Figure 4-5. X-ray crystal structure of the 2 : 1 host/guest inclusion complex between calix[4]arene **2** and *N*-methyl pyridinium chloride (guest methyl group is statistically disordered over two equivalent different orientations and an ethanol molecule is co-crystallized around the chloride ion).

3. CALIX[6]ARENE WHEELS FOR ROTAXANE SYNTHESIS

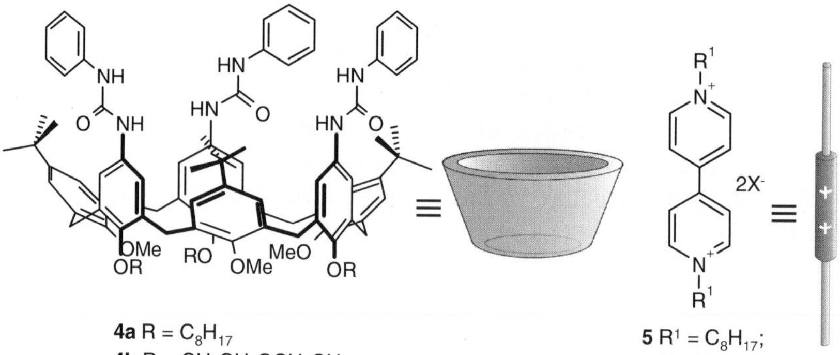

4a R = C_8H_{17}
4b R = $CH_2CH_2OCH_2CH_3$

5 $R^1 = C_8H_{17}$; $X^- =$ Ts, I, Br

Figure 4-6. Calix[6]arene wheels **4a-b** and dialkylviologen-based "axles" **5** used for the formation of calixarene-based pseudorotaxanes compounds.

One of the major contributions to the growth of nanoscience and the development of nanotechnology through the bottom-up approach could come from the transfer of the concepts that are the basis of host-guest chemistry to the synthesis of working devices.[2a] Calixarenes, that can be functionalized at both rims in a selective and stereo-controlled manner, have already demonstrated their potential within this context. In these macrocycles however, the two rims have been usually considered as two distinct and separate domains and examples of axial complexation where all sites of these hosts are simultaneously involved in binding of the guest have been reported only very recently. In this section, a rational approach to the utilization of calixarene receptors for the synthesis of pseudorotaxanes and rotaxanes will be discussed. To this end, the calix[6]arene platform was chosen since it possesses an annulus that is large enough to allow a guest having the size of an aromatic ring to cross the two rims.

Based on our systematic investigations on the parameters that affect the efficiency of calix[4]arene derivatives as host for QUATS,[12d,22] the triphenylureidocalix[6]arene derivative **4a** (see Fig. 6) was selected as suitable host because of its "restricted" conformational flexibility and because of the presence of three phenyureido moieties at the upper rim that extend the calixarene cavity and could act as additional binding sites.

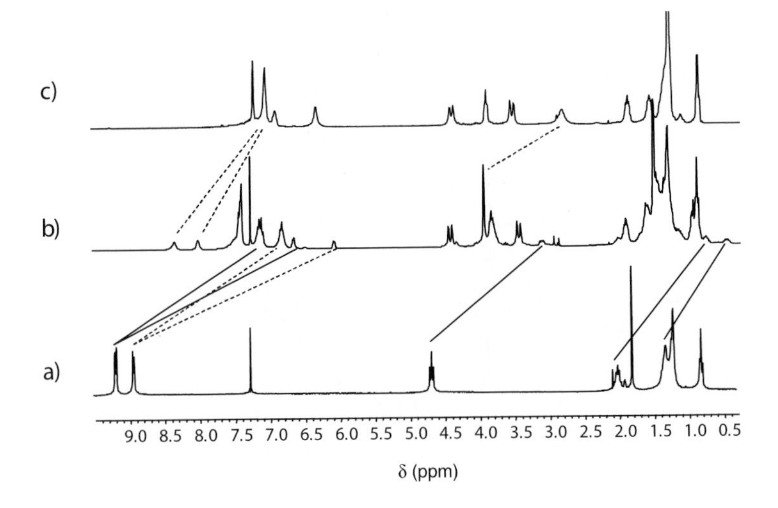

Figure 4-7. ^{1}H NMR spectra (300 MHz, T = 300 K) of a) **5**•2I in 10% $CD_3CN/CDCl_3$ solution, b) pseudorotaxane **4a** ⊃ **5**•2I$^-$ in $CDCl_3$, and c) triphenylureidocalix[6]arene **4a** in $CDCl_3$.

The propensity of **4a** to act as receptor for organic cations was tested using dialkyviologen salts as guests.[23] It was thus found that **4a** is able to dissolve the colorless crystals of dioctylviologen ditosylate (**5**•2TsO$^-$) in chloroform, affording a deep red solution. The ^{1}H NMR analysis of the solution obtained by equilibrating a suspension of dioctylviologen diiodide (**5**•2I$^-$) with a $CDCl_3$ solution of **4a** showed the formation of a 1:1 complex (see Fig. 7). The extensive upfield shift (1 to 3 ppm) of the aromatic CH and NCH_2 hydrogen signals of the guest suggests that this portion of the viologen dication is positioned inside the calixarene cavity. Quite interestingly, the three methoxy groups that in the free host resonate, as a broad signal, at δ = 2.8 ppm, in the complex undergo a downfield shift of about 1 ppm. This indicates that their orientation has changed to a situation where they are no longer oriented inside the calixarene cavity. In addition, the six protons of the host ureido NH groups suffer a downfield shift of about 1.5 ppm, suggesting their involvement in hydrogen bonding with the two iodide anions of the guest.

These chemical shift variations, together with 2D NMR experiments are in agreement with a supramolecular complex belonging to the class of pseudorotaxane. The structure of this complex was unambiguously confirmed through X-Ray diffraction studies (see Fig. 8) that showed, for the first time that a calixarene can be threaded by a suitable guest and behaves as a heteroditopic three-dimensional wheel.[23]

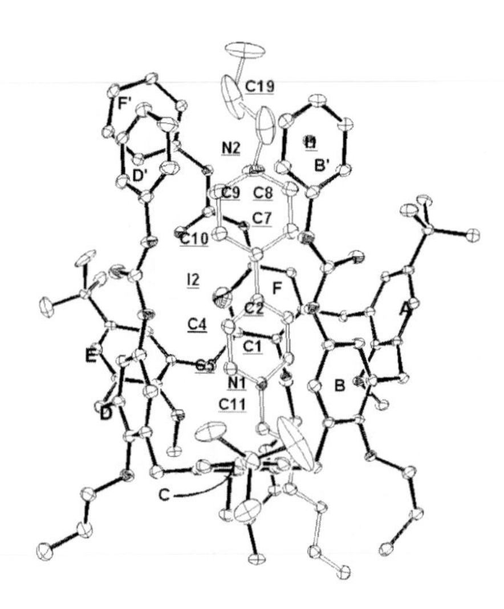

Figure 4-8. X-ray structure of pseudorotaxane **4a** ⊃ **5**•2I (part of the structure is omitted for clarity).

Particularly informative was the observation that all sites of the calixarene wheel participate in the stabilization of the complex through a combination of several host-guest intermolecular interactions. For example, the N 1 atom of lower pyridinium of the axle is located *ca.* 0.4 Å above the least-squares plane through the bridging methylene groups of the calixarene; this pyridinium unit is sandwiched and stacked by rings F and B of the host and points its octyl chain across the macrocycle annulus, expelling the three methoxy groups from the aromatic cavity. These latter moieties are involved in hydrogen bonding with protons in C1, C5 of the lower pyridinium fragment and C11 of the lower octyl chain of the axial component. The aromatic rings D' and B' of the phenylureido moieties also participate in binding through CH-π interactions with the hydrogen atoms in C19 of the upper octyl chain and hydrogen atoms in C8 and C9 of the upper pyridinium system (distance *ca.* 2.641 and 2.639 Å, respectively). The two iodide anions are involved in the stabilization of the complex and are hydrogen bonded by the three ureido groups that are present at the upper rim of the wheel (see Fig. 8).

The pseudorotaxane formed by **4a** and **6**•2TsO$^-$ was stoppered with diphenylacetyl chloride to afford the corresponding calixarene-based rotaxane **7**•2TsO$^-$ (see Scheme 2).

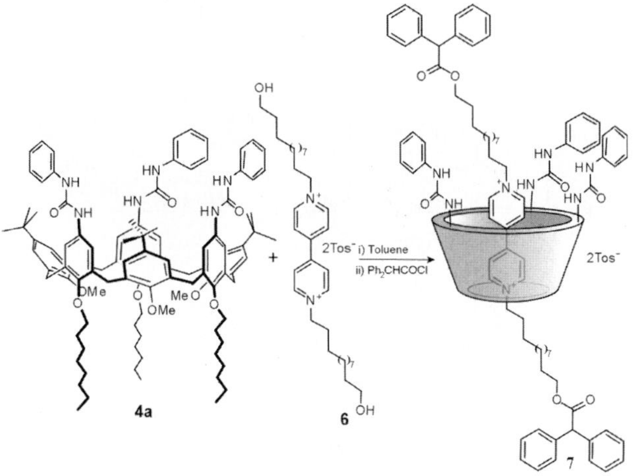

Scheme 4-2.

In CH_2Cl_2, the energy associated with this self-assembly process is around –8 kcal mol^{-1} and is strongly dependent upon the nature of the counteranion of the bipyridinium component.[24] In fact, the use of either tosylate or hexafluorophosphate salts of the guest affects both the stability of the complexes and the rate of the threading process. Such effects have been interpreted in terms of ion-pair recognition, suggesting that coordination of the counteranions of the viologen thread by the ureido groups of the calixarene wheel is crucial for the breaking of tight ion pairs prior to threading. These pseudorotaxanes can be disassembled in a fast and reversible manner by one-electron reduction of their axle components. In addition, it was also found that the heterogeneous electron-transfer kinetics for the reduction of the bipyridinium unit is slowed upon encapsulation into the calixarene cavity. The pseudorotaxane species undergo fast dethreading (submicrosecond time scale) on electrochemical reduction of the guest.[24]

In principle, an axle having the size of the bipyridyl could thread the calixarene wheel either from the wider or the narrower rim. Nevertheless, in **4a-b**, the wider rim possesses remarkable hydrogen-bond donor ability because of the presence of the three ureido moieties, while the narrow rim, that bears alkyl groups, is hydrophobic. This observation prompted us to verify whether these differences could be exploited as control elements to guide the direction of an axle threading, selectively from one of the two calixarene rims.

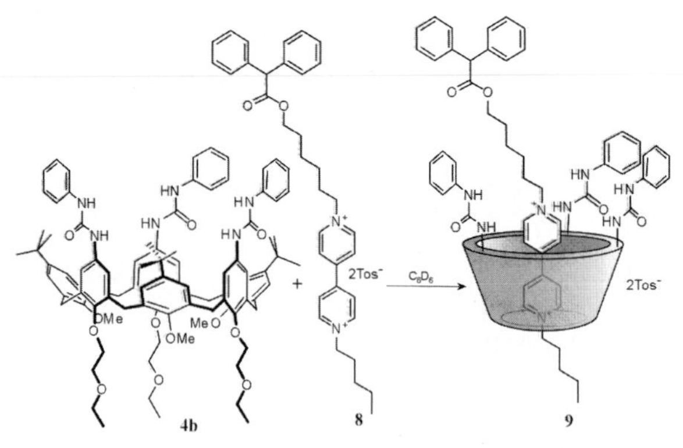

Scheme 4-3.

To address this issue we synthesized the asymmetrical axle **8**•$2TsO^-$ (see Scheme 3), derived from 4,4'-bipyridyl functionalized with two alkyl chains, one of which bears a diphenylacetyl stopper, and studied the structure of the complexes formed with the calixarene wheel **4b**.[25]

The ^{1}H NMR spectrum of the deep red solution obtained after filtration of the suspension of **4b** and an excess of axle **8**•2Tos in C_6D_6, appears as a single set of signals consistent with a 1:1 host-guest complex that is in slow exchange in the NMR timescale. As expected, the spectrum of this complex (**9**•2Tos), shows that wheel **4b** undergoes a substantial conformational rearrangement as evidenced, for example by the downfield shift (about 1 ppm) of the three methoxy groups that, because of threading, are expelled from the cavity. The six ureido NH protons suffer a downfield shift of about 2 ppm, indicating their involvement in hydrogen bonding with the two counteranions of the axial component.

The orientation of the unsymmetrical axle toward the two calixarene rims was inferred through 2D ^{1}H NMR techniques. Particularly informative were the ROE cross peaks between the pentyl chain of the axle and the alkyl groups present at the narrower rim of the wheel and those between the phenyl protons of the phenylureido groups and the hexyl chain of the thread. These ROESY features, together with the chemical shift variation experienced by the protons of both components, are consistent with the hypothesis of a structure in which the axle points its pentyl chain towards the lower rim, while the stopper is positioned in the region of the phenylureido moieties. This structure suggests that the cationic portion of the axle has threaded the calixarene cavity only through the wider rim and that the complex formed possesses the structure of an oriented pseudorotaxane.[25]

To further verify this issue and also to exclude the presence of the parent pseudorotaxane that could have formed from the threading of this axle from the narrower rim of **4b**, the new rotaxane **11**, obtained by stoppering the pseudorotaxane derived from **4b** and the symmetrical axle **10**•2TsO$^-$ with diphenylacetyl chloride was synthesized (see Scheme 4).

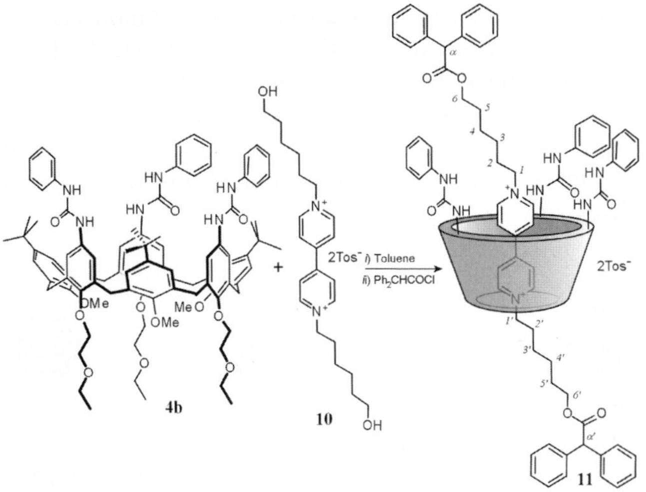

Scheme 4-4.

In the ^{1}H NMR spectrum of **11**•2TsO$^-$, the aliphatic protons of the axle portion that reside at the wide rim show chemical shift variations that can be related and are very similar to those observed in pseudorotaxane **9**•2TsO$^-$. In fact, in rotaxane **11**•2TsO$^-$, protons α and 1, which resonate at δ = 5.23 and 4.17 respectively, experience chemical shifts that are almost identical to those of the corresponding methylene protons (δ = 5.24 and δ = 4.15) present in the C_6 spacer of pseudorotaxane **9**•2TsO$^-$. On the other hand, those that protrude from the narrow rim of **4b** could not be found in the spectra of the oriented pseudorotaxane **9**•2TsO$^-$. Particularly informative was also the observation that, because of their proximity to the narrow rim, protons α' and 1' that resonate at δ = 5.28 and 4.44 ppm respectively, are distinguishable from those at the wider rim and could thus be utilized to establish the orientation of the two diphenylacetyl stoppers with respect to the two calixarene rims.

This selective threading could be explained by considering the concomitant action of the following factors: a) in C_6D_6, the three methoxy groups that are present at the narrow rim of the calixarene are oriented inward, occupying the cavity and thus disfavoring, by repulsive intermolecular interactions, the access of the axle from this rim; b) the size and binding properties of the wheel inner volume are suitable only for the inclusion of the cationic portion of the axle; c) in apolar media, the axles employed are present as tight ion pairs and therefore a process that implies the partial separation of the dication from its counteranions should take place before threading; d) the phenyl ureido moieties that are at the wide rim of the wheel are potent hydrogen bond donor groups and participate in the overall binding process by ligating the two anions of the axle, thus favoring the formation of a ligand-separated ion pair. It may thus be reasonable to assume that the NH groups force the cationic axle to thread from the wide rim of the calix [6]arene wheel.

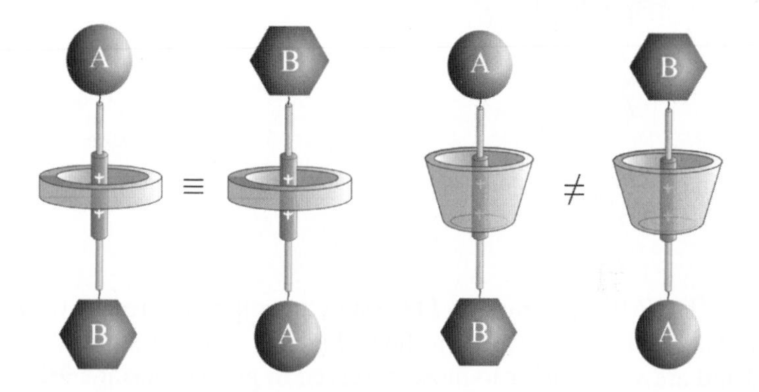

Figure 4-9. Constitutional isomers obtained by threading the unsymmetrical calix[6]arene wheel with axle having different stoppers A and B, respectively.

This piece of information was then applied to the synthesis of constitutionally isomeric calix[6]arene based rotaxanes (see Fig. 9). In fact, in spite of the remarkable achievements gained in the synthesis of rotaxane as well as on the control of the movement of their cyclic component along the dumbbell, the synthesis of rotaxanes endowed with a unique and programmable relative orientation of the wheel/dumbbell systems has been ignored so far. This can be understood considering that the structure of the cyclic components often employed in rotaxane syntheses is related to that of a two-dimensional macrocycle such as that of crown ethers, or three-dimensional symmetrical systems such as cucurbiturils, that can be considered as degenerate wheels in the sense of having identical portals that face the two identical[26] or different[27] stoppers of the dumbbell.

In the case of cyclodextrin-based rotaxanes, because of the lack of control elements on their two different rims, an axle can enter these three-dimensional nonsymmetrical wheels without preference from both rims, thus leading to a mixture of oriented pseudorotaxane isomers[28] in which the orientation of the cyclodextrin along the axle is virtually unpredictable.[29]

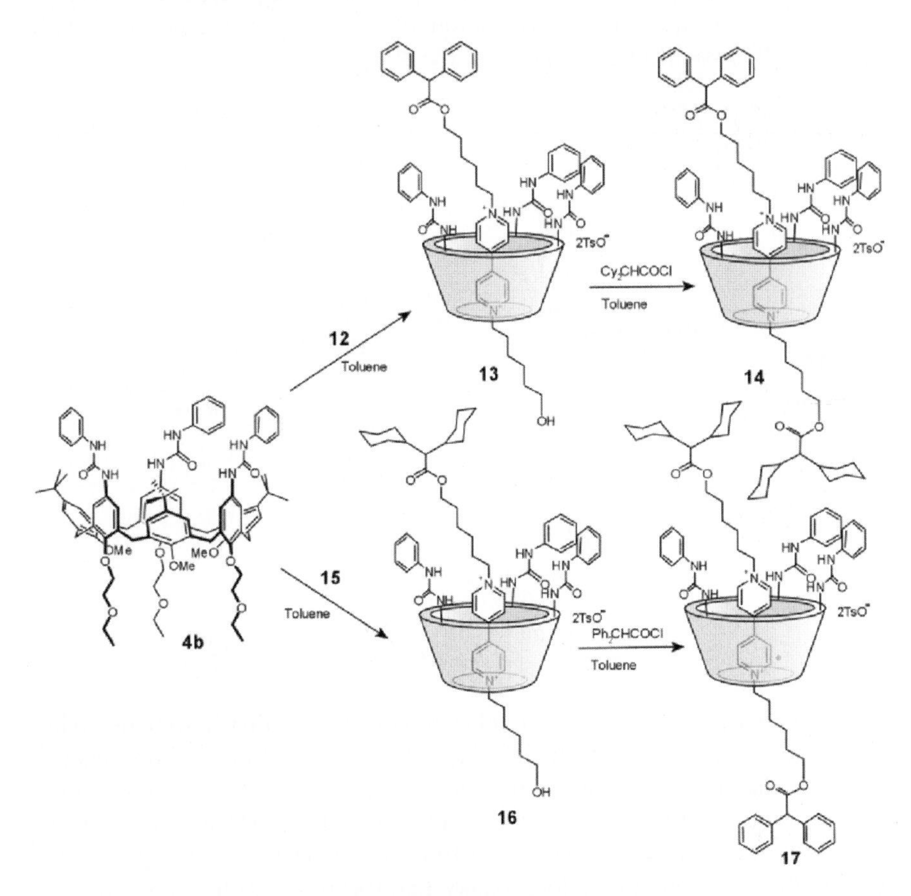

Scheme 4-5.

In order to test the possibility of exploiting the unique binding properties of **4b** for the synthesis of this new type of rotaxane, axles **12**•2TsO$^-$ and **15**•2TsO$^-$ were prepared.[30] Initially, the unsymmetrical axle **12**•2TsO$^-$ was submitted to threading reaction with the calixarene wheel **4b** using toluene as the solvent. To the deep red solution thus obtained, dicyclohexyl chloride was added to stopper the OH group of the thread. This reaction gave, after workup, only the oriented rotaxane **14**. By applying the same reaction sequence and adding diphenylacetyl chloride to the complex formed by **4b** and **15**•2TsO$^-$, the rotaxane isomer **17** was obtained (see Scheme 5).

The two rotaxanes **14** and **17** are two constitutionally isomeric rotaxanes, characterized by the unique orientation of the two calixarene rims with respect to the two different stoppers of the dumbbell.[30]

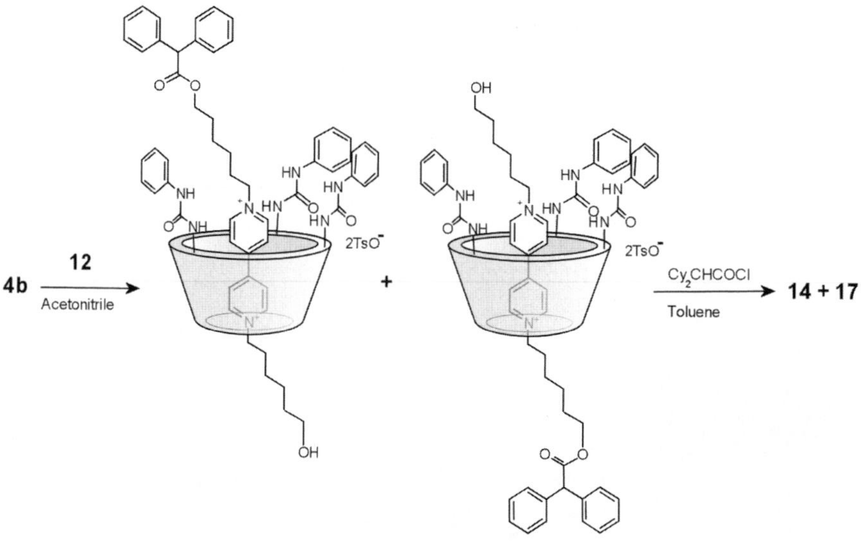

Scheme 4-6.

These data can be tentatively explained considering that in toluene, axles **12** and **15** are present as tight ion pairs during the initial pseudorotaxane formation that as such cannot fit into the calixarene cavity. These conditions maximize the pivoting action of the ureido NHs toward the cationic portion of the axle that thus thread the wheel from the upper rim. The high kinetic stability of the oriented pseudorotaxanes that forms favors the stoppering of the OH group that protrudes from the lower rim of the wheel, without detectable axle-wheel isomerization. The question that needed to be addressed was whether these axles could thread the same wheel also from the narrow rim or, because of the insufficient size of this rim, this direction of threading would remain forbidden. To answer this, axle **12** was equilibrated with **4b** in acetonitrile. This more polar solvent should in fact decrease the extent of the axle ion-pairing and also decrease the magnitude of the pivoting role exerted by the ureido groups of the wheel during threading. After removal of acetonitrile from the adduct formed by **12**•2TsO$^-$ and **4b** and subsequent addition of toluene, dicyclohexylacetylchloride was added. After workup of the reaction mixture, a mixture of both rotaxane isomers **14** and **17** was obtained as unique chromatographic fraction. This clearly indicates that indeed the calix[6]arene annulus is large enough to allow the passage of axle derived from 4,4'-bipyridine also from the lower rim (see Scheme 6).

4. CALIX[4]ARENES AS PROTECTIVE LAYERS ON GOLD NANOPARTICLES

Atoms and molecules at surfaces represent a fourth state of matter, where the gradients in properties are greatest. 2D-SAMs (Self-Assembled Monolayers on flat surfaces) are organic assemblies formed by the adsorption of molecular constituents from solution or the gas phase onto the surface of solids. Often the adsorbates organize spontaneously into crystalline (or semicrystalline) structures. The molecules or ligands that form SAMs possess a functional group, or "head-group", with specific and high affinity for the material present on the surface.[31]

There are many head-groups that bind to specific metals. The most extensively studied class of SAMs is derived from the adsorption of alkanethiols on noble metals and in particular on gold. The high affinity of thiols for these metal surfaces makes it possible to generate well-defined organic coatings with useful and highly alterable chemical functionalities displayed at the exposed interface.[31]

3D-SAMs (Self-Assembled Monolayers on Nanoparticles) are an important class of nanometer-scale materials having a typical diameter of 1-40 nm, usually built up from a metal core. The more extensively studied have been those derived from gold, coated with an organic monolayer. The small dimension of these materials, which are a bridge between molecular and bulk materials, give them unique physical properties.[32,33]

Initially, alkanethiols were employed as protective and stabilizing shell around the metal core to obtain relatively monodisperse nanoparticles. Their easy availability was soon seen as a new tool to study the structure, properties and reactivity of organic species arranged in monolayers. In this regard, in the last decade, 3D-SAMs have been employed both as a possible new systems to study the supramolecular aspect of functionalized monolayers and as new building blocks for self- assembly processes.[34] These nanoparticles, which are stable in several conditions, show long shelf life, can be easily prepared and stabilized by reduction of a metal ion in the presence of functionalized organic molecules and can be obtained with controllable metal core size.[32,35] Unlike 2D-SAMs, they are soluble so that their characterization and the study of their properties can be carried out in solution.

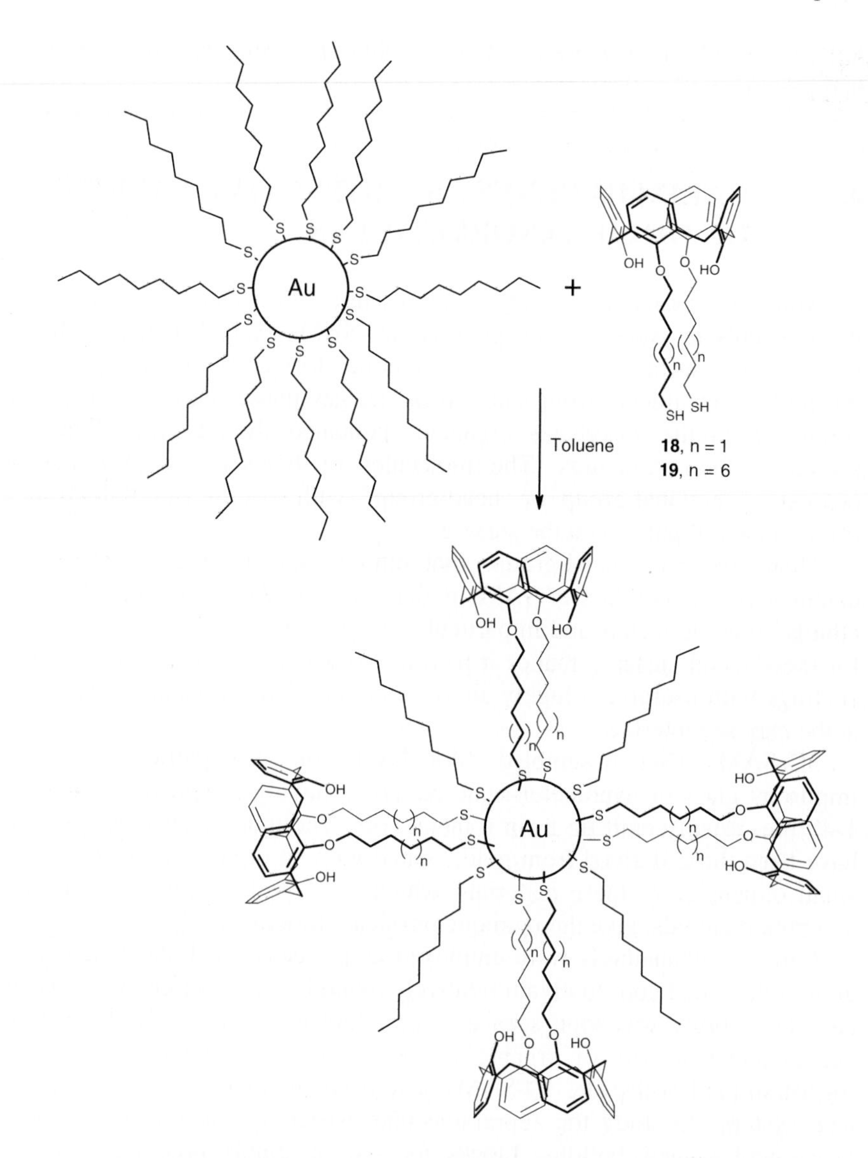

Figure 4-10. Schematic representation of synthesis of calix[4]arene-based gold nanoparticles.

Displacement of one ligand for another is a second strategy for modifying the organic surface of nanoparticles after their formation. These so-called "ligand-exchange" methods are particularly useful if the desired ligand is not compatible with the highly reductive environment required by nanoparticle synthesis or if the desired ligand is particularly valuable (or simply not commercially available) and cannot be used in the large excess necessary for stabilization during synthesis. Mixed-monolayer systems greatly enhance the versatility of these nanomaterials, allowing for multiple functionalities to be appended onto their outer layer.[36]

Another interesting characteristic of 3D-SAMs is the high radius of curvature of the substrate onto the metal core. An important consequence of this curvature is the decrease of the chain density moving away from the metal surface (radial nature of the ligand), that minimize steric hindrance that can be present in 2D-SAMs.[37]

A very attractive topological property of MPNs (monolayer-protected nanoparticles) is therefore the possibility of anchoring onto the surface of the metal a discrete number of suitable receptors in a radial three-dimensional arrangement. In addition, the recognition properties of the cluster of receptors thus obtained can be studied in solution. In spite of the remarkable results obtained so far on the use of these systems as receptors for ion pairs and for neutral molecule recognition, only few attempts to compare their binding abilities with those of the monomeric model host have been carried out.[38] It however appeared us that a possible approach to the rational design of MPC (monolayer-protected core) supported hosts should be based on the understanding of the several factors introduced by the clustering of receptors.[39]

On this base the synthesis of gold nanoparticles coated with calix[4]arene derivatives was undertaken, with the perspective of manufacturing nanoscale devices with potential applications as sensors, switches and new materials endowed with tunable properties. The hosts chosen for these studies were 1,3-dialkoxycalix[4]arenes, which are known to form endo cavity inclusion complexes with quaternary ammonium cations. To disclose a possible effect of the distance of the calixarene cavities from the gold core on the recognition efficiency, calix[4]arenes bearing two alkanethiol chains of six **18** or eleven carbon **19** atoms were synthesised.[40] The anchoring of these hosts onto gold nanoparticles was performed by an exchange reaction starting from dodecanethiol MPNs prepared according to the Brust–Schiffrin method, to obtain gold MPCs having a core diameter of about 2 nm (see Fig. 10).[35]

The "average binding constants" of these different gold nanoparticles ("polyreceptors") toward N-methylpyridinium tosylate were evaluated in chloroform, and compared with that of 1,3-dipropoxycalix[4]arene that was employed as reference "monoreceptor". The most efficient nanoparticle is that protected with the calix[4]arene ligand having the longer C11 chains. With this study it was verified that the efficiency in the recognition of ion pairs by hosts anchored onto a radial monolayer is very high.[40]

As previously reported mixed-monolayer protected nanoparticles greatly enhance the versatility of MPNs, allowing for multiple functionalities to be appended onto the monolayer. In particular, in view also of future bioanalytical applications of calixarene-coated nanoparticles, it could be necessary to have stable MPNs able to recognize efficiently selected guests in aqueous media.[41] To this end, the high lipophilicity of calixarenes must be overcome. Very stable, yet chemically versatile water soluble MPNs can be obtained when the sulfanylalkyl oligo(ethyleneglycol) **20** is used as a stabilizing ligand.[42] On this basis, gold MPNs having an average diameter of 14-nm, stabilized with **20** and having the calix[4]arene **19** in their ligand shell were prepared and studied (see Fig. 11).[41]

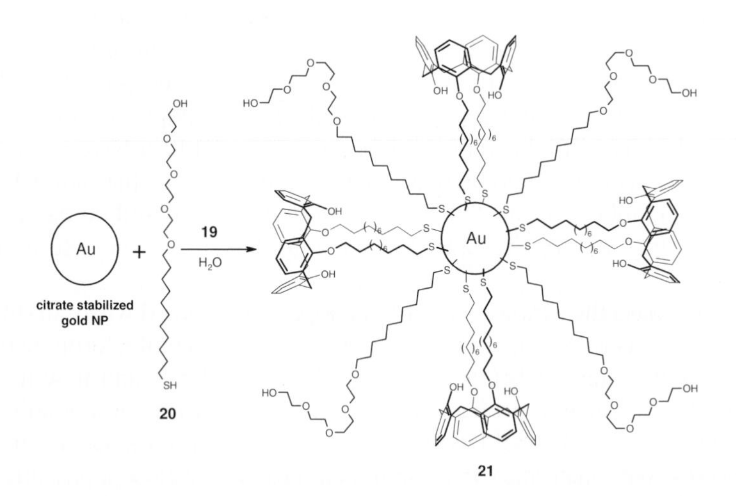

Figure 4-11. Schematic representation of water soluble calix[4]arene-based gold nanoparticles **21** obtained from calix[4]arene **19** and sulfanylalkyl oligo(ethyleneglycol) **20**.

The molecular recognition properties of these new nanoparticles, tested in water, have been qualitatively demonstrated with specific binding studies using the pyridinium ions **22** [TsO^- = (4-methylphenyl)sulfonyl] supported on a flat gold surface as the substrate. As demonstrated by the AFM images of Fig. 12, the calixarene modified water soluble nanoparticles were found to bind selectively to this surface. In contrast, when these nanoparticles were equilibrated with clean gold surfaces, their binding occurred only as a non-specific and adventitious event. Water soluble MPNs without calixarene ligands in their shell did not show any significant binding to the SAMs or to clean gold surfaces.[41]

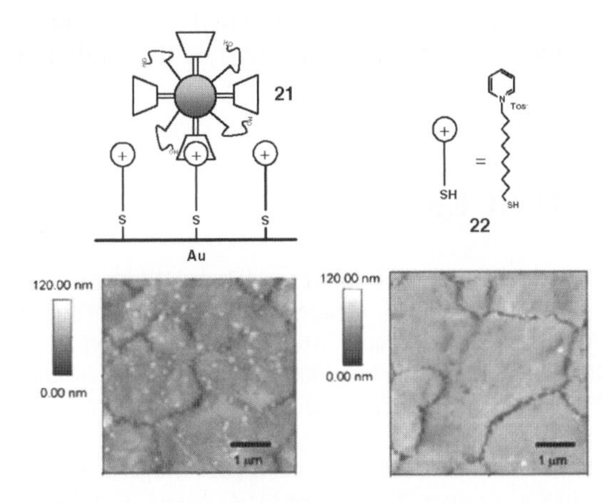

Figure 4-12. Specific binding of the calixarene-modified gold nanoparticles **21**, from aqueous solution to a self-assembled monolayer (SAM) of the pyridinium ions of **22** [TsO^-= (4-methylphenyl)sulfonyl] on a gold surface shown by AFM.

We have thus introduced a very simple and promising strategy for the preparation of water-soluble calixarenes. Importantly, it has also been demonstrated that in aqueous solution the calixarene cavity maintains its molecular recognition properties, which have been utilized to bind the MPNs selectively to chemically modified substrates. These preliminary experiments could be the starting point for the development of new self assembled 3D-SAMs.

5. OUTLOOK

The calix[n]arenes have been one of the most convenient platforms for the development of supramolecular chemistry and, together with resorcinarenes and cyclodextrins, have facilitated the transition of this discipline from the planar to the three-dimensional world. Given the efforts carried out during these last decades for their regio- and stereoselective selective functionalization, it can be foreseen that the development of targeted and efficient procedures for the synthetic manipulation of all members of the calixarene series will contribute to the expansion of nanoscience and nanotechnology. The construction of calixarene-based working devices able to transfer their unique recognition properties to the nanometre scale and respond, in a programmable manner, to external stimuli will be the challenge of calixarene chemistry.

6. REFERENCES

1. a) J.-M. Lehn, *PNAS* **99**, 4763-4768 (2002); b) G. M. Whitesides, and M. Boncheva, *PNAS* **99**, 4769-4774 (2002); c) F. M. Menger, *PNAS* **99**, 4818-4822 (2002).
2. a) V. Balzani, A. Credi, and M. Venturi, *Molecular Devices and Machines - A Journey into the Nano World* (VCH, Weinheim, 2003); b) *Molecular Catenanes, Rotaxanes and Knots*, edited by J.-P. Sauvage, C. O. Dietrich-Buchecker (Wiley-VCH, Weinheim, 1999).
3. (a) *Calixarenes in Action*, edited by L. Mandolini, R. Ungaro (Imperial College Press, London, 2000); (b) *Calixarenes 2001*, edited by Z. Asfari, V. Böhmer, J. Harrowfield, J. Vicens (Kluwer: Dordrecht, 2001).
4. a) A. Ikeda, H. Tsuzuki, and S. Shinkai, *J. Chem. Soc., Perkin Trans. 2*, 2073-2080 (1994); b) A. Arduini, M. Fabbi, M. Mantovani, L. Mirone, A. Pochini, A. Secchi, and R. Ungaro, *J. Org. Chem.* **60**, 1454-1457 (1995).
5. A. Arduini, W. M. McGregor, D. Paganuzzi, A. Pochini, A. Secchi, F. Ugozzoli, and R. Ungaro, *J. Chem. Soc. Perkin Trans. 2* 839-846 (1996).
6. S. Smirnov, V. Sidorov, E. Pinkhassik, J. Havlicek, and I. Stibor, *Supramol. Chem.* **8**, 187-196 (1997).
7. W. P. Van Hoorn, M. G. H. Morshuis, F. C. J. M. van Veggel, and D. N. Reinhoudt, *J. Phys. Chem. A* **102**, 1130-1138 (1998).
8. T. Kusano, M. Tabatabai, Y. Okamoto, and V. Böhmer, *J. Am. Chem. Soc.* **121**, 3789-3790 (1999).
9. M. Orda-Zgadzaj, V. Wendel, M. Fehlinger, B. Ziemer, and W. Abraham, *Eur. J. Org. Chem.* 1549-1561 (2001).
10. See *e.g.* a) P. R. Ashton, P. J. Campbell, E. J. T. Chrystal, P. T. Glink, S. Menzer, D. Philip, N. Spencer, J. F. Stoddart, P. A. Tasker, and D. J. Williams, *Angew. Chem. Int. Ed. Engl.* **34,** 1865-1869 (1995); b) V. P. Solov'ev, N. N. Strakhova, O. A. Raevsky, V. Rüdiger, and H.-J. Schneider *J. Org. Chem.* **61**, 5221-5226 (1996).
11. See *e.g.* J. W. Jones, and H. W. Gibson, *J. Am. Chem. Soc.* **125**, 7001-7004 (2003).
12. a) S. Roelens, and R. Torriti, *J. Am. Chem. Soc.* **120**, 12443-12452 (1998); b) S. Bartoli, and S. Roelens, *J. Am. Chem. Soc.* **121**, 11908-11909 (1999); c) V. Böhmer, A. Dalla Cort, and L. Mandolini, *J. Org. Chem.* **66**, 1900-1902 (2001); d) A. Arduini, E. Brindani, G. Giorgi, A. Pochini, and A. Secchi *J. Org. Chem.* **67**, 6188-6194 (2002); e) S. Bartoli, and S. Roelens, *J. Am. Chem. Soc.* **124**, 8307-8315 (2002).
13. K.-S. Jeong, K.-M. Hahn, and Y. L. Cho, *Tetrahedron Lett.* **39**, 3779-3782 (1998).
14. a) P. D. Beer, and P. A. Gale, *Angew. Chem. Int. Ed. Engl.* **40**, 486-516 (2001); b) G. J. Kirkovits, J. A. Shriver, P. A. Gale, and J. L. Sessler, *J. Incl. Phenom. Macrocyc. Chem.* **41**, 69-75 (2001); c) P. A. Gale, *Coord. Chem. Rev.* **240**, 191-221 (2003).
15. For more recent papers on ion-pairs recognition, see *e.g.* : a) M. R. Sambrook, P. D. Beer, J. A. Wisner, R. L. Paul, A. R. Cowley, F. Szemes, and M. G. B. Drew, *J. Am. Chem.*

Soc. **127**, 2292-2302 (2005); b) J. M. Mahoney, G. U. Nawaratna, A. M. Beatty, P. J. Duggan, and B. D. Smith, *Inorg. Chem.* **43**, 5902-5907 (2004); c) J. M. Mahoney, J. P. Davis, A. M. Beatty, and B. D. Smith, *J. Org. Chem.* **68**, 9819-9820 (2003).

16. a) J. Gong, B. C. Gibb, *Chem. Commun.* 1393-1395 (2005); b) K.-S. Jeong, K. H. Shin, and S.-H. Kim, *Chem. Lett.* 1166-1167 (2002).
17. For heteroditopic receptors based on calix[6]arenes see Part 3 of this Chapter.
18. A. Arduini, A. Secchi, and A. Pochini, *J. Org. Chem.* **65**, 9085-9091 (2000).
19. R. A. Lalancette, W. Furev, J. N. Costanzo, P. R. Hemmes, and F. Jordan, *Acta Cryst.* **B34**, 2950-2953 (1978).
20. S. Tsuzuki, H. Tokuda, K. Hayamizu, and M. Watanabe, *J. Phys. Chem. B* **109**, 16474-16481 (2005).
21. A. Arduini, A. Pochini, A. Secchi, and F. Ugozzoli *unpublished results*.
22. A. Arduini, A. Pochini, and A. Secchi, *Eur. J. Org. Chem.* 2325-2334 (2000).
23. A. Arduini, R. Ferdani, A. Pochini, A. Secchi, and F. Ugozzoli, *Angew. Chem Int. Ed.* **39**, 3453-3456 (2000).
24. A. Credi, S. Dumas, S. Silvi, M. Venturi, A. Arduini, A. Pochini, and A. Secchi, *J. Org. Chem.* **69**, 5881-5887 (2004).
25. A. Arduini, F. Calzavacca; A. Pochini, and A. Secchi, *Chem. Eur. J.* **9**, 793-799 (2003).
26. H. R. Tseng, S. A. Vignon, P. C. Celeste, J. Perkins, J. O. Jeppesen, A. Di Fabio, R. Ballardini, M. T. Gandolfi, M. Venturi, V. Balzani, and J. F. Stoddart, *Chem. Eur. J.* **10**, 155-172 (2004).
27. a) G. W. H. Wurpel, A. M. Brouwer, I. H. M. Stokkum, A. Farran, and D. A. Leigh, *J. Am. Chem. Soc.* **123**, 11327-11328 (2001); b) J. O. Jeppesen, K. A. Nielsen, J. Perkins, S. A. Vignon, A. Di Fabio, R. Ballardini, M. T. Gandolfi, M. Venturi, V. Balzani, J. Becher, and J. F. Stoddart, *Chem. Eur. J.* **9**, 2982-3007 (2003). c) Q.-C. Wang, X. Ma, D.-H. Qu, H. Tian, *Chem. Eur. J.* **12**, 1088-1096 (2006).
28. B. C. Gibb, *J. Supramol. Chem.* **2**, 123-131 (2002).
29. Examples of cyclodextrins based oriented pseudorotaxanes exist: a) H. Yonemura, M. Kasahara, H. Saito, H. Nakamura, T. Matsuo, *J. Phys. Chem.* **96**, 5765-5770 (1992); b) M. R. Craig, M. G. Hutchings, T. D. W. Claridge, H. L. Anderson, *Angew. Chem. Int. Ed.* **40**, 1071-1074 (2001); c) R. Isnin, A. E. Kaifer, *J. Am. Chem. Soc.* **113**, 8188-8190 (1991).
30. A. Arduini, F. Ciesa, M. Fragassi, A. Pochini, and A. Secchi, *Angew. Chem. Int. Ed.* **44**, 278-281 (2005).
31. J. C. Love, L. A. Estroff, J. K. Kriebel, R. G. Nuzzo, and G. M. Whitesides, *Chem. Rev.* **105**, 1103-1169 (2005).
32. a) M.-C. Daniel, and D. Astruc, *Chem. Rev.* **104**, 293-346 (2004); b) M. Brust, and C. J. Kiely, *Colloids Colloid Assem.* 96-119 (2004); c) R. Shenhar, and V. M. Rotello, *Acc. Chem. Res.* **36**, 549-561 (2003); d) A. C. Templeton, W. P. Wuelfing, and R. W. Murray, *Acc. Chem. Res.* **33**, 27-36 (2000).
33. U. Drechsler, B. Erdogan, and V. M. Rotello, *Chem. Eur. J.* **10**, 5570-5579 (2004).
34. See *e.g.* a) G. Schmid, and U. Simon, *Chem. Commun.* 697-710 (2005); b) M. Brust, *Nature Materials* **4**, 364-365 (2005).
35. a) M. Brust, M. Walker, D. Bethell, D. J. Schiffrin and R.Whyman, *J. Chem. Soc., Chem. Commun.* 801-802 (1994); b) M. J. Hostetler, J. E. Wingate, C.-J. Zhong, J. E. Harris, R. W. Vachet, M. R. Clark, J. D. Londono, S. J. Green, J. J. Stokes, G. D. Wignall, G. L. Glish, M. D. Porter, N. D. Evans, and R.W.Murray, *Langmuir* **14**, 17-30 (1998).
36. M. J. Hostetler, S. J. Green, J. J. Stokes, and R. W. Murray, *J. Am.Chem. Soc.* **118**, 4212-4213 (1996).
37. A. K. Boal, V. M. Rotello, *J. Am. Chem. Soc.* **124**, 5019-5024 (2002).
38. P. D. Beer, D. P. Cormode, and J. J. Davis, *Chem. Commun.* 414-415 (2004).

39. J. D. Badjicä , A. Nelson, S. J. Cantrill, W. B. Turnbull, and J. F. Stoddart, *Acc. Chem. Res.* **38**, 723-732 (2005).
40. A. Arduini, D. Demuru, A. Pochini, and A. Secchi, *Chem. Comm*, 645-647 (2005).
41. T. R. Tshikhudo, D. Demuru, Z. Wang, M. Brust, A. Secchi, A. Arduini, and A. Pochini, *Angew. Chem. Int. Ed.* **44**, 2913-2916 (2005).
42. A. G. Kanaras, F. S. Kamounah, K. Schaumburg, C. J. Kiely, and M. Brust, *Chem. Commun.* 2294-2295 (2002).

Chapter 5

CALIXDENDRIMERS

Design of multi-calixarenes

Najah Cheriaa,[a,b] Mouna Mahouachi,[a,b] Amel Ben Othman,[a,b] Lassaad Baklouti,[a,b] Rym Abidi,[b] Jong Seung Kim,[c] Yang Kim,[d] Jack Harrowfield,[e] and Jacques Vicens [a]

[a]*Ecole Chimie Polymères Matériaux, UMR 7512 du CNRS, 67087 Strasbourg, France* [b]*Facultés des Sciences de Bizerte, 7021 Zarzouna-Bizerte, Tunisie,* [c]*Department of Chemistry, Dankook University, Seoul 140-714, Korea,* [d]*Department of Chemistry and Advanced Materials, Kosin University, 149-1, Dongsam-dong, Yeongdo-gu, Busan, 606-701, Korea,* [e]*Institut de Science et d'Ingénierie Supramoléculaires, UMR 7006 du CNRS, 67083 Strasbourg, France*

Abstract: This chapter places our investigations of calixdendrimers within the context of extant literature.

Key words: Dendrimers, convergent synthesis, divergent synthesis, complexation.

1. INTRODUCTION

"Nanoscience" is a term used to describe the study of properties of materials composed of particles, "nanoparticles", with dimensions of the order of tens to hundreds of nanometres. The properties of such materials are neither those of their atomic or molecular constituents nor those of the bulk materials they may form, and are usually dependent upon the exact particle size. "Nanotechnology" refers to the design, characterisation, production and application of functional materials, structures, devices and systems based on nanoparticles.[1-3] In some senses, nanoscience and nanotechnology are not new. As exemplified by polymer chemistry, chemists have long been concerned with the aggregation of subnanometre particles (atoms and most molecules) to give larger structures with at least nanometre dimensions. In the natural world, colloidal solutions such as milk can be considered to be nanoparticle suspensions and proteins, polysaccharides and polynucleotides generally are sufficiently large molecules to be considered as nanoparticles. Computer technology has engendered enormous pressure for the preparation of

J. Vicens and J. Harrowfield (eds.), Calixarenes in the Nanoworld, 89–107.

structures of ever-diminishing dimensions, in an engineering/physical approach to objects of multinanometre dimensions which may be termed the "top down" method. In contrast, the essence of increasingly sophisticated chemical approaches to nanostructure synthesis is the "bottom up" or "modular" method that consists of combining molecular segments to reach a larger molecular structure with desired properties.[4] One application of new synthetic methods[5] is the preparation of biomolecule analogues[6] which may undergo self-organisation[7] into complex structures with properties which may mimic or even extend those of living systems.[8-11]

An important group of new molecular nano-objects is that of "dendrimers". Dendrimers (greek *dendron* = tree) are macromolecules with a tree-like structure.[12] That the branched structures of three-dimensional polymers should engender unique physical and chemical properties was recognised, in theory, by Flory as early as 1941.[13-15] Such molecules were not, however, actually synthesised until the early 70s, when Vögtle and co-workers[16] reported discrete, branched, polyamines prepared by an iterative, step-wise 'cascade synthesis'. In 1981, Denkewalter *et al.*[17] reported in a patent the preparation of polylysine-based dendrimers. In 1985, Newkome *et al.*[18] and Tomalia *et al.*[19] published different routes to polyamido *arborols* and to polyamino *dendrimers*. Since these seminal publications, dendrimers have been widely developed and find many applications, such as in receptor molecules, sensors and drug-transport agents,[12d] metallo-dendrimeric catalysts,[20] redox and optical (light conversion and nonlinear optics) nanodevices,[21] as well as in MRI contrast and EPR imaging agents in gene therapy and prion research.[22]

The special properties of dendrimers come from their very special molecular structures.[12] They are unusual polymers because of their necessarily three-dimensional form and in particular because they are monodisperse. Their essential components are a central core, branches, and end groups. The preparation of such branched structures demands the use of particular building blocks with the appropriate stereochemistry and multiple, equivalent reaction centres. Calixarenes,[23] with their multiple sites for functionalisation on a conformationally restricted, macrocyclic scaffold, are attractive substrates for such modular syntheses. Their chemistry is well-established and has engendered extensive research not only because of their capacity for forming complexes with a variety of guests, both charged and neutral, but also because of their ease of functionalisation, as reflected in their use in the construction of sophisticated derivatives such as calixcrowns[24,25] calixcryptands[26] and calixspherands.[27] The ease of linking calixarene units together raises particularly interesting prospects not only for the formation of calixdendrimers, *viz.* dendrimers in which both the core and the branches contain calixarene units, but also for their use simply as dendrimer cores.[28]

2. CALIXDENDRIMERS

The first calixarene-based dendrimers were prepared in 1991 by Newkome and co-workers,[29] who attached nine hydroxyl groups *via* three amide bonds to each phenyl ring of a calix[4]arene to form "silvanols" **1**. These silvanols are water soluble.

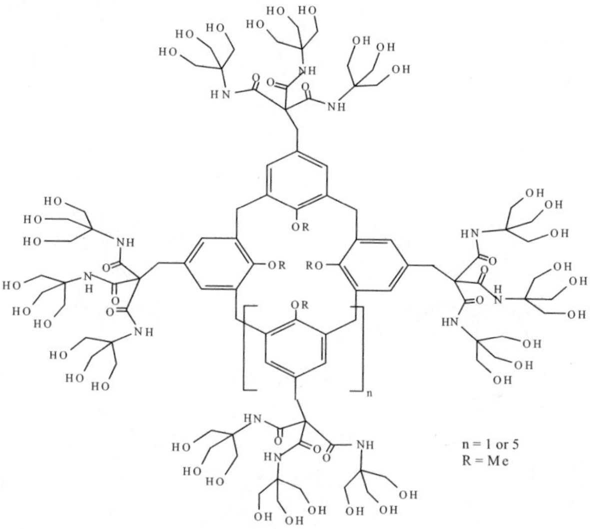

n = 1 or 5
R = Me

1

In 1995, Lhotak and Shinkai[30] reported a series of oligo-calixarenes linked through phenolic oxygens via aliphatic chains (lower rim-lower rim connections). Monobromoalkyl derivatives **2** (n = 2, 3 and 6) of O-tripropyl-substituted calix[4]arene were used as starting molecules. Reactions of **2** with suitable differently substituted calixarenes gave double-, triple- and penta-calixarenes **3-5**.

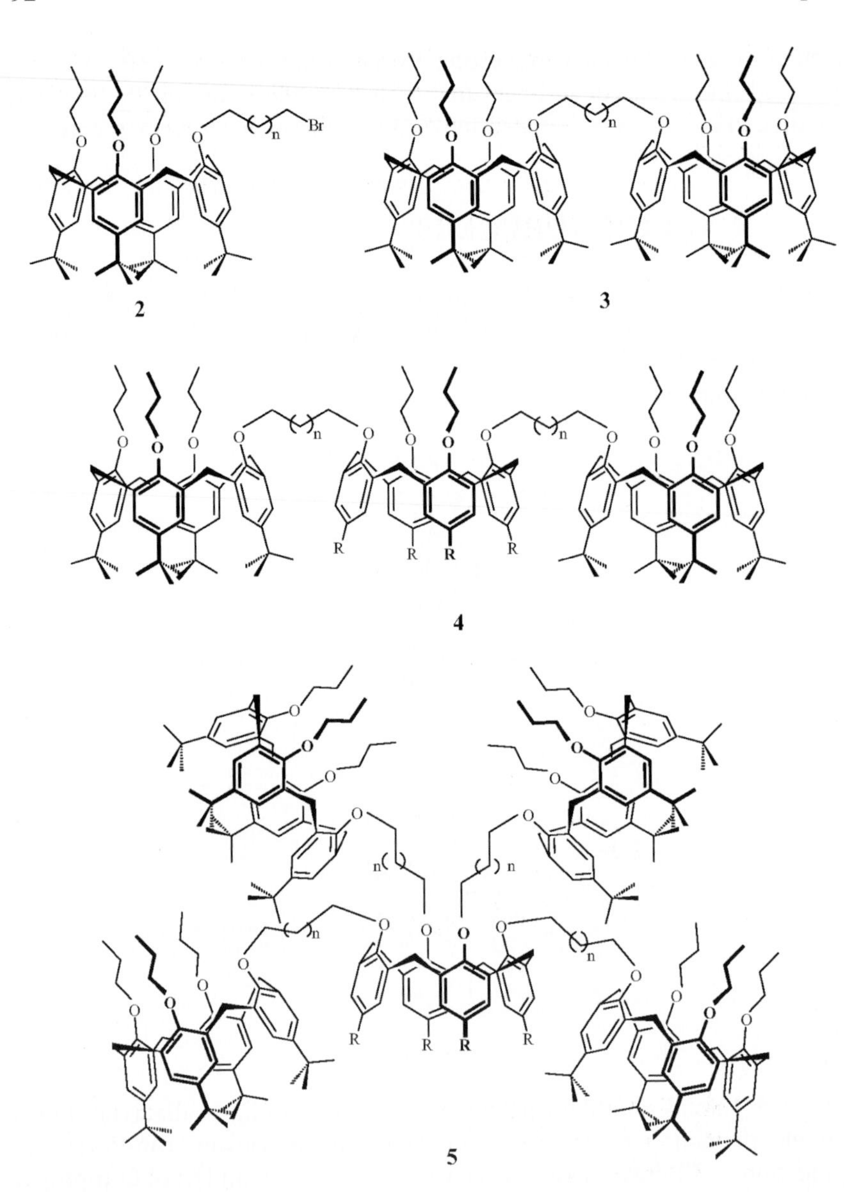

All the calix[4]arene moieties were shown to be in the cone conformation. These oligo-calixarenes were claimed to represent the first step towards purely calixarene-based dendrimers. ^{1}H NMR studies showed that they could bind a number of alkali metal cations (Na^{+} and Li^{+} as perchlorates in $CDCl_3$:CD_3CN = 4:1 v/v) at least equal to the number of calix[4]arene units in the molecule.

In 1997, Nagasaki and co-workers[31] synthesized photochromic dendrimers **6** which involve a 1,3-alternate conformer of a calix[4]arene as a core and azobenzene units as branches. The typical photo-switching of the azobenzene units was maintained in the interior of the dendrimers, allowing the regulation of the nanoparticle size and leading to potential development of drug delivery systems.

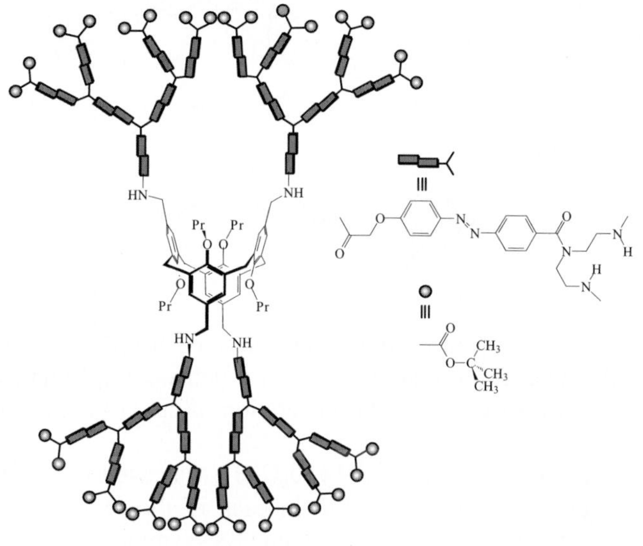

6

In 1998, Mogck *et al.*[32] reported the synthesis multi-calixarenes based on the formation of multiple amide links. Mono *ipso*-nitration of *p-tert*-butylcalix[4]arene tetraethers and subsequent reduction provided an easy access to *p*-monoaminocalix[4]arenes. Reactions with various di- and triacid chlorides led to double- and triple-calix[4]arenes **7** and **8**.

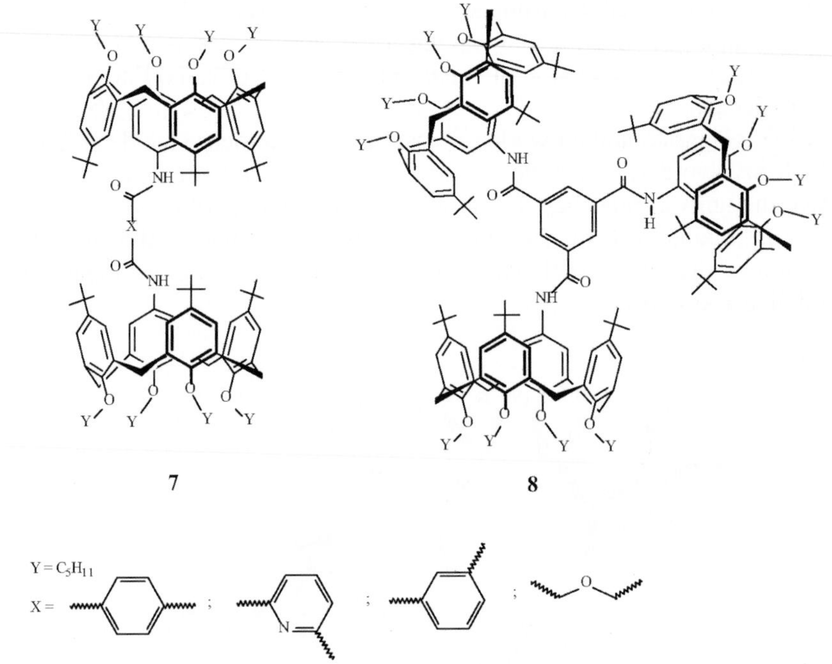

7 **8**

Y = C_5H_{11}

X =

Similar reactions of tetraacid chlorides derived from calix[4]arenes in the cone 1,3-alternate conformations provided penta-calix[4]arenes **9** and **10**, which can be regarded as the first generation of calix[4]arene-based dendrimers.

Penta-calix[4]arene **10** (Y = $-CH_2CO_2C_2H_5$) was used to complex NaSCN in $CDCl_3$. Complexation of one Na^+ in each of the four tetraester cavities was demonstrated by observing the retained S_4-symmetry of the ligand after complexation. Addition of free ligand leads to a spectrum corresponding to a superimposition of that of the complex and that of the free ligand, indicating exchange of the cation to be very slow.

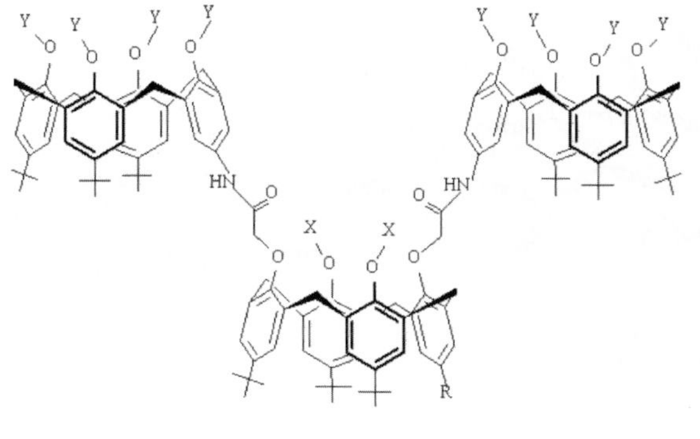

cal =

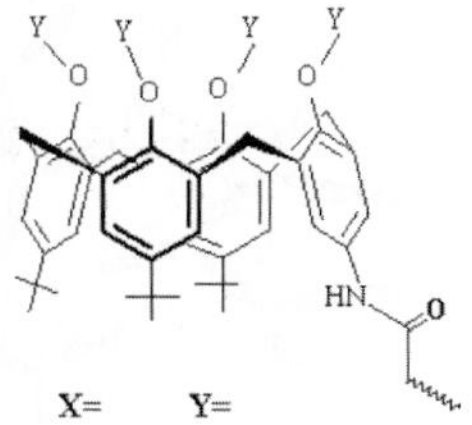

R=	X=	Y=
tBu	H	C_5H_{11}
tBu	H	$CH_2COOC_2H_5$
tBu	cal	C_5H_{11}
tBu	cal	$CH_2COOC_2H_5$
NO_2	cal	$CH_2COOC_2H_5$

9

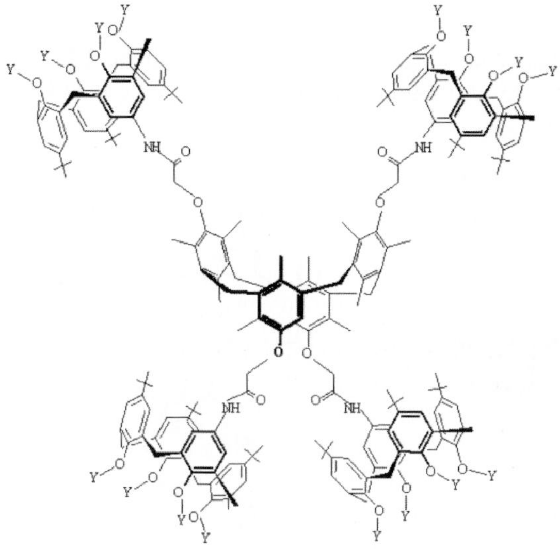

10

In 1999, Roy and Kim[33] described the first synthesis of a dendrimeric water-soluble, *p-tert*-butyl calix[4]arene **11** with sugar residues at its periphery. Lectin binding properties were observed. The lipophilic, t-butyl rim of the calixarene unit in **11** provides a site for surface adhesion, so that the oppositely-directed hydrophilic carbohydrate units mimic the saccharide rich surface of cells. The lectin binding ability indicates that bioanalytical applications based on the adsorption of the dendrimers on polystyrene-coated microtitre plates could be developed.

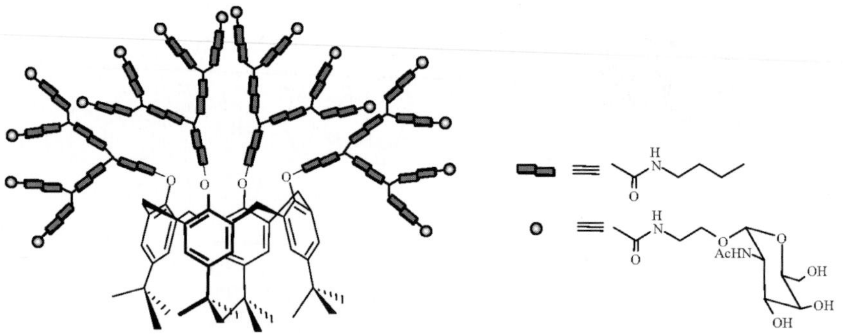

11

In 2002, Szemes *et al.*[34] reported the synthesis, from **12**, of calix[4]arene-based dendrimers, **13** and **14**, containing up to seven calix[4]arene moieties.

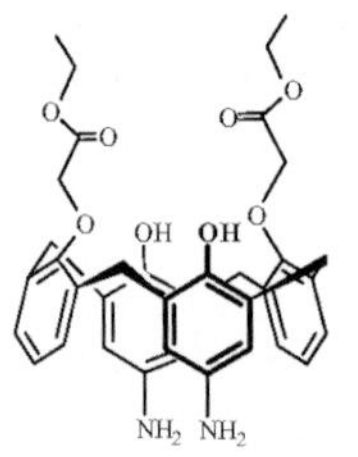

12

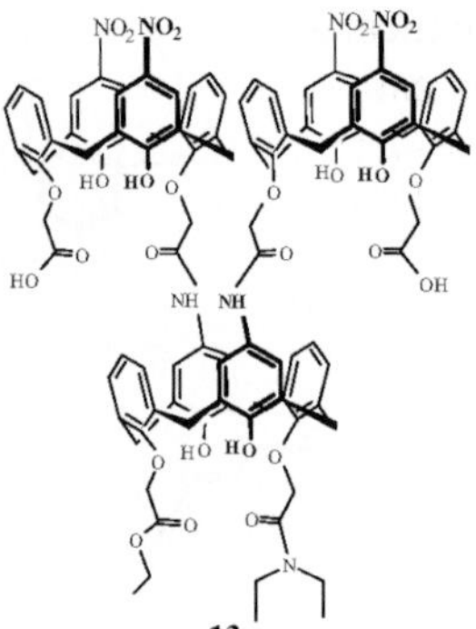

13

The construction takes advantage of the selective 1,3-O-dialkylation of calix[4]arene and subsequent selective p-nitration of the remaining phenolic rings. The linkage of the calix[4]arenes is made after hydrogenation of the nitro functions (see calixarene **12**), giving amines which can be reacted with acyl chlorides. Tricalix[4]arene **13** forms strong complexes with La^{3+}, Gd^{3+} and Lu^{3+}, as shown by UV-Vis titrations.

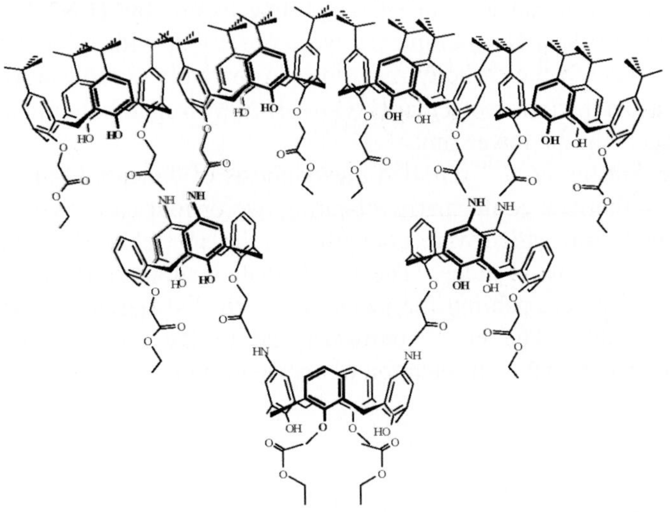

14

In 2003, Xu *et al.*[35] described a modular strategy towards macromolecules which combines the methods of peptide synthesis with functionalized calixarene chemistry. Various calix[4]arene amino acids were used to construct multi-valent calix-peptide-dendrimers. The first generations of two calix[4]arene peptide dendrimers **15** have been described.

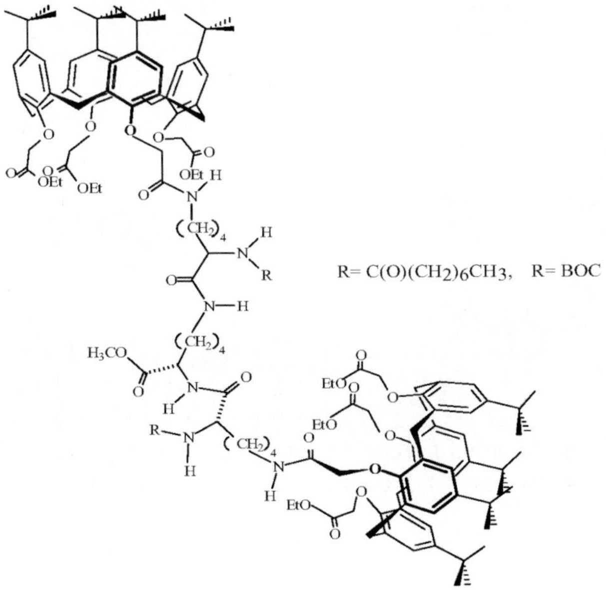

R= $C(O)(CH_2)_6CH_3$, R= BOC

15

Complexation of $NaClO_4$ in $CDCl_3$ was investigated by ^1H NMR spectroscopy. In both cases, Na^+ complexes were observed, with a localisation of the cation close to the carbonyl functions. Apparently complexation of Na^+ cation disrupts the intramolecular C=O·······H-N hydrogen bonding present in the free ligand at the lower rim.

In 2004, Šťastný *et al.*[36] reported the synthesis of thiacalix[4]arenes in the cone or 1,3-alternate conformations bearing two or four carboxylic functions on the lower rim which were converted to the acyl chlorides and reacted with para amino calix[4]arenes. The 1,3-alternate conformer **16** was observed to be less reactive, enabling the isolation of tricalix[4]arene **17** possessing inherent chirality. The cone conformer led to the corresponding penta calixarenes **18** and **19** with retention of configuration.

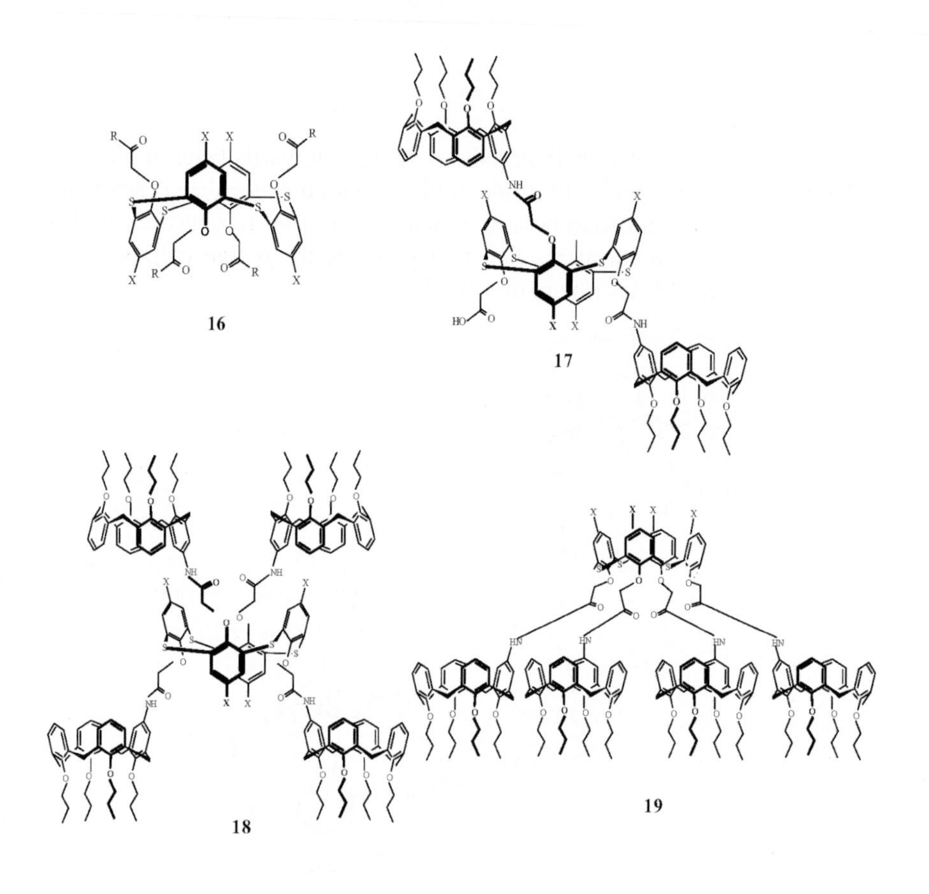

The same year, Appelhans and co-workers[37] used similar thiacalix[4]arene carboxylic acids **20** in the 1,3-alternate conformation for the design of dendritic cores with amino surface groups. To minimise steric effects, a phenyl spacer was added, leading to the formation of novel thiacalixarenes **21** bearing protected lysine groups.

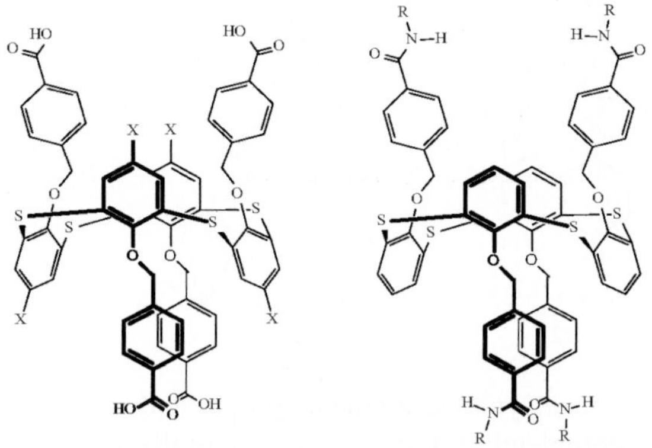

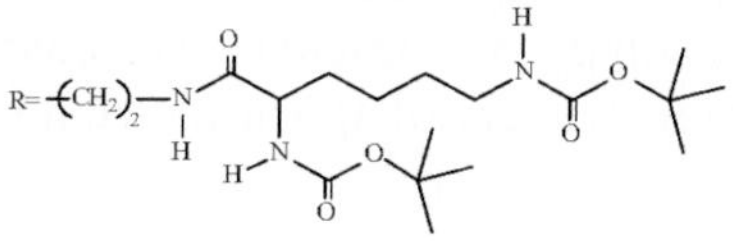

These compounds were claimed to be the first example of thiacalix[4]arene derivatives potentially useful as dendritic cores for subsequent branching derivatization.[37a] Recently, this team enlarged this family of molecules, including derivatives of the cone isomer of **20**, to form cone-type and spherical lysine dendrimers up to generation 3.[37b]

Wang *et al.*[38] synthesized calix[4]arenes **22** and **23** substituted at the narrow or wide rim by eight carbamoylmethyl-phosphine oxide (CMPO) functions in a dendritic manner. Extractions of Eu^{3+} and Am^{3+} were carried out from water to *o*-nitrophenylhexyl ether. ^{1}H-NMR relaxivity titrations for wide rim octa-CMPO derivatives revealed the formation of solvent-free 1:2 ligand/metal complexes, while the wide rim tetra-CMPO derivative formed oligomeric complexes under similar conditions.

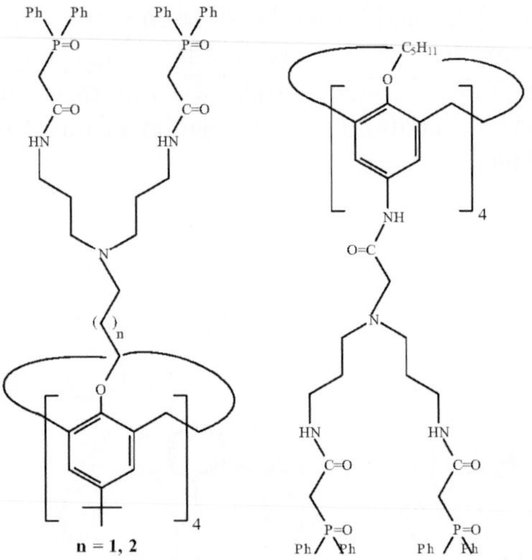

n = 1, 2

22 **23**

Kellermann *et al.*[39] have synthesized the amphiphilic dendro-calixarene **24** that assembles into completely uniform and structurally persistent micelles formed by the arrangement of seven molecules maintained together by hydrogen bonding. A particular use envisaged for such micelles is as vehicles for the delivery of apolar molecules.

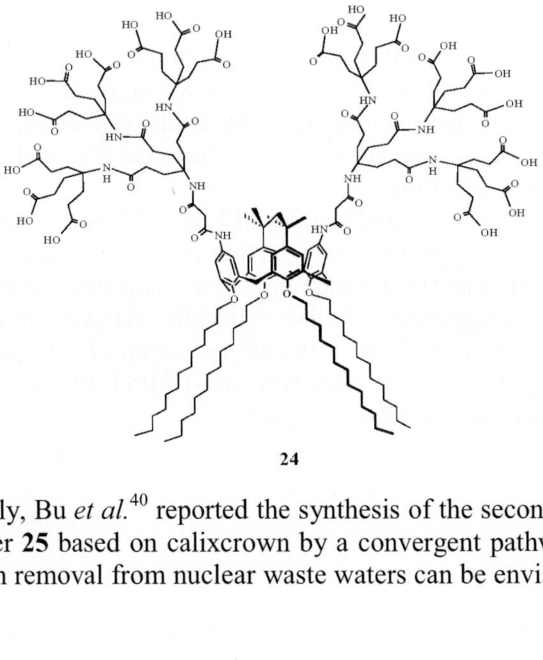

24

Recently, Bu *et al.*[40] reported the synthesis of the second generation of the dendrimer **25** based on calixcrown by a convergent pathway. Potential uses in cesium removal from nuclear waste waters can be envisaged.

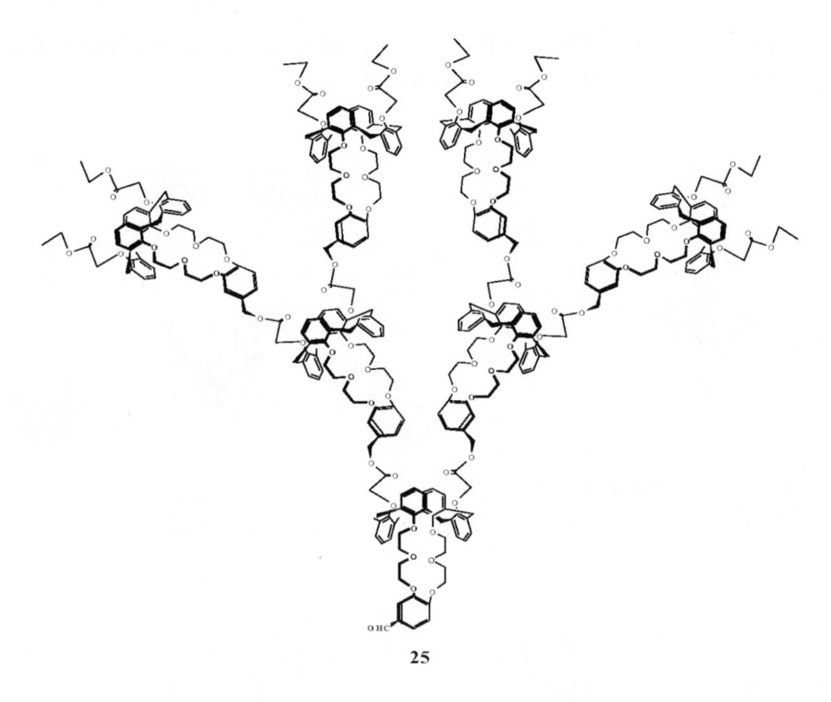

25

In work associated with that of the present authors, Cheriaa *et al.*[41] synthesized a diamido calix[4]arene **26** derivative from 'tren' and mono-carboxymethylcalix[4]arene which can be used for the preparation of a variety of hyperbranched molecules by acylation of the remaining amino group.

26 **27**

The molecule **26** is a dendron suitable for convergent dendrimer syntheses whereby the Y-shaped bis(calixarene) unit is added multiple times to a polyfunctional core. Where this core is also a calixarene derivative, molecules such as **27-30** may be obtained.[42]

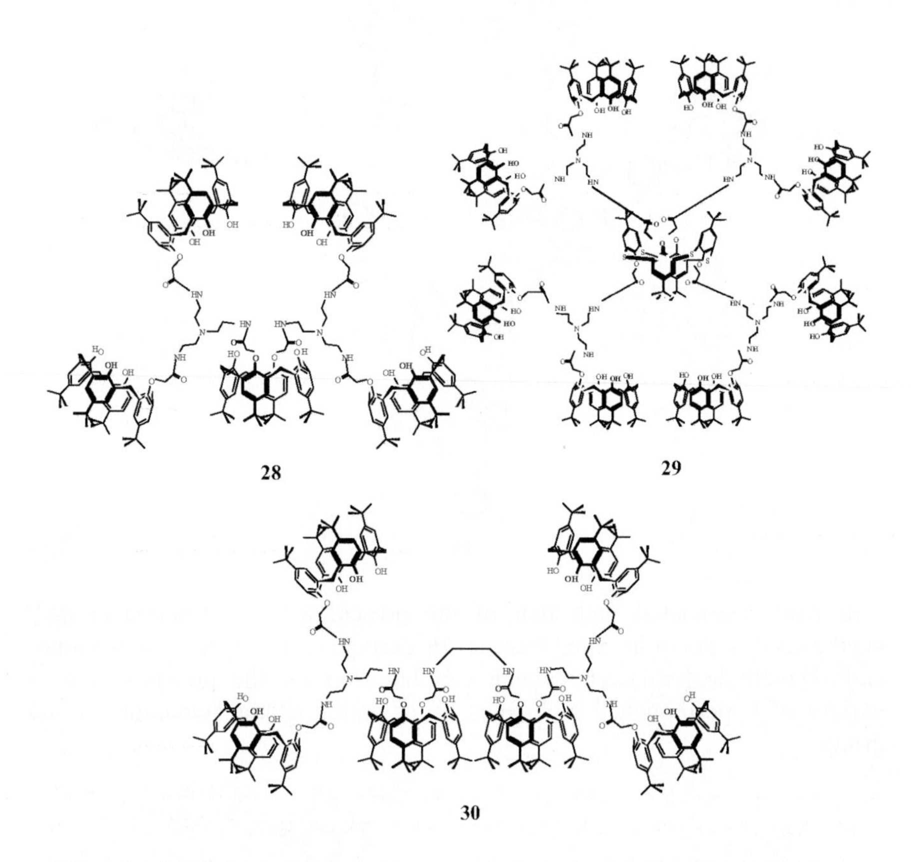

The tricalix[4]arene **27** may be regarded as a first generation dendrimer derived from the tetramine "tren" (tris(aminoethyl)amine) and can be used in a divergent synthesis of the second generation dendrimer **31**, which can also be obtained by a convergent procedure based on similar chemistry.[42]

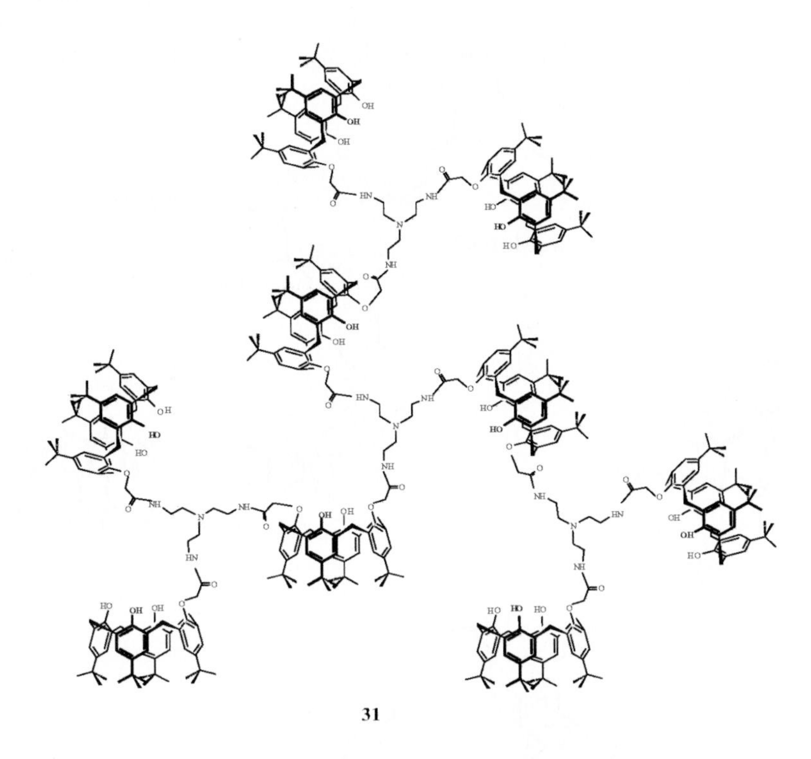

31

Another example of calixdendrimer formation via amide formation from alkoxycarbonylmethyl derivatives of calix[4]arene is provided in the synthesis of the tricalixarene **34** from the triaminotrithia-alcohol "hyten" and monocarboxymethylcalix[4]arene **32**.[43] Here, the core of the dendrimer contains an N_3S_3 donor atom array potentially useful for binding a six-coordinate metal ion, whereas **27** can provide only a site for four-coordination. The bis(calix) dendron **33** was also isolated from the reactions of hyten.

These molecules, designed as dendrimer precursors, retain a core capable of acting as a multidentate ligand for metal ions. Studies of complex formation with Zn(II) and Co(III) by these **27, 33, 34** showed that indeed there is preferential binding to the core (rather than the phenolic sites of the calixarene units), suggesting new mechanisms for the control of the structure and stereochemistry of dendrimer species.

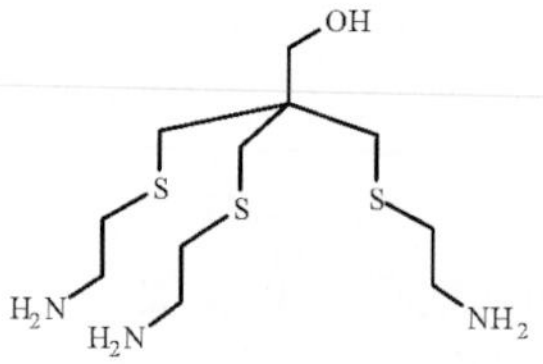

"hyten"

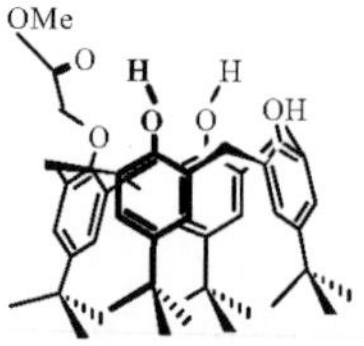

32

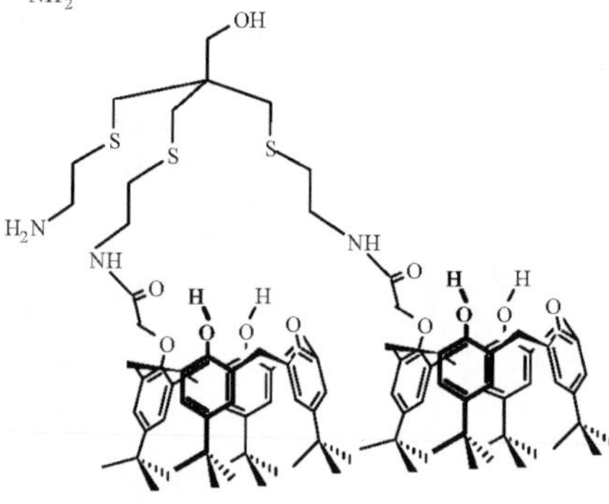

33

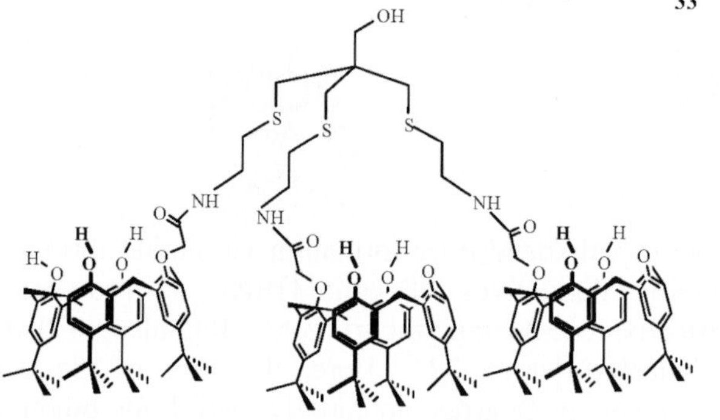

34

Reactions of other tripodal amines, "tame" ($CH_3C(CH_2NH_2)_3$) and "sen" ($CH_3C(CH_2NHCH_2CH_2NH_2)_3$) with **32** result in the tricalix derivatives **35** and **36**, respectively.[44] The formation of **36** demonstrates the selective acylation of the primary amino groups of sen. Studies of Zn(II) complexation by **35** and **36** indicate that coordination to O-donor sites of the calixarene units is preferred but prospects exist nonetheless, for **36** in particular, to bind other metals at the N-donor sites and so control the dendrimer stereochemistry.

The examples discussed above are all of calixdendrimers obtained by the formation of kinetically inert bonds. One recent example has been published of a calixdendrimer built by labile self-assembly (see chapter 2). Rudzevich *et al.*[45] demonstrated that a self-sorting process of different tri- and tetraurea derivatives can give rise to structurally uniform dendrimers by dimerization. The uniform size of the dendritic assemblies was shown by ^{1}H DOSY NMR experiments (diffusion coefficients of 5.4×10^{-11} m s^{-2} for all ^{1}H signals with an effective hydrodynamic radius of 2.1 nm) and confirmed by dynamic light scattering (which led to a hydrodynamic radius of 2.4 nm).

Similar chemistry has been exploited with a series of penta-calixarenes **39** where a 1,3-alternate calix[4]arene is linked to four cone calix[4]arene units, each with four urea units attached to the wide rim. Via H-bonding with simple tetra-aryl- or tetra-tosyl-urea-calix[4]arenes, **39** provides tetracapsular derivatives.[46]

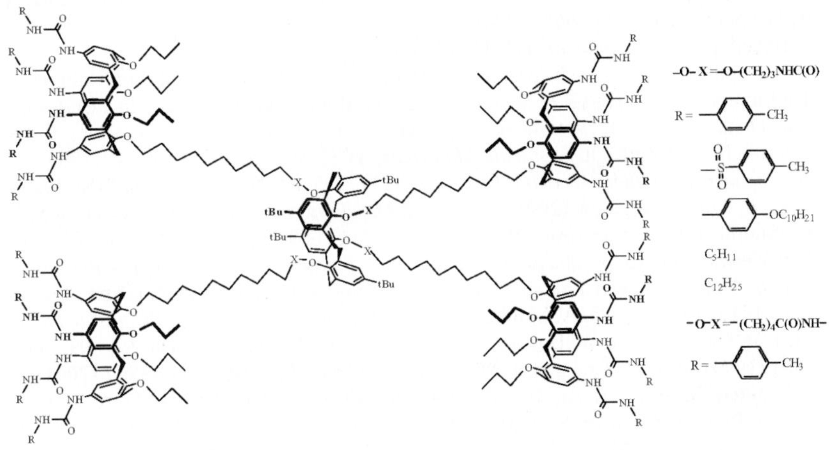

37

3. CONCLUSION

In many areas, calixarene and dendrimer chemistries have real and important applications[12,23] and the meeting of two fields both approaching true maturity may lead to a new field of applications. The chemistry discussed above shows the ease with which calixdendrimers that are certainly of nanometre dimensions may be synthesised. It is clear that there are many attractive prospects remaining to be exploited in this area and, given already known developments of calixarenes as molecular machines, transport agents, molecular reactors and sensors, considerable optimism is justified for the future of calixdendrimers.

4. REFERENCES

1. H. S. Nalwa (Ed.) *Encyclopedia of Nanoscience and Nanotechnology*, American Scientific Publishers, Stevenson Ranch (California), (2004), Vols 1 - 10.
2. D. Astruc, *C. R. Chimie* **6**, 709-711 (2003).
3. (a) G. M. Whitesides, *Small* **1**, 172-179 (2005); (b) V. Balzani, *Small* **1**, 278-283 (2005).
4. G. M. Whitesides, *Angew. Chem. Int. Ed.* **43**, 3632-3641 (2004).
5. C. A. Schalley, A. Lützen, M. Albrecht, *Chem Eur. J.* **10**, 1072-1083 (2004).
6. B. Bensaude-Vincent, H. Arribart, Y. Bouligand, C. Sanchez, *New J. Chem.* **26**, 1-5 (2002).
7. J. Skär, *Phil. Trans. R. Soc. Lond. A* **361**, 1049-1056 (2003).
8. P. A. Gale, *Phil. Trans. R. Soc. Lond. A* **358**, 431-453 (2000).
9. J.-M. Lehn, *PNAS* **99**, 4763-4768 (2002).
10. S Rasmussen, L. Chen, D. Deamer, D. C. Krakaeur, N. H. Packard, P. F. Stadler, M. A. Bedau, *Science*, **303**, 963-965 (2004).
11. B. Imperiali, *Chem. Commun.* 445-447 (2003).
12. Recent publications which provide reviews of the synthesis and applications of dendrimers include: (a) G. R. Newkome, C. N. Moorefield, F. Vögtle: In *Dendrimers and Dendrons*, VCH, Weinheim, Germany, (2001); (b) G. R. Newkome, C. N. Moorefield, F. Vögtle: In *Dendritic Molecules*, VCH, Weinheim, Germany, (1996); (c) *Dendrimers and Other Dendritic Polymers*, J. M. Fréchet, D. A. Tomalia, Eds., Wiley: London, United Kingdom, (2001); (d) *Recent Developments in Dendrimer Chemistry*, D. K. Smith, Ed., *Tetrahedron Symposium in Print* **59**, pp 3787-4024 (2003).; (e) *Dendrimers and Nanosciences*, D. Astruc, Ed. Numéro spécial *Comptes Rendus de Chimie*, **Vol. 6**, Elsevier France, (2003); (f) F. Zeng, S. C. Zimmerman, *Chem. Rev.* **97**, 1681-1712 (1997); (g) O. A. Matthews, A. N. Shipway, J. F. Stoddart, *Prog. Polym. Sci.* **23**, 1-56 (1998); (h) A. W. Bosman, H. M. Janssen, E. W. Meijer, *Chem. Rev.* **99**, 1665-1688 (1999) (i) S. M. Grayson, J. M. J. Fréchet, *Chem. Rev.* **101**, 3819-3867 (2001); (j) B. Klajnert, M. Bryszewska, *Acta Bio. Polonica* **48**, 199-208 (2001); (k) D. A. Tomalia, J. M. J. Fréchet, *J. Polym. Science, Part A : Polym. Chem.* **40**, 2719-2728 (2002); (l) A.-M. Caminade, J.-P. Marjoral, *Acc. Chem Res.* **37**, 341-348 (2004); (m) D. A. Tomalia, *Aldrichimica Acta* **37**, 39-57 (2004)
13. P. J. Flory, *J. Am. Chem. Soc.*, **63**, 3083-3090 (1941).
14. P. J. Flory, *J. Am. Chem. Soc.*, **63**, 3091-3096 (1941).
15. P. J. Flory, *J. Am. Chem. Soc.*, **63**, 3096-3100 (1941).
16. E. Buhleier, W. Wehmer, F. Vögtle, *Synthesis*, 155-162 (1978).
17. R. G. Denkelwalter, J. Kolc, W. J. Lukasavage, *U. S. Pat.* 4,289,872 (Sept. 15, 1981).
18. G. R. Newkome, Z.-Q. Yao, G. R. Baker, V. K. Gupta, *J. Org. Chem.* **50**, 2003-2017 (1985).
19. D. A. Tomalia, H. Baker, J. R. Dewald, M. Hall, G. Kallos, S. Martin, J. Roeck, J. Ryder, P. Smith, *Polym. J.* **17**, 117-131 (1985).
20. (a) C. B. Gorman, J. C. Smith, *Acc. Chem. Res.* **34**, 60-71 (2001); (b) D. Astruc, F. Chardac, *Chem. Rev.* **101**, 2991-3023 (2001); (c) R. M. Crooks, M. Zhao, L. Sun, V. Chechik, L. K. Yeung, *Acc. Chem Res.* **34**, 181-190 (2001); (d) R. van Heerbeek, P. C. J. Kamer, P. W. N. M. van Leeuwen, J. N. H. Reek, *Chem. Rev.* **102**, 3717-3756 (2002); (e) L. J. Twyman, A. S. H. King, I. K. Martin, *Chem. Soc. Rev.* **31**, 69-82 (2002); (f) M. Castagnola, C. Zuppi, D. V. Rossetti, F. Vincenzoni, A. Lupi, A. Vital, E. Meucci, I. Messana, *Electrophoresis* **23**, 1769-1778 (2002).
21. (a) V. Balzani, P. Ceroni, M. Maestri, V. Vicinelli, *Curr. Opin. Chem. Biol.* **7**, 657-665 (2003); (b) T. G. Goodson III, *Acc. Chem. Res.* **38**, 99-107 (2005).
22. (a) P. Veprek, J. Jezek, *J. Peptide Sci.* **5**, 203-220 (1999); (b) Y. Kim, S. C. Zimmerman, *Curr. Opin. Chem. Biol.* **2**, 733-742 (1998); (c) A. K. Patri, I. J. Majoros, J. R. Baker Jr, *Curr. Opin. Chem. Biol..* **6**, 466-471 (2002); (d) M. J. Cloniger, *Curr. Opin. Chem. Biol..*

2, 742-748 (2002); (e) U. Boas, P. M. Heegaard, *Chem. Soc. Rev.* **33**, 43-63 (2004); (f) E. R. Gillies, J. M. J. Fréchet, *Drug Discovery Today* **10**, 35-43 (2005).

23. For books on calixarene chemistry cf. the following: (a) *Calixarenes 2001*, Z. Asfari, V. Böhmer, J. Harrowfield, J. Vicens Eds., Kluwer Academic Publishers: Dordrecht, Netherlands, (2001); (b) C. D. Gutsche: In *Calixarenes Revisited*, J. F. Stoddart, Ed., Monographs in Supramolecular Chemsitry, Royal Society of Chemistry: London, (1998); (c) *Calixarenes: a Versatile Class of Macrocyclic Compounds*, J. Vicens, V. Böhmer Eds.; Kluwer Academic Publishers, Dordercht, Netherlands, (1991); (d) C. D. Gutsche: In *Calixarenes*, J. F. Stoddart, Ed., Monographs in Supramolecular Chemistry, No. 1, Royal Society of Chemistry : London, (1989).
24. A. Casnati, R. Ungaro, Z. Asfari, J. Vicens, see reference 2(a), chapter 20.
25. B. Pulpoka, V. Ruangpomvisuti, Z. Asfari, J. Vicens, in *Cyclophane Chemistry for the 21st Century* : H. Takemura, Ed., Research Signpost, Kerala, India (2002), chapter 3.
26. D. N. Reinhoudt, P. J. Dijkstra, P. J. In't Veld, K. E. Bugge, S. Harkema, R. Ungaro, E. Ghidini, *J. Am. Chem. Soc.* **109**, 4761-4762 (1987).
27. L. C. Groenen, J. A. J. Brunink, W. I. Iwema-Bakkler, S. Harkema, S. S. Wijmenga, D. N. Reinhoudt, *J. Chem. Soc. Perkin Trans. 2*, 1899-1906 (1992).
28. (a) M. Saadioui, V. Böhmer, see reference 2(a) chapter 7. (b) J.-M. Liu, Y.-S. Zheng, Q.-Y. Zheng, J. Wie, M.-X. Wang, Z.-J. Huang, *Tetrahedron* **58**, 3729-3736 (2002).
29. G. R. Newkome, Y. Hu, M. J. Saunders, F. R. Fronczek, *Tetrahedron* **55**, 1133-1137 (1991).
30. P. Lhotak, S. Shinkai, *Tetrahedron* **51**, 7681-7696 (1995).
31. T. Nagasaki, S. Tamagaki, K. Ogino: *Chem. Lett.*, **19**, 717-718 (1997).
32. O. Mogck, P. Parzuchowski, M. Nissinen, V. Böhmer, G. Rokicki, K. Rissanen, *Tetrahedron* **54**, 10053-10068 (1998).
33. R. Roy, J. M. Kim: *Angew. Chem. Int. Ed.*, **38**, 369-372 (1999).
34. F. Szemes, M. G. B. Drew, P. D. Beer: *Chem. Commun.* 1228-1229 (2002).
35. H. Xu, G. R. Kinsel, J. Zhang, M. Li, D. M. Rudkevich: *Tetrahedron* **59**, 5837-5848 (2003).
36. V. Štastný, I. Stibor, H. Dvořáková, P. Lhoták, *Tetrahedron* **60,** 3383-3391 (2004).
37. (a) D. Appelhans, V. Stastny, H. Komber, D. Voigt, B. Voit, P. Lhoták, I. Stibor, *Tetrahedron Lett.* **45**, 7145-7149 (2004).(b) D. Appelhans, M. Smet, G. Khimich, H. Komber, D. Voigt, P. Lhotàk, D. Kucklinge, B. Voit: *New J. Chem.* **29**, 1386-1389 (2005).
38. P. Wang, M. Saadioui, C. Schmidt, V. Böhmer, V. Host, J. F. Desreux, J.-F. Dozol, *Tetrahedron* **60**, 2509-2515 (2004).
39. M. Kellermann, W. Bauer, A. Hirsch, B. Scade, K. Ludwig, C. Böttcher: *Angew. Chem. Int. Ed.*, **43**, 2959-2962 (2004).
40. J.-H. Bu, Q.-Y. Zheng, C.-F. Chen, Z.-T. Huang, *Tetrahedron* **61**, 897-902 (2005).
41. (a) N. Cheriaa, R. Abidi, J. Vicens: *Tetrahedron Lett.* **45**, 7795-7799 (2004).(b) N. Cheriaa, R. Abidi, J. Vicens: *Tetrahedron Lett.* **46**, 1533-1536 (2005).
42. N. Cheriaa, M. Mahouachi, A. Ben Othman, L. Baklouti, Y. Kim, R. Abidi, J. Vicens: *Supramolecular Chem.* in press.
43. M. Mahouachi, Y. Kim, S.H. Lee, R. Abidi, J. Harrowfield, J. Vicens: *Supramolecular Chem.* **14**, 323-330 (2005).
44. M. Mahouachi, R. Abidi, Y. Kim, S. H. Lee, Jong H. Jung, J. S. Kim, J. Vicens: *Org. Chem. Lett.* submitted
45. Y. Rudzevich, V. Rudzevich, C. Moon, I. Schnell, K. Fischer, V. Böhmer: *J. Am. Chem. Soc.* **127**, 14168-14174 (2005).
46. Y. Rudzevich, K. Fischer, M. Schmidt, V. Böhmer: *Org. Biomol. Chem.* **3**, 3916-3925 (2005).

Chapter 6

CALIX[4]TUBES AND CALIX[4]SEMITUBES

A new class of ionophore

Susan E. Matthews[a] and Paul D. Beer[b]

[a]*School of Chemical Sciences and Pharmacy, University of East Anglia, Norwich, NR4 7TJ, UK.* [b]*Inorganic Chemistry Laboratory, University of Oxford, South Parks Road, Oxford, OX1 3QR, UK.*

Abstract: The development of a novel class of ionophore based on bis-calix[4]arene molecular structures, the calix[4]tubes and the structurally related calix[4] semitubes, is reviewed in this chapter. The parent system exhibits remarkable selectivity for potassium over all group 1 metal cations and barium. Metal cation selectivity and rates of complexation can be further tuned *via* upper-rim substituent variation and by the nature of the lower-rim bridging linkages. The incorporation of quinone redox-active groups into the tubular structural design has produced potassium, and the first rubidium and caesium selective redox sensing ionophores.

Key words: Calix[4]tube, calix[4]semitube, cation binding.

1. INTRODUCTION

The selective recognition and extraction of Group 1 metals has garnered considerable interest in many fields including the preparation of mimics for biological sodium and potassium channels and for the development of selective extractants for caesium for nuclear waste remediation[1]. Since the pioneering work of Charles Pedersen[2], the macrocyclic crown ether derivatives have proved attractive as selective hosts for group 1, group 2 and ammonium cations. These hosts feature multiple hard donor oxygen atoms and can be readily synthetically modified to provide complementary cation to cavity size, cavity pre-organisation and additional π-cation binding interactions. In recent years, the calixcrowns have been developed which, through variation

J. Vicens and J. Harrowfield (eds.), Calixarenes in the Nanoworld, 109–133.

of calixarene conformation and crown cavity size have shown selectivity for sodium[3], potassium[4] and caesium.[5]

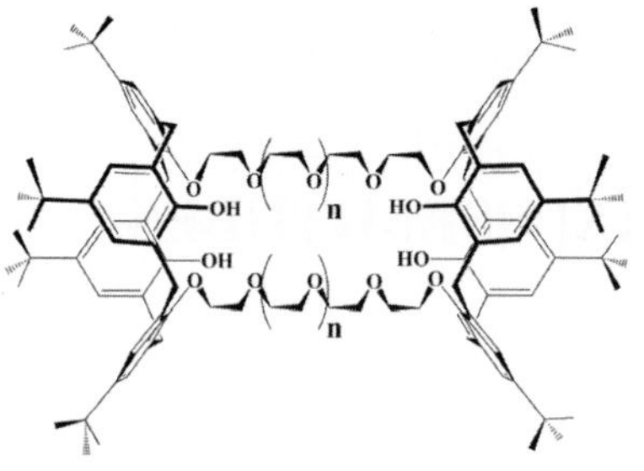

1

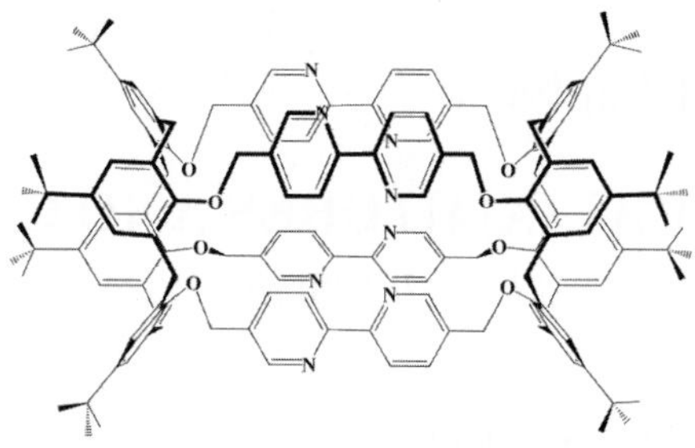

2

1a n = 0; **1b** n = 1; **1c** n = 2

Figure 6-1. Calix[4]polyether (partial) tubes[7] (left) and a calix[4]barreland[8,9] (right).

Following the structural elucidation of the potassium ion channel of *Streptomyces lividans*[6] and the identification of a potential four-tyrosine selectivity filter we were interested in developing potassium selective ionophores based on calixarenes where the calixarene cavity would restrict entry to the binding array in a size-selective manner. Building on the work of Shinkai and co-workers, who demonstrated shuttling of a sodium metal cation between two binding sites in a polyether partialtube **1**[7] and that of the group of Ziessel,[8,9] who developed a series of bis-calixarene barrelands **2** (Fig. 1), we were interested in exploiting bis-calixarene systems. To this end, we developed a new class of ionophore, the calix[4]tube system, which displays exceptional selectivity in potassium binding. In this chapter, we review the synthesis and binding capabilities of the calix[4]tubes and the development of structurally related novel systems such as the calix[4]semitubes (Fig. 2).

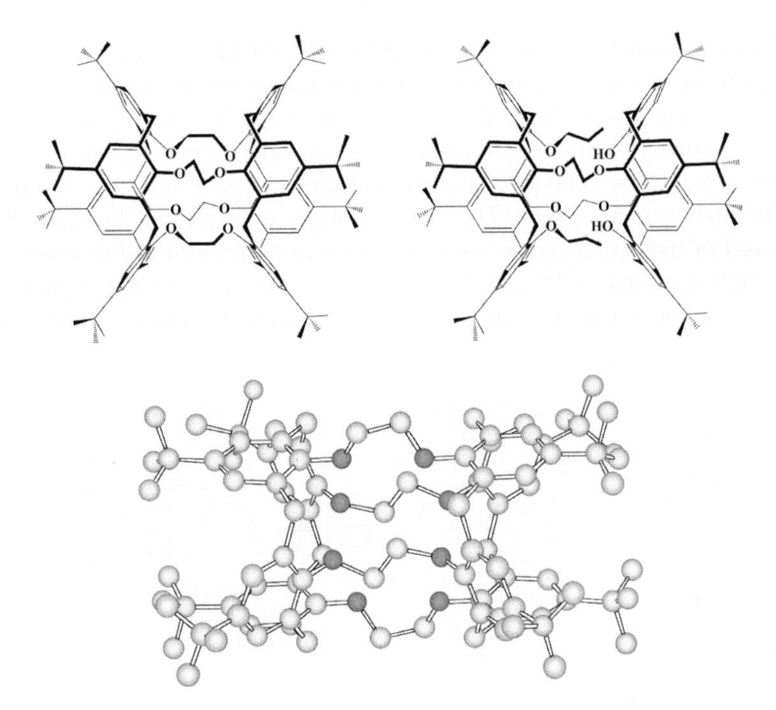

Figure 6-2. Representations of a calix[4]tube (upper left) and a calix[4]semitube (upper right), and the *p*-*t*-butylcalix[4]tube molecule as found in the crystal (disorder not shown).

2. CALIXTUBES

2.1 Calix[4]tubes

2.1.1 Synthesis and Structure of Calix[4]tubes

First reported in 1997,[10] the synthesis of the parent calix[4]tube **5** was achieved in 51% yield on treatment of *p*-*t*-butyl calix[4]arene **3** with *p*-*t*-butyl calix[4]arene tetratosylate **4** in the presence of K_2CO_3 in refluxing acetonitrile (Fig. 3). This yield is exceptionally high for such a multiple intermolecular condensation reaction and suggests a templation role for the potassium ion. This hypothesis was confirmed by repeating the synthesis using Na_2CO_3 and Cs_2CO_3, where, in both cases, no tube product was isolated. ^1H NMR studies on the calix[4]tube **5** indicated that the structure was highly rigid, with each calixarene adopting a flattened cone conformation, as evidenced by a doubling of all signals excluding those of the methylene bridges, which was unchanged on heating up to 328 K. The crystal structure (Fig. 2) shows that this C_{2v} symmetry of the calixarenes is maintained in the solid-state and that the ethylene linkages adopt a trans-gauche-trans-gauche arrangement (*tgtg*). This torsion angle arrangement has also been observed in molecular dynamics simulations to be present in the lowest energy conformations of the calix[4]tube whereas the *gggg* arrangement necessary for complexation of a cation is seen in the highest energy conformers.

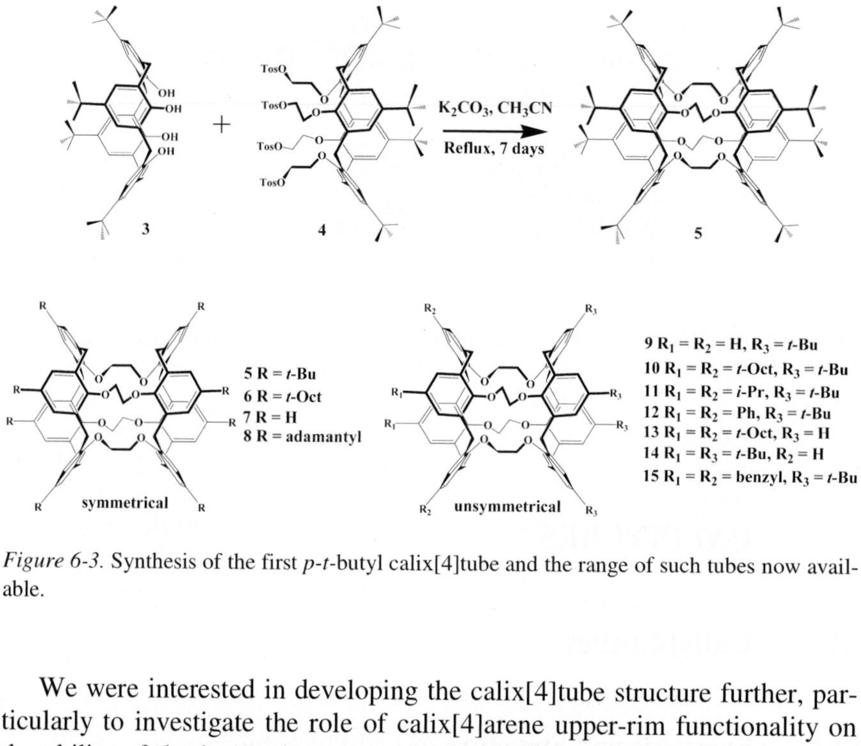

Figure 6-3. Synthesis of the first *p*-*t*-butyl calix[4]tube and the range of such tubes now available.

We were interested in developing the calix[4]tube structure further, particularly to investigate the role of calix[4]arene upper-rim functionality on the ability of the lower-rim cryptand like array to complex cations and the potential for achieving binding selectivity.[11] On extending the series for the preparation of calix[4]tubes **6**,**7**,**9-14** further insights into the synthetic approach were elucidated (Fig. 3). Whilst acetonitrile is an effective solvent for the synthesis of the parent tube, with other calixarenes, for example calix[4]arene and *p*-*t*-octyl calix[4]arene, the reactions proved lower-yielding and gave more complex mixtures of products. However, by changing the solvent for the tube-forming step to the higher-boiling xylene, 50-60% yields of pure material could be achieved. A comparison can be made for the calix[4]tube **9**. Using the initial method, Stibor and co-workers reported a

14% yield[12] whereas this yield can be increased to 66% using the xylene method.[11] With asymmetric tubes the choice of synthetic route can greatly affect the yield observed; for example enhanced yields are achieved using tetratosylate derivatives of *p*-*t*-butylcalix[4]arene or *p*-*t*-octylcalix[4]arene rather that calix[4]arene as a consequence of the greater solubility of both reagents in xylene (*e.g.* for **13**, *p*-*t*-octylcalix[4]arene tetratosylate 61%, calix[4]arene tetratosylate 20%). Two additional calix[4]tubes have been reported, **15**, which was prepared from *p*-benzylcalix[4]arene tetratosylate using the acetonitrile method,[13] and the adamantyl derivative **8**.[14]

All of the calix[4]tubes adopt the same flattened cone structure as the parent tube in solution, which is maintained when heated between 278 and 328 K. However, the rate of exchange between the two extreme flattened cone conformations through the C_{4v} symmetrical intermediate varies between the tubes. Using 1D-EXSY NMR spectroscopy a rate of 0.94 s^{-1} was calculated for **5** and 2.95 s^{-1} for **6** in $CDCl_3$:CD_3OD 4:1 at 328 K. With calix[4]tube **14**, the two flattened cone conformers are no longer equivalent as a consequence of the 1,3-derivatisation. The two conformers are observed separately in the NMR spectrum and a preference (1.6:1) is observed for the conformer in which the aryl residues bearing hydrogen adopt the upright position, a result which can be related to the lesser steric requirements of a hydrogen atom compared to that of a *t*-butyl group.[11]

2.1.1.1 Derivatisation of Calix[4]tubes

A particular area of interest within the calix[4]tube field is upper-rim functionalisation to enable the preparation of potential ion-pair heteroditopic receptors[15]. Such receptors would enable the counter anion to be tightly bound by hydrogen bonds at the same time as a cation is bound in the cryptand-like array. Most efforts have focused on the introduction of halo-functionality. For example, bromination could allow access into a carboxylic acid derivative and iodination to an amino derivative *via* Gabriel methodology.

Two routes are available for the preparation of tetra functionalised derivatives; either functionalisation of an asymmetric tube *i.e.* after the tube has been prepared, or functionalisation of one of the calixarene sub-units before forming the tube. The latter convergent approach is more attractive for the preparation of a large family of functionalised tubes. However, despite introducing the bromo functionality at the remote site, to minimise interference in the tube-forming step, no functionalised tubes could be prepared by this route using our standard conditions (K_2CO_3/xylene).[16]

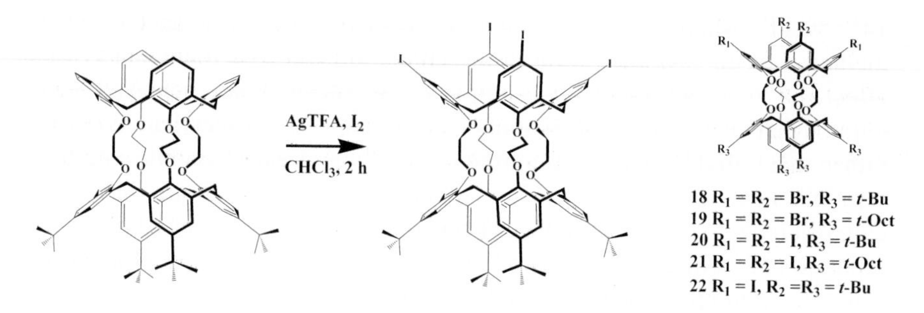

Figure 6-4. Synthesis of an iodotube, and the range of known halotubes.

In contrast, bromo functionalised tubes can be prepared from the asymmetric tubes **9** and **13** using a large excess of N-bromosuccinimide in butan-2-one, with the initially low yields being improved slightly by heating the reaction at reflux. Iodination proved considerably more successful as a functionalisation route, with yields of above 60% of the functionalised tubes **20-22** after two hours treatment with iodine in the presence of silver trifluoracetate (Fig. 4). Solid-state structures were obtained for both bromo and iodo derivatives and these are similar to that of the parent tube, with a *tgtg* arrangement of the linkers and the calixarenes adopting even more pronounced flattened cone conformations. ^{1}H NMR studies showed once again that the C_{2v} arrangement is adopted in solution and that exchange between the two conformations is slow. Interestingly, 1D-EXSY measurements show that the interconversion is faster in these functionalised derivatives than for the alkyl calix[4]tubes. Again, with the 1,3-derivatised calix[4]tube **22**, a preference is observed for the derivative in which the iodo aryl group adopts the upright position but this is less marked than for the unsubstituted calix[4]tube **14** (1.3:1 *cf* 1.6:1), due to the increased steric requirements of iodine.

2.1.2 Complexation of Cations

2.1.2.1 Complexation of Potassium

The selective and strong binding of potassium by calix[4]tubes is remarkable for such a simple structure.[10,11] ^{1}H NMR solid-liquid extraction studies with a 20 fold excess of potassium iodide showed that complexation is slow on the NMR time scale. This enables analysis of complexation through a direct comparison of the integration of the NMR peaks relating to complexed and uncomplexed calix[4]tubes. On complexation the NMR spectrum simplifies from the C_2 symmetrical structure of the free calix[4]tube to a time-averaged C_4 structure with spectroscopic shifts that are consistent with potassium binding within the cryptand-like binding array. This solution behaviour is consistent with the solid-state structure of the potassium complex of **5.KI**, (Fig. 5), where the cation has approximately four-fold rotational symmetry. The flattened cone structures of the uncomplexed tube become regular cone conformations on binding and the ethylene linkers rearrange to give all gauche torsion angles (*gggg*), enabling the potassium to be bound to the eight oxygen donor atoms.

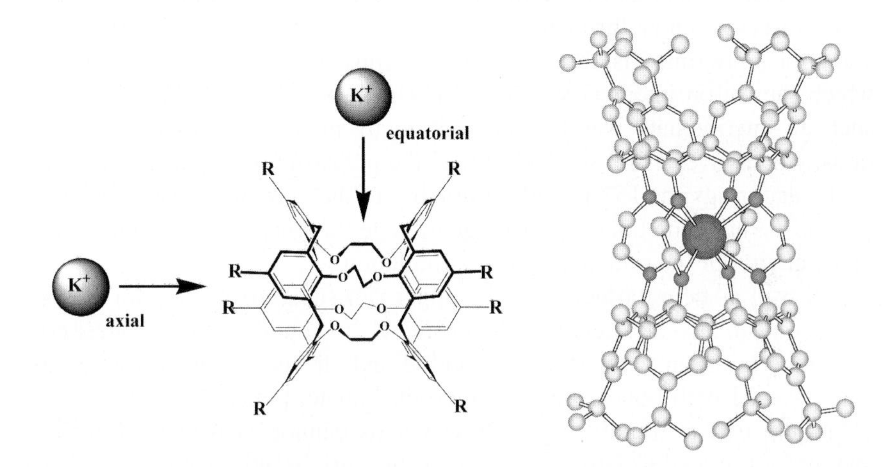

Figure 6-5. Possible modes of entry of potassium ion into a calix[4]tube and the form of the complex of the *t*-butyl tube found in the crystal.

When considering the range of calix[4]tubes with different upper rim substituents, three different rates of uptake of potassium are observed, those which are fast and exponential, those with intermediate kinetics, often exhibiting a sigmoidal profile for the binding, and those which are very slow. With the four related calix[4]tubes **5, 14, 9** and **7** which differ in the degree of *t*-butyl substitution, calix[4]tube **5** is at equilibrium within 2 h (k_{obs} = 3.9 x 10^{-3} s^{-1} at 298 K). Removal of just two *t*-butyl groups results in significant slowing of complexation (k_{obs} = 6.8 x 10^{-3} s^{-1} at 298 K for **14**) and further modification leads to further rate decreases, with calix[4]tube **9** showing 10% complexation at 2 h and with **7** showing 11% complexation after 15 h. The *p*-*t*-octyl derivatives **6** and **10** exhibit sigmoidal binding profiles, whereas **11** and **12** exhibit very slow binding kinetics which may be a result of enhanced π-π interactions preventing entry into the binding array or of a diminished ability of the linkers to change conformation and re-arrange to accommodate the cation.

This large range of the rate of potassium cation complexation with variation of the upper-rim substituent of calix[4]tubes suggests that the cation enters the cryptand-like cavity through the aromatic filter of the calixarene cavity. The route of entry of a cation into the cavity (Fig. 6), either axially through the calixarene or equatorially through direct re-arrangement of the ethylene linkers, has been investigated using molecular modelling simulations.[11,17] A potassium cation was placed at different distances from the binding array of **5** either along the main axis of the molecule or perpendicular to it. In all cases, entry through the axial route is the lowest energy pathway and a local minimum is observed which is consistent with an intermediate in which the cation interacts with the aryl rings of the filter. The presence of such an intermediate was further verified by molecular dynamics simulations, where, over a 500 ps time course, the potassium ion remains at the top of the aryl rings for 152 ps before moving to the central cavity at the same time as the ethylene linkers re-arrange to *gggg* torsion angles; it then remains in the cavity for the remainder of the simulation.

As well as being affected by the upper-rim substituent, the rate of complexation can also be altered by the nature of the counter-anion. Fastest rates of uptake are seen with potassium iodide and slower, yet measurable rates are observed with potassium hexafluorophosphate, potassium benzoate and potassium triflate. Interestingly, these results cannot easily be rationalised and are not related to lattice energies of the salts, as no uptake is seen over two hours with potassium nitrate, potassium perchlorate and potassium picrate. The strength of binding within the calix[4]tube **5** has been probed through competitive binding studies with the well-known potassium ionophore 2.2.2-cryptand. The calix[4]tube complex proves highly stable, with only 28% conversion over 15 h of a pre-complexed solution of the calix[4]tube to the potassium complex of 2.2.2-cryptand.

The calix[4]tubes are exceptionally selective for potassium over other Group 1 metals and the similarly sized barium cation. Qualitative competitive electrospray mass spectrometry experiments give evidence for only the potassium complex. This kinetic selectivity is confirmed with ^{1}H NMR studies which show, for example, that after 48 h, **5** has complexed less than 5% rubidium compared with 100% complexation of potassium at the same time-point. Molecular modelling studies of the complexation process show that smaller cations such as Na^+ and Li^+ enter the cavity but also leave rapidly, while the larger cations Cs^+ and Rb^+ preferentially remain at the upper aryl rim of the calixarene.[10]

The introduction of halogen substituents to the upper-rim of the calix[4]tubes results in a slowing in the rate of potassium complexation but does not affect selectivity for potassium over the rest of the Group 1 metals.

After 15 h in a ^{1}H NMR solid-liquid extraction study, the potassium uptake ranged from 3-42%, the greatest uptake being for calix[4]tube **22** in which only two iodo substituents are present. Molecular modelling with these derivatives indicates a preference for the cation to enter axially through the *p*-alkyl calix[4]arene annulus rather than through that of the *p*-halo calix[4]arene.[16]

2.1.2.2 Complexation of Silver

The ability of calix[4]tubes to complex the smaller silver cation in one of two modes, either in the cryptand polyether array or the tetraaryl calixarene cavity, depending on the counter ion present is one of the most interesting aspects of tube complexation. Our initial binding studies with silver iodide (20 eq. AgI in $CDCl_3$: CD_3OD 4:1) and calix[4]tubes **5** and **6** gave no indication of binding. However, on changing the counter-ion to hexafluorophosphate, structural changes consistent with binding in the cryptand array were observed.[17] The ^{1}H NMR spectrum simplified, as with potassium binding, indicating a change in the structure from C_2 to C_4 symmetry. In contrast to potassium binding, complexation was rapid (Ag^+ 5 min, K^+ 60 h) and gave rise to broad peaks for the complex. Molecular modelling studies have also been undertaken for the binding of silver and show that silver enters into the cryptand cavity more readily than potassium but that once the silver is bound the polyether linkers are still able to change conformation indicating that the smaller silver cation is a less good fit for the binding array.

The importance of counter anion on silver binding was confirmed additionally by studies using silver trifluoroacetate with the iodo functionalised calix[4]tubes.[16] Silver trifluoroacetate is used during the synthetic conversion to the iodo tube but the product is isolated as the uncomplexed structure. This observation is accounted for by binding studies which show no changes when the silver trifluoroacetate salt is used.

The group of Stibor has reported a completely different binding mode for silver with the asymmetric tube **9**.[12] When silver triflate is used in binding studies the silver is complexed within the arene array of the calix[4]arene under the same conditions. The binding of silver does not result in a simplification of the spectrum but in ^{1}H NMR spectroscopic shifts which are particularly marked in the calix[4]arene region. The importance to silver binding of the calix[4]arene adopting a C_{2v} structure was demonstrated through a competition study with a pre-complexed solution of **9** with potassium. Once potassium is bound the silver can no longer bind, although the use of two different counter-ions somewhat complicates this result. A solid-state structure was obtained (Fig. 6), confirming arene-site binding and showing a dimer of calix[4]tubes with a silver ion occupying the *p*-H-calix[4]arene cavity of both tubes and two triflate counter-ions acting as bridges between the dimers. In a later account[18] they have also reported the extension of this work into crystal engineering. By complexation of **7** with silver triflate an infinite linear array of *p*-H-calix[4]tubes can be prepared in which a silver ion is bound in each aryl cavity of the calix[4]tube.

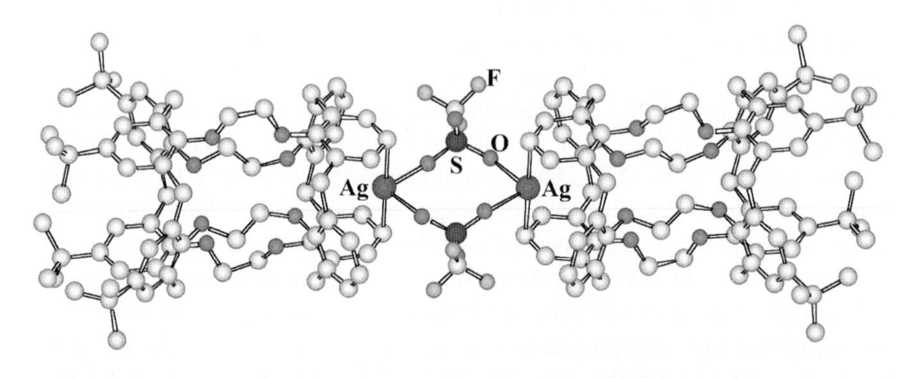

Figure 6-6. The triflate-bridged, binuclear Ag(I) complex of the asymmetric calix[4]tube **9**.

2.1.2.3 Complexation of Thallium

As has been discussed the nature of both the upper-rim aryl substituent and the counter anion can greatly affect the rate of binding and binding mode, be it within the polyether cryptand-like array or within the calixarene aryl cavity. Following our studies with silver we were interested in investigating the binding of thallium (I), a cation of similar size to potassium.[19] ^{1}H NMR solid-liquid extraction studies (20 eq. $TlPF_6$, $CDCl_3$:CD_3OD 4:1) showed that complexation was slow and that whilst structural changes were occurring on complexation, the lack of perturbation in the cryptand polyether region and the loss of symmetry in the aryl region pointed towards binding in the calix[4]arene cavity. ^{205}Tl NMR studies gave a more direct measure of the process that occurs. Both symmetrical tubes **5** and **6** behave similarly and an intermediate, which was not observed in the ^{1}H NMR studies, was identified as arising from the binding of one cation in one calixarene aryl cavity before the formation of a second stable complex in which two cations are bound (Fig. 7). The shifts observed (δ-599 to δ-634 ppm) for both the intermediate and the final complexes are not consistent with binding in the polyether array (expected shift Δδ~60 ppm). With the asymmetric calix[4]tube **10**, a more complex situation is observed; the final complex shows two thallium peaks which can be directly related to the symmetric structures and indicate binding in each of the different cavities. Interestingly, only one intermediate is observed, consistent with binding of one cation in the *p-t*-octyl calix[4]arene. This may be a result of preferential binding in the *p-t*-octyl cavity or of fast binding of a second cation from the *p-t*-butyl monocation intermediate. These solution studies are consistent with the unprecedented structure of the bis thallium (I) complex obtained with the *p-t*-octyl calix[4]tube (Fig. 7). Here two discrete cations are bound within the aryl calixarene cavities and whilst these cavities have become less flattened on binding this occurs without alteration of the conformation of the polyether cryptand recognition site.

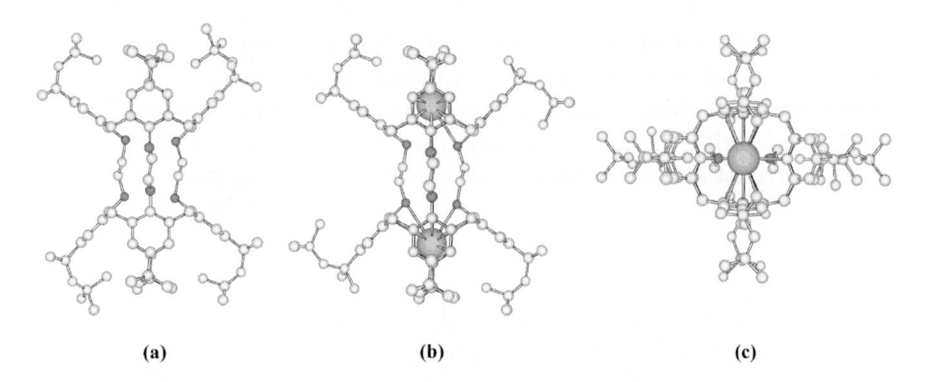

Figure 6-7. (a) The molecular structure of the ligand **6** and (b), (c) orthogonal views of its *bis*-Tl(I) complex.

2.2 Calix[5]tubes and Related Derivatives

2.2.1 Calix[5]tubes

The successful preparation of the calix[4]tube family led us to investigate the development of extended systems in which larger cryptand binding arrays might prove selective for cations other than potassium. Initially, we investigated the preparation of calix[5]tubes using our standard tube-forming conditions but failed to isolate any product.[20] Raston and co-workers[13] have also thoroughly investigated the formation of calix[5]tubes from the pentatosylate derivative of calix[5]arene under a variety of different conditions using a range of templates, temperatures and dilution approaches. They also explored the direct reaction of calix[5]arene with chloroethyl tosylate and isolated a crowned derivative which suggests that an initial trialkylation is followed by an intramolecular alkylation rather than the desired intermolecular reaction.

2.2.2 Calix[4]-calix[n]tubes

The groups of Pappalardo[21] and Raston[13] have both investigated the synthesis of hybrid tubes focusing on the reaction of tetratosylate derivatives of calix[4]arenes with calix[5],[6] or [8] arene, but to-date these efforts have not been extended to an evaluation of cation binding. Pappalardo and coworkers[21] reported that the reaction of *p-t*-butyl calix[4]arene tetratosylate **4** with two equivalents of *p-t*-butyl calix[5]arene in the presence of K_2CO_3 in acetonitrile gave, along with the expected majority trimeric product **23** (65% yield), the hybrid tube **24** in 7% yield (Fig. 8). The group of Raston have also prepared **24** in a higher yield of 40% using a similar method.[13]

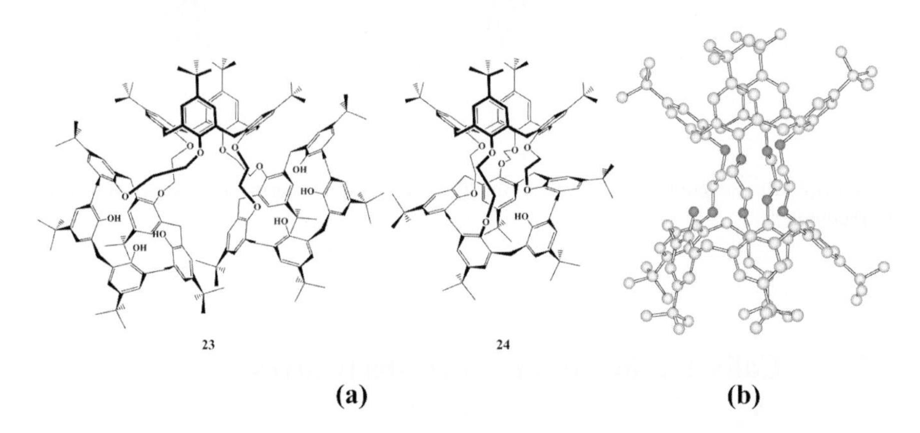

Figure 6-8. (a) The hybrid tubes **23** and **24** incorporating calix[4] and calix[5] entities. (b) The molecular structure of **24** determined by X-ray crystallography.

This tube **24** is more flexible than the calix[4]tube and, at temperatures between 233K and 393K, the ethylene linkers interchange between trans and gauche arrangements, resulting in an overall C_s symmetry.[21] In the solid-state, the calix[4]arene adopts a flattened cone conformation, and the linkers have a *tgtg* geometry similarly to the calix[4]tube. The calix[5]arene unit, however adopts a more distorted cone conformation.

In contrast, reaction of *p-t*-butyl calix[4]arene tetratosylate **4** with *p-t*-butyl calix[6]arene results only in the formation of a trimeric structure, identified by mass spectrometry, in 26% yield.[13] Two isomeric tubes **25** and **26** have been formed in the reaction between **4** and *p-t*-butyl calix[8]arene.[13] These were prepared in equivalent yields (15% each) but, unfortunately, could not be separated from each other. The two isomers have been proposed to have the symmetrical structure **25,** in which an enlarged cavity for binding is available, and the asymmetric structure **26** which is similar to the calix[4]-calix[5]tube (Fig. 9).

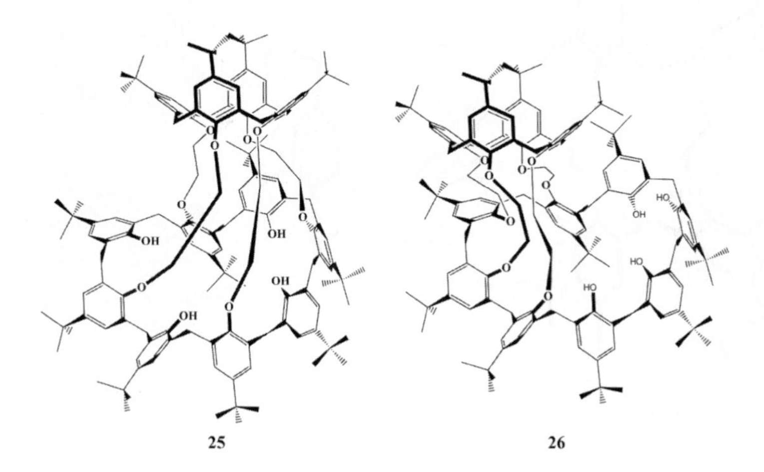

Figure 6-9. Probable isomeric forms of the calix[4]-calix[8] partial tubes **25** and **26**.

2.3 Thiacalix[4]tubes

2.3.1 Synthesis and Structure of Thiacalix[4]tubes

Whilst attempts to prepare the calix[5]tube have failed to date, an expanded cavity tube has been isolated: the thiacalix[4]tube **27** (Fig. 10).[22] Using our standard approach of reaction of the tetratosylate derivative with thiacalix[4]arene in the presence of K_2CO_3 as both base and template, the tube was isolated in 10% yield and was the first example of a dimeric thiacalix[4]arene. This low yield may be the result of intramolecular cyclisation or a consequence of potassium being an unsuitable template for the larger cavity cryptand. The work of the group of Hosseini on calix[5]arene derivatisation points to a competitive cyclisation reaction.[23] They investigated treatment of the tetratosylate with thiacalix[4]arene in the presence of Cs_2CO_3, isolating a 65% yield of **28,** and proposed a mechanism for the formation of this cyclised material, which was also obtained when thiacalix[4]-arene was treated with 1,2-dibromoethane.

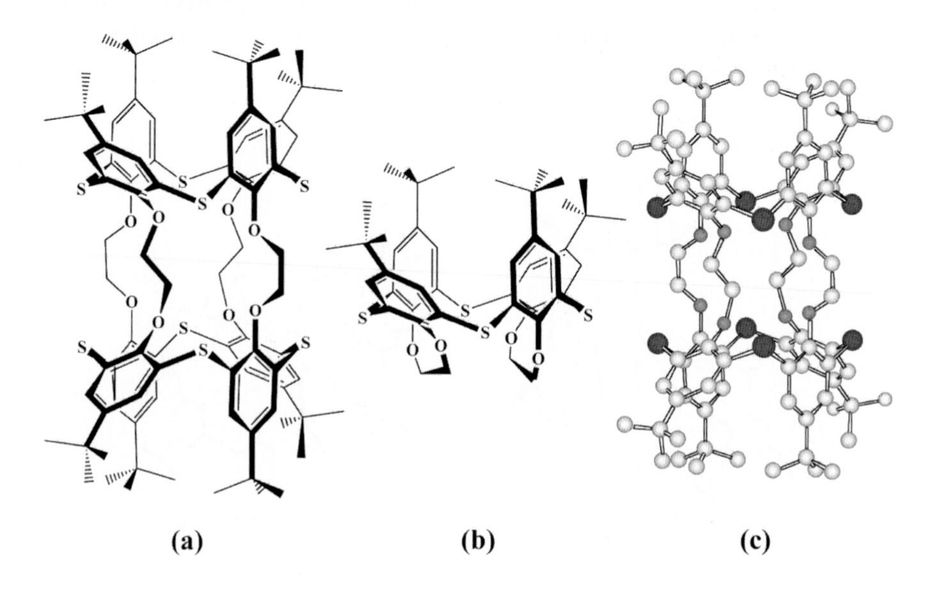

Figure 6-10. Representations of (a) the thiacalix[4]tube **27** and (b) the major side product, **28**, of its synthesis. (c) The molecular structure of **27**, determined by X-ray crystallography.

Whilst ^{1}H NMR studies in 4:1 $CDCl_3$:CD_3OD show that the thiacalix[4]-tube structure is similar to that of the parent calix[4]tube at room temperature, the structure is more flexible, with the rate of exchange between the two conformers being much higher (9.25 s^{-1}). From variable temperature NMR studies there is evidence for additional conformations between 303 K and 323 K before signals coalescence to give broad peaks consistent with a C_{4v} structure occurs above 328 K. The solid-state structure (Fig. 10) is similar to that seen for calix[4]tubes and is consistent with the lower temperature NMR studies.

Recently, the group of Kovalev have conducted preliminary studies on hybrid calix[4]-thiacalix[4]tubes **29-31** which show intermediate flexibility between calix[4]tubes and thiacalix[4]tubes.[14] Interestingly, they have also prepared the sulfinyl derivative **32**, through treatment of the hybrid tube with

$NaNO_3$ in acetic acid, the first calix[4]tube in which the methylene bridges are functionalised.

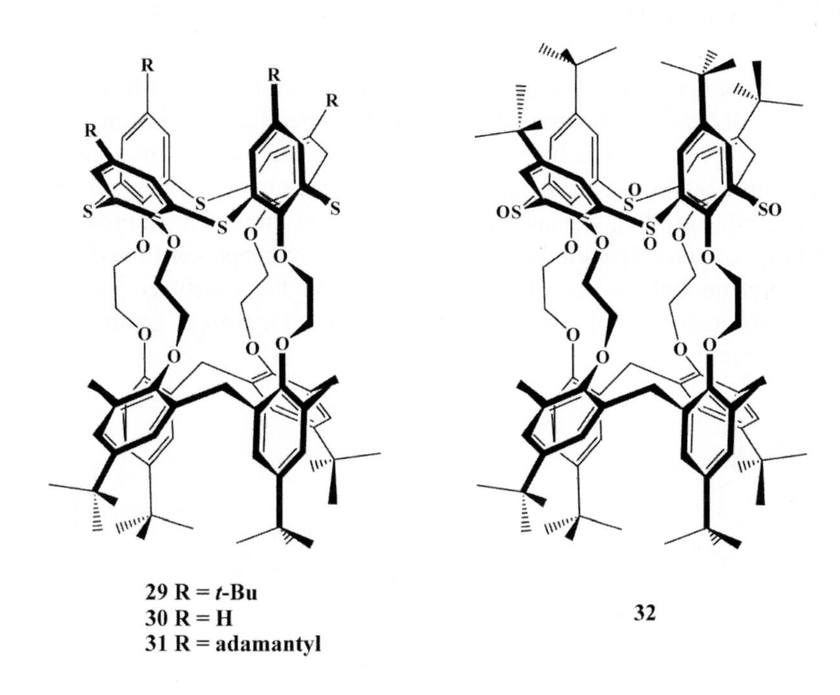

Figure 6-11. Representations of hybrid tubes derived from thiacalix[4]arene.

2.3.2 Complexation of Cations

Qualitative electrospray mass spectrometry cation binding studies on the novel thiacalix[4]tube **27** showed this receptor to be less discriminatory than the calix[4]tube, with significant binding of both sodium and potassium and the appearance of smaller peaks for the binding of rubidium and caesium.[22] This poor ionophoric behaviour was confirmed by ^{1}H NMR studies where very small shifts ($\Delta\delta \leq 0.05$) were observed in the presence of potassium. Molecular modelling studies showed that whilst axial entry of the potassium cation into the thiacalix[4]tube was comparable to that with the calix[4]tube, the size of the metal ion co-ordination sphere is significantly larger, which suggests that the metal will not be retained in the polyether cavity and reduced binding can be expected.

3. CALIXSEMITUBES

3.1 Calix[4]semitubes

As the calix[4]tube is a particularly rigid structure with little flexibility, *i.e.* a pre-organised spherand with exceptional selectivity in binding but slow binding kinetics, there is interest in the development of structural analogues that would retain the remarkable potassium cation selectivity but offer a more flexible binding site and faster kinetics of binding. One approach is the synthesis of calix[4]semitubes, in which the calixarenes are linked by only two ethylene linkers in a 1,3-arrangement and the cavity is additionally closed through the alkylation of the remaining hydroxyl groups of one calix[4]arene.

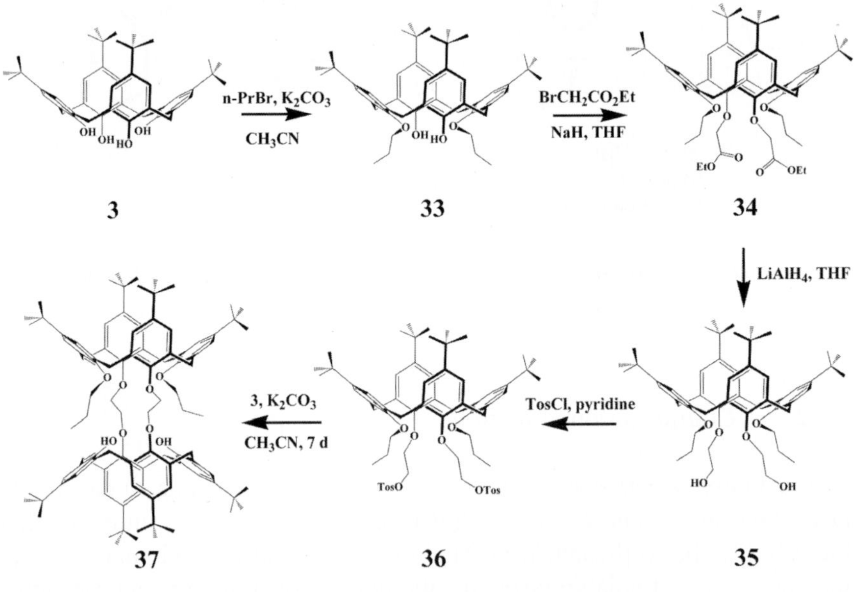

Figure 6-12. Synthesis of calix[4]semitubes.

3.1.1 Synthesis and Structure of Calix[4]semitubes

A range of semitubes **37-40** have been prepared, through a stepwise approach in which calix[4]arene is first 1,3-dialkylated with 1-bromopropane before introduction of the tosyl functionality and subsequent condensation with calix[4]arene (Fig. 12).[24] This final tube-forming step is achieved in yields of between 50-70%. Interestingly, using a similar method, Tuntulani and co-workers have prepared the more flexible methyl derivative **40**[25,26] and demonstrated the importance of initial alkylation to avoid intramolecular cyclisation during tube-formation. Treatment of the 1,3 di(tosyloxyethyl) derivative of *p-t*-butylcalix[4]arene with p-*t*-butyl calix[4]arene in the presence of base resulted in the preferential intramolecular formation of the crowned derivative **41** (Fig. 13) in which the calixarene adopts a 1,2-alternate conformation, whereas the same reaction with the 2,4-methyl derivative provides the desired tube product.

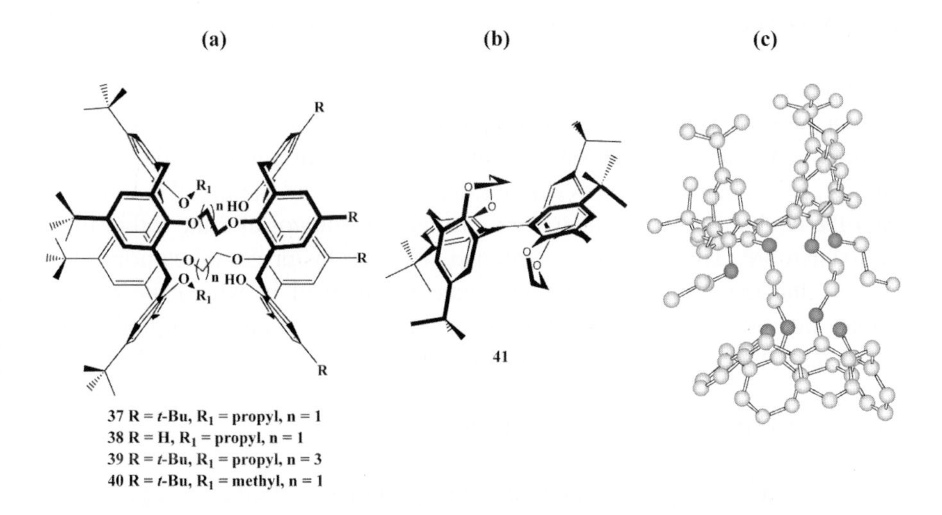

Figure 6-13. (a) The range of known calix[4]semitubes; (b) the side product resulting from intramolecular cyclisation during semitube synthesis. (c) The molecular structure of the semitube **38**, showing the different conformations of the links.

Unlike the calix[4]tubes, the solid-state structure of the calix[4]semitube **38**[24] (Fig. 13) indicates that the two linkers adopt different torsion angles, one being *trans* and the other being *gauche*. Both calix[4]arenes have a cone conformation but whilst the unsubstituted calix[4]arene is a regular cone, the *p-t*-butyl calix[4]arene is more distorted, with the aryl residues bearing the linkers being flattened.

3.1.1.1 Complexation of Cations by Calix[4]semitubes

The complexation behaviour of calix[4]semitubes has been investigated using ^{1}H NMR studies with 1 eq. of $NaClO_4$, KSCN or $RbPF_6$ in $CDCl_3$:CD_3OD (1:1).[24] Although complexation is slow on the NMR timescale, equilibration occurs considerably faster than for the calix[4]tubes. Thus, within 4 minutes, 100% complexation of potassium is observed for **37** and **38** with a stability constant of greater than 10^5 M^{-1}. The binding observed with sodium and rubidium is weaker (K_{ass} Na^+ 20 M^{-1} Rb^+ 30 M^{-1}) and no binding is seen for the larger caesium cation. Analysis of the ^{1}H NMR spectrum indicates that complexation results in a structural change within the molecule with the propyl residues moving towards the cavity on binding. For the butylene-linked derivative **39**, no binding of Group 1 cations was observed which confirms the importance of cavity size on achieving both strong complexation and selectivity.[24]

Molecular modelling studies have been undertaken to determine the route by which the potassium cation is complexed. Unlike the calix[4]tubes, where the proposed route of complexation is the axial route through the calix[4]arene annulus, molecular mechanics calculations indicate that the metal cation can enter the cavity at lower temperature through the equatorial route rather than the axial route. These results, combined with the faster complexation rates suggest that the semitubes operate a gate mechanism in which the propyl groups move to allow entry of the cation before undergoing a reorganisation and closing the cavity to encapsulate the cation.[24]

3.1.2 Derivatisation of Calix[4]semitubes

The unique calix[4]semitube structure provides a synthetic opportunity for the selective introduction of functionality to one calixarene subunit and for the preparation of novel ditopic receptors that can potentially bind anions and cations.[27] Selective 1,3-dinitration at the para position of the free phenolic residues can be achieved under standard conditions, albeit in the low yield of 19%, and the product can be readily reduced to the amine. For the formation of an anion binding cavity, the amine was converted to the hexyl urea **42** (Fig. 14) through treatment with hexyl isocyanate.

The presence of free phenolic residues in the semi[4]tube structures also offers a route into highly selective electrochemical sensors for cations. Through treatment with thallium (III) trifluoracetate, the phenols were successfully oxidised to quinones **43** and **44** (Fig. 14).[25,26,28] The solid-state structure of the propyl derivative **43**[28] (Fig. 14) shows that both calixarenes adopt a flattened cone conformation and, in the case of the calixquinone, the aryl rings are flattened whilst the quinones are upright and inclined towards each other at the upper-rim. This conformation is also evident in solution.

Variable temperature NMR studies over the range 193-288 K show that despite the ability of the quinones to rotate through the annulus, no partial cone conformations can be observed.

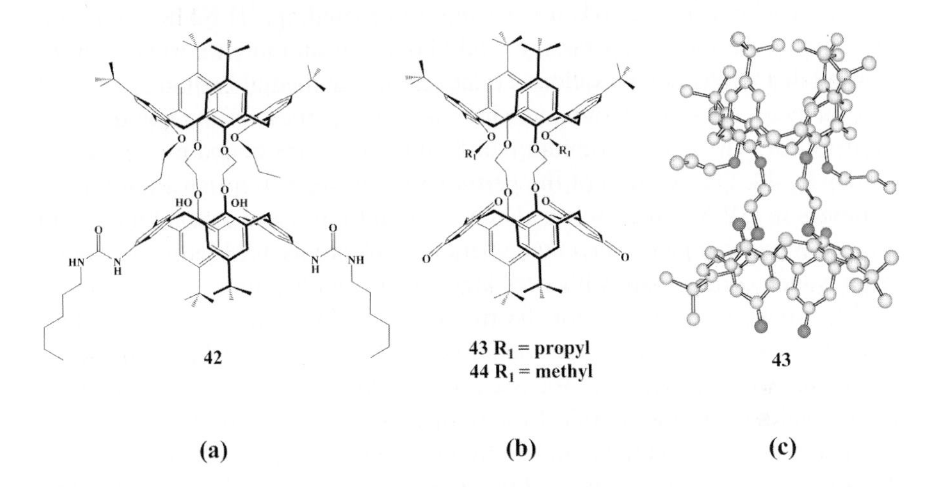

Figure 6-14. Calix[4]semitube (a) urea and (b) quinone derivatives, and (c) the molecular structure of the diquinone **43**, as determined by X-ray crystallography.

3.1.2.1 Complexation of Group 1 Salts by Ditopic Calix[4]semitubes

The anion-binding behaviour of the ditopic calix[4]semitube **42** has been investigated both in the presence and absence of a complexing cation.[27] Whilst binding of a range of anions from their tetrabutylammonium salts is weak, significant increases are observed in the presence of sodium and potassium cations. This is particularly marked for bromide and iodide where enhancements of up to thirty fold can be seen (*e.g.* K_{ass} Br^- 20 M^{-1}, K_{ass} K^+ Br^- 550 M^{-1} in $CDCl_3$: CD_3CN 2:1). These increases can be attributed to a number of factors including the increased acidity of the anion binding urea protons, the rigidification of the receptor and pre-organisation of the anion-binding site and an electrostatic interaction between the cation and anion. In the case of solid-liquid extractions with an excess of potassium or sodium chloride or iodide, the extent of extraction is directly related to lattice energy of the salt with the most efficient extraction into a chloroform solution, after 48 h mechanical shaking, being for KI (95%).

3.1.2.2 Complexation of Cations by Quinone Derivatives

A range of techniques have been used to investigate the binding of cations by calix[4]semitube quinones. In particular the presence of the quinone moieties offers the advantage of using both UV-Vis spectroscopy and electrochemical methods for sensing cation binding. ^{1}H NMR studies on the binding of sodium, potassium, rubidium and ammonium ions with **43** showed that whilst complexation is relatively fast and equilibration is reached in 15 minutes for sodium and potassium, it is slower than for the non-quinone derivatives.[28] In contrast, rubidium and ammonium are bound exceptionally slowly by **43**. Observation of the perturbation of the $\pi^* \leftarrow n$ transition for the quinones in UV-Vis studies has also been used to measure cation binding. In a range of polar organic solvents remarkable selectivity for binding of potassium is seen concomitant with slow kinetics of complexation.

Electrochemical studies for the free receptor **43** show two major reduction redox waves in the cyclic voltammogram, the first of which can be assigned to two one-electron processes for the formation of the semiquinone radical anions and the second to a further two one-electron processes and formation of the tetraanion. On addition of 0.5 eq. of KSCN, a current diminution in these waves is seen and two new waves appear which are substantially anodically shifted ($\Delta E = 340$ mV); with further addition of 0.5 eq., the original waves disappear and only the new waves are observed. With $NaClO_4$, which binds more weakly to the receptor, even with 5 eq., the original waves are retained whilst the emerging waves remain small.

With the more flexible methyl derivative **44**, lithium, sodium and potassium are bound by the receptor in ^{1}H NMR studies but caesium is too large to bind.[25,26] However, the structure of the complex appears to vary, depending on the metal ion. For the smaller lithium and sodium the original cone conformation of the calixarene changes to a mixture of conformations due to the ease of rotation of the methyl residues through the annulus. In the case of potassium, the cone conformation is maintained and a binding constant can be calculated (K_{ass} 458 M^{-1} in $CDCl_3 : CD_3CN$ 4:1).

A structurally analogous series of tetra quinones which show exceptional selectivity for the larger alkali metal cations, rubidium and caesium, have also been prepared (Fig. 15). The initial tube molecules were easily prepared by reaction of calix[4]arene with an appropriate difunctional alkane.[25,26,29,30,31] A solid-state structure has been obtained of the propylene derivative **46**[31] which showed that, despite the expected ease of rotation of the phenolic residues through the calixarene annulus, regular cone conformations are adopted by both calixarenes. This calixarene conformation may be a consequence of hydrogen bonding between the oxygens at the lower-rim.

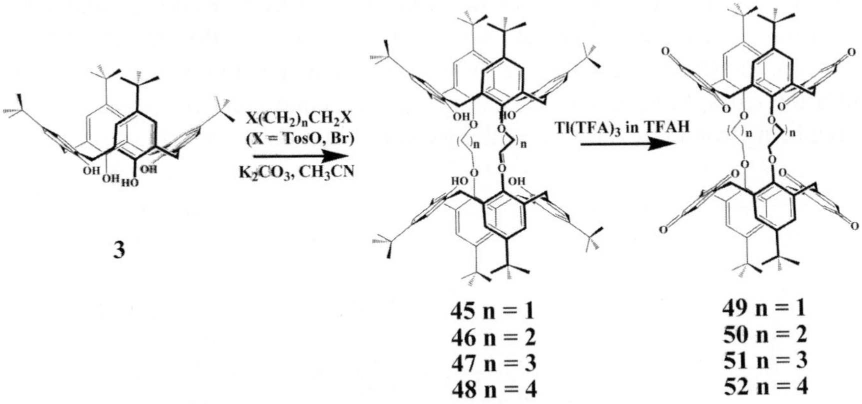

Figure 6-15. Synthesis of tetraquinone calix[4]partialtubes.

Using standard oxidation conditions ($Tl(OCOCF_3)_3$ in trifluoroacetic acid) the novel tetraquinones **49-52** were prepared in yields of between 8 and 28%. A solid-state structure has been obtained for the pentylene-linked derivative **52**[31] (Fig. 16(a)) and shows that interestingly both calixarenes adopt a 1,3-alternate conformation with the linkers forming a closed cage which is too small to enable cation complexation without re-organisation. In solution at room temperature, ^1H NMR studies indicate that both calixarenes adopt a cone conformation. On reducing the temperature, the spectrum for the propylene derivative **50** broadens and splitting is observed in the *t*-butyl region suggesting that other conformations may be adopted through rotation of the quinone moieties.

Extensive alkali metal co-ordination studies have been undertaken on the tetra quinone derivatives **50-52** using a variety of techniques including mass spectrometry, UV-Vis spectroscopy, ^1H NMR spectroscopy, electrochemistry and solid-state determinations. These investigations indicate that linker length and, thus, cavity size are crucial for determining selectivity within the Group 1 metal cations.

The propylene linked receptor **50** showed, in competitive electrospray mass spectrometry studies, a preference for binding rubidium over sodium, potassium and caesium.[30,31] This qualitative result was confirmed by UV-Vis studies, where a stable complex of rubidium (K_{ass} 6.3 x 10^4 M^{-1} in DMSO : H_2O 99:1) is observed with a selectivity profile of Rb^+>Cs^+>K^+>>Na^+. The structure of the cation complex has been probed by variable temperature NMR studies over a temperature range of 193 to 318 K. For the rubidium complex, no changes are seen over the whole temperature range, showing that the cation is bound tightly to all donor atoms. With the smaller sodium cation, the NMR spectra differ with temperature. At 273 K, a time-averaged spectrum is observed where the sodium is in motion within the cavity; on reducing the temperature further the spectrum changes to be consistent with both the calix[4]diquinones adopting a partial cone conformations and sodium being bound by only six of the possible eight donor atoms.

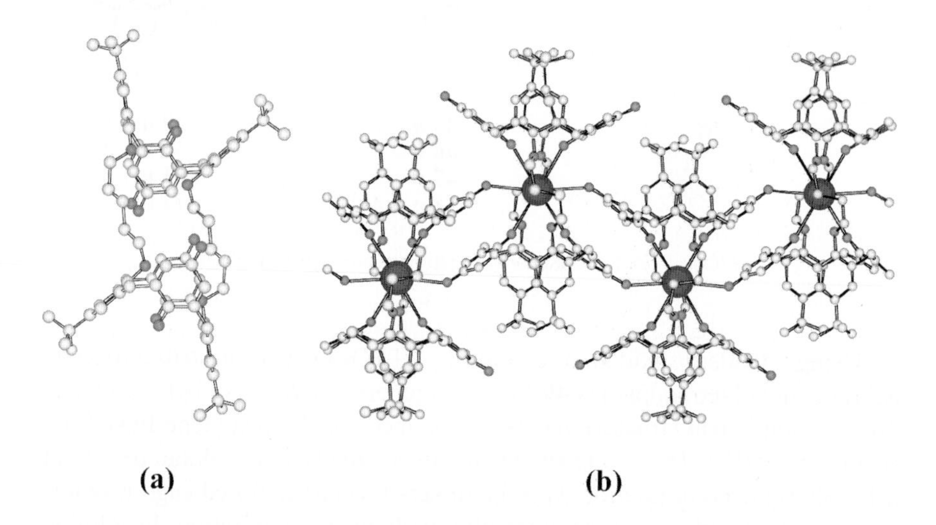

Figure 6-16. (a) The molecular structure of the tetraquinone **52**, showing the 1,3-alternate conformation of the calixarene units. (b) Part of the polymeric chain present in the lattice of the Cs(I) complex of tetraquinone **51**. (For clarity, Cs-O(ether) interactions are not shown.)

When the cavity size is extended further, as in the case of the butylene linked tube **51**, both electrospray mass spectrometry and UV-Vis studies, reveal remarkable selective binding of caesium (K_{ass} $1.6x10^3M^{-1}$ in DMSO: H_2O 99:1) over rubidium and an absence of binding of sodium and potassium is seen in this competitive solvent mixture.[30,31] The solid-state structure of the complex shows the caesium cation bound to all eight oxygens (Fig. 16(b)). The calix[4]diquinones adopt a flattened cone conformation with the aryl residues perpendicular to the plane of the methylene bridges whilst the quinones are flattened. Interestingly, the caesium is additionally bound to two upper-rim quinone oxygens on adjacent molecules, leading to a zig-zag polymeric structure. This binding mode was confirmed in solution by variable temperature NMR studies, where the symmetrical spectrum is unchanged between 193 and 291K. Molecular modelling studies confirm the selectivity of **51** for caesium and indicate that the caesium ion is the only alkali metal cation that is of sufficient size to bind to all eight oxygen atoms of the ionophore.

Preliminary studies have been reported on the ethylene linked receptor **49** which indicate that a 1:1 binding stoichiometry is observed and that the receptor is selective for binding of sodium over lithium and potassium while caesium is not bound (K_{ass} Na^+ 4.4 x 10^4 M^{-1}, Li^+ 3.0 x 10^4 M^{-1}, K^+ 1.4 x 10^3 M^{-1} in $CDCl_3$: CD_3CN 4:1).[25,26]

Electrochemical studies have been undertaken on the free receptors[25,26,30,31] and in the presence of cations. The complex CV waves, featuring significant anodic shifts of the redox couples, observed on cation binding have been assigned to a multi-step reduction process.[31] Initially, the cation is bound symmetrically; one calixquinone is then preferentially reduced to the semiquinone, which attracts the cation to give an asymmetric structure. The subsequent reduction of the second calixquinone results in a tetra-semiquinone in which the cation is bound symmetrically before the structure is reduced further to the octaanion. Electrochemical competition experiments reveal **51** is capable of selectivity sensing caesium in the presence of excess amounts of sodium and potassium cations.

4. CONCLUSIONS

The calix[4]tube systems are highly effective ionophores, where metal ion selectivity can be tuned by linker length and cavity size, and rate of complexation controlled through the use of upper-rim substituents or through modifications to the binding array. The development of calix[4]tubes has introduced a unique class of highly potassium selective ionophores whilst the preparation of the calix[4]semitube family has enabled the design of selective potassium ionophores with fast complexation kinetics. The ability to introduce redox sensing moieties, in the form of quinones, into the tubular structural design has resulted in the development of the first rubidium and caesium selective redox-active ionophores. Additionally the introduction of functionality to the upper-rim has resulted in the development of a ditopic receptor which, through combining the highly cation selective ionophore with an anion binding motif, has enabled binding and extraction of alkali metal salts. Recent advances such as the hybrid calix[4]-thiacalix[4]tube further extend the range of cations that may be complexed by these systems and offer pointers to the development of this field in the future.

5. REFERENCES

1. (a) A. Ikeda, S. Shinkai, *Chem. Rev.*, **97**, 1713-1734 (1997); (b) P. D. Beer, P. A. Gale, Z. Chen, *Adv. Phys. Org. Chem.*, **31**, 1-90 (1998); (c) X. X. Zhang, R. M. Izatt, J. S. Bradshaw, K. E. Krakowiak, *Coord. Chem. Rev.*, **174**, 179-189 (1998); d) G. W. Gokel, W. M. Leevy, M. E. Weber, *Chem Rev*, **104**, 2723-2750 (2004).
2. (a) C. J. Pedersen, *J. Am. Chem. Soc.*, **89**, 2495-2496 (1967); (b) C. J. Pedersen, *J. Am. Chem. Soc.*, **89**, 7017-7036 (1967).
3. H. Yamamoto, S. Shinkai, *Chem Lett,* 1115-1118 (1994).
4. A. Casnati, A. Pochini, R. Ungaro, C. Bocchi, F. Ugozzoli, R. J. M. Egberink, H. Struijk, R. Lugtenberg, F. de Jong, D. N. Reinhoudt, *Chem. Eur. J.*, **2**, 436-445 (1996).
5. (a) A. Casnati, A. Pochini, R. Ungaro, F. Ugozzoli, F. Arnaud, S. Fanni, M. J. Schwing, R. J. M. Egberink, F. de Jong, D. N. Reinhoudt, *J. Am. Chem. Soc.*, **117**, 2767-2777 (1995). (b) F. Arnaud-Neu, R. Arnecke, V. Böhmer, S. Fanni, J. L. M. Gordon, M. J. Schwing-Weill, W. Vogt, *J. Chem Soc., Perkin Trans. 2*, 1855-1860 (1996).
6. D. A. Doyle, J. M. Cabral, R. A. Pfuetzner, A. Kuo, J. M. Gulbis, S. L. Cohen, B. T. Chait, R. MacKinnon, *Science*, **280**, 69-77 (1998).
7. F. Ohseto, S. Shinkai, *J. Chem. Soc., Perkin Trans.*, 1103-1109 (1995).
8. G. Ulrich, R. Ziessel, *Tetrahedron Lett.*, **35**, 6299-6302 (1994).
9. G. Ulrich, R. Ziessel, I. Manet, M. Guardigli, N. Sabbatini. F. Fraternali, G. Wipff, *Chem. Eur. J.,* **3**, 1815-1822 (1997).
10. P. Schmitt, P. D. Beer, M. G. B. Drew, P. D. Sheen, *Angew. Chem. Int. Ed. Engl.,* **36**, 1840-1842 (1997).
11. S. E. Matthews, P. Schmitt, V. Felix, M. G. B. Drew, P. D. Beer, *J. Am. Chem. Soc.*, **124**, 1341-1353 (2002).
12. J. Budka, P. Lhoták, I. Stibor, V. Michlová, J. Sykora, I. Cisarová, *Tetrahedron Lett.*, **43**, 2857-2861 (2002).
13. M. Makha, P. J. Nichols, M. J. Hardie, C. L. Raston, *J. Chem. Soc., Perkin Trans 1*, 354-359 (2002).
14. V. Kovalev, E. Shokova, I. Vatsouro, E. Khomich, A. Motornaya, Proceedings of the 8th International Conference on Calixarenes (Calix 2005), Prague, Czech Republic.
15. B. D. Smith, Ion-Pair Recognition by Ditopic Macrocyclic Receptors in Macrocyclic Chemistry: Current Trends and Future Perspectives, Ed. K. Gloe, Springer, 137-151 (2005).
16. S. E. Matthews, V. Felix, M. G. B. Drew, P. D. Beer, *Org. Biomol. Chem.*, **1**, 1232-1239 (2003).
17. V. Felix, S. E. Matthews, P. D. Beer, M. G. B. Drew, *Phys. Chem., Chem. Phys.*, **4**, 3849-3858 (2002).
18. J. Budka, P. Lhoták, I. Stibor, J. Sykora, I. Cisarová, *Supramol. Chem.*, **15**, 353-357 (2003).
19. S. E. Matthews, N. H. Rees, V. Felix, M. G. B. Drew, P. D. Beer, *Inorg. Chem.*, **42**, 729-734 (2003).
20. P. Schmitt, P. D. Beer, unpublished results.
21. A. Notti, S. Occhipinti, S. Pappalardo, M. F. Parisi, I. Pisagatti, A. J. P. White, D. J. Williams, *J. Org. Chem.*, **67**, 7569-7572 (2002).
22. S. E. Matthews, V. Felix, M. G. B. Drew, P. D. Beer, *New J. Chem.*, **25**, 1355-1358 (2001).
23. H. Akdas, L. Bringel, V. Bulach, E. Graf, M. W. Hosseini, A. De Cian, *Tetrahedron Lett.*, **43**, 8975-8979 (2002).

24. P. R. A. Webber, A. Cowley, M. G. B. Drew, P. D. Beer, *Chem. Eur. J.*, **9**, 2439-2446 (2003).
25. K. Tantrakarn, C. Ratanatawante, T. Pinsuk, O. Chailapakul, T. Tuntulani, *Tetrahedron Lett.*, **44**, 33-36 (2003).
26. N. Kerdpaiboon, B. Tompatanaget, O. Chailapakul, T. Tuntulani, *J. Org. Chem.*, **70**, 4797-4804 (2005).
27. P. R. A. Webber, P. D. Beer, *Dalton Trans.*, 2249-2252 (2003).
28. P. R. A. Webber, A. Cowley, P. D. Beer, *Dalton Trans.*, 3922-3926 (2003).
29. B. Tompatanaget, B. Pulpoka, T. Tuntulani, *Chem. Lett.*, 1037-1038 (1998).
30. P. R. A. Webber, G. Z. Chen, M. G. B. Drew, P. D. Beer, *Angew. Chem. Int. Ed. Engl.*, **40**, 2265-2268 (2001).
31. P. R. A. Webber, P. D. Beer, G. Z. Chen, V. Felix, M. G. B. Drew, *J. Am. Chem. Soc.*, **125**, 5774-5785 (2003).
26. N. Kerdpaiboon, B. Tompatanaget, O. Chailapakul, T. Tuntulani, *J. Org. Chem.*, 2005, 70, 4797-4804.
27. P. R. A. Webber, P. D. Beer, *Dalton Trans.*, 2003, 2249-2252.
28. P. R. A. Webber, A. Cowley, P. D. Beer, *Dalton Trans.*, 2003, 3922-3926.
29. B. Tompatanaget, B. Pulpoka, T. Tuntulani, *Chem. Lett.*, 1998, 1037-1038.
30. P. R. A. Webber, G. Z. Chen, M. G. B. Drew, P. D. Beer, *Angew. Chem. Int. Ed. Engl.*, 2001, 40, 2265-2268.
31. P. R. A. Webber, P. D. Beer, G. Z. Chen, V. Felix, M. G. B. Drew, *J. Am. Chem. Soc.*, 2003, 125, 5774-5785.

Chapter 7

1,3-ALTERNATE CALIX TUBES

Shuttling of cations

Buncha Pulpoka[a], Lassaad Baklouti[b], Jong Seung Kim[c], and Jacques Vicens[d]

[a]Supramolecular Chemistry Research Unit and Organic Synthesis Research Unit, Department of Chemistry, Faculty of Science, Chulalongkorn University, Phyathai Road, Bangkok 10330, Thailand. E-mail: Buncha.P@chula.ac.th; [b]Faculté des Sciences de Bizerte, Laboratoire de Chimie des Interactions Moléculaires, 7021 Zarzouna, Tunisie; E-mail: bakloutilassaad@yahoo.fr; [c]Department of Chemistry, Dankook University, Seoul, 140-714 Republic of Korea; E-mail: jongskim@dankook.ac.kr; [d]Ecole Chimie Polymères et Matériaux, Laboratoire de Conception Moléculaire, UMR 7512, 25 rue Becquerel, F-67087 Strasbourg, France. E-mail: vicens@chimie.u-strasbg.fr

Abstract: Calix[4]arenes in their 1,3-alternate conformation are particularly convenient scaffolds for the construction of extended calix tubes. These molecules can act as polytopic cation receptors and have the fascinating property of allowing cation transport ("shuttling") along the tube by passing through the "π-basic tube" formed by the macrocyclic rings of the calixarene units. Although calix tubes can also influence anion transport, this does not appear to involve anion shuttling.

Key words: Calixarenes, nanotubes, calix tubes, 1,3-alternate conformation.

1. INTRODUCTION

The desire to mimic and thereby understand the functions of biological systems has been a powerful stimulus in many areas of chemistry. The inherently three-dimensional nature and nanoscale dimensions of biological systems[1] place enormous demands upon the ability of the synthetic chemist to create "supermolecules"[2] of the appropriate stereochemistry and function. As described in the preceding chapter 6, tube-like molecules derived from calix[4]arenes in their cone conformation have provided fascinating insights into the nature of selective ion binding and transport. There is, however, some difficulty in extending the syntheses used for such molecules to create

J. Vicens and J. Harrowfield (eds.), Calixarenes in the Nanoworld, 135–149.

longer tubes of dimensions such that they might be considered analogues of membrane-spanning ionophores. This difficulty is much diminished by the use of calix[4]arenes in their 1,3-alternate conformation, where the basic functional groups are now oriented in a manner which facilitates oligomer and polymer formation. Thus, the following discussion is a survey of the exploitation of this approach to the production of synthetic ionophores. The mimicry of trans-membrane ion transport[3] using synthetic ion channels[4] has, of course, been explored in various ways, the earliest successful system being based on a functionalised cyclodextrin which in fact proved more efficient (for Co(II) transport) than its natural analogue.[5]

2. THE FIRST 1,3-ALTERNATE CALIX TUBES

An essential feature of a membrane-spanning ion channel is the presence of ion-binding sites at both termini of the channel. These sites may be considered portals to the membrane space and a relatively simple design,[6] readily open to variation, of a molecule incorporating portals for cation binding (or, when protonated, for anion binding) is that shown in Fig. 1. Molecules of this type have been shown to function as cation transporters in phospholipid bilayers[7] and their form, with multiple tails attached to a hollow central unit (relay), structurally analogous to that of polyps of the genus *Hydra*, has led to them being referred to as "hydraphiles".[8] The design of hydraphiles may be said to embody a "three macrocycles" concept,[8,9] with a central macrocyclic ligand serving to pass a cation from one macrocyclic portal to the other. Hydrophobic spacers linking the macrocycles provide a means of adjusting the channel length to the thickness of a given membrane.

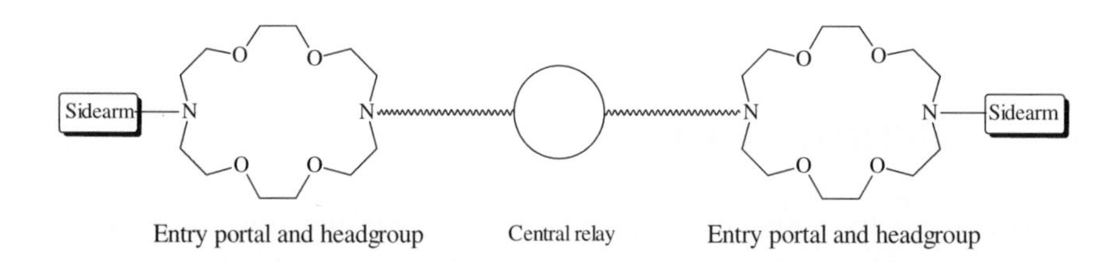

Figure 7-1. 'Hydraphiles', designed cation-conducting channels.

An idealised representation of a "tunnel" form of a tris(macrocycle) **1** based on the linking of three diaza-18-crown-6 units is shown below. Such a configuration should also be realisable with a calix[4]arene unit as the central macrocycle and the first such derivatives, **2a** and **2b**, in which the spacer chains and headgroups were those that had proved efficacious in the hydraphile family, were prepared with the calixarene in its cone and 1,3-alternate conformations, respectively.[10] Conductance measurements showed that while cone **2a** appeared to be inactive as a cation transporter, 1,3-alternate **2b** was active and remained so even with t-butyl substituents on the para positions of the phenyl rings. Since such substitution was expected to block passage of a cation through the calixarene annulus, the intriguing question arose as to how a cation might be transferred from one portal to the other.

1

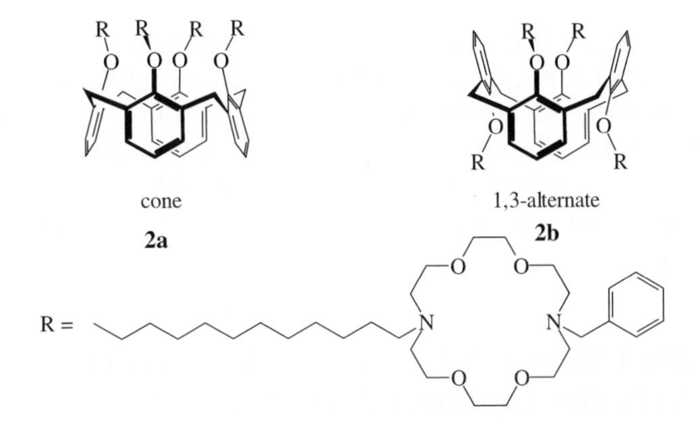

By analogy with a synthetic sterol-based ion channel[11] consisting a tartrate-derived crown ether supporting six steroids, the calix[4]arene-cholic acid conjugates **3-7**, again incorporating both cone and 1,3-alternate conformers, were prepared recently.[12] Molecular modeling of the fully extended conformations of these species indicated that the 1,3-alternate calix derivatives **3-5** should have the ability to span a membrane 35 ± 2 Å thick, whereas cone-form derivatives **6** and **7** could span only 25 ± 2 Å.

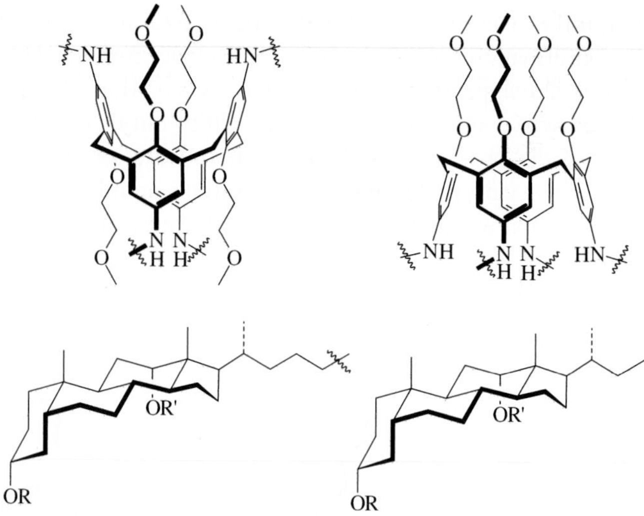

3 **R = R' = Ac**
4 **R = H, R' = Ac**
5 **R = R' = H**

6 **R = R' = Ac**
7 **R = H, R' = Ac**

Measurements of both the H^+ and Na^+ transporting abilities of these compounds showed that, once again, the 1,3-alternate calixarene derivatives were more efficient. The crucial factor in these cases may be the difference in the length of the channel possibly formed by the ionophore.[12] While the activity of these synthetic ionophores is comparable to that of natural systems, the exact mechanism of their ion transport remains to be established.

3. METAL OSCILLATION THROUGH THE Π-BASE TUNNEL OF THE 1,3-ALTERNATE CONFORMATION

The 1,3-alternate conformation of calix[4]arene can be considered a 'smart' building block for constructing original structures directed towards designed properties.[13] Its utility stems from its ditopic form, with two divergent binding sites connected by orthogonal pairs of parallel phenyl rings forming a "π-base tunnel". The development of a facile synthesis of tetra-O-alkylated 1,3-alternate calixarenes **8** based on the use of Cs_2CO_3 as a base catalyst in dimethylformamide (dmf)[14] led to extensive studies of their metal-ion binding capacity and the early discovery[15] that stability constant values for many metal ions exceeded those for the analogous cone conformer complexes. This enhanced ionophoricity was attributed to the involvement of the π electrons of the phenyl groups as donor centres,[16] various crystallographic studies, *e.g.* [17], providing evidence in support of this proposal.

8

Thus, a possible solution to the enigma of the means of passage of a cation via the central calixarene unit of a hydraphile was to propose that the "π-base tunnel" could function as a temporary binding site to allow transit of the cation through the macrocyclic cavity (Fig. 2).

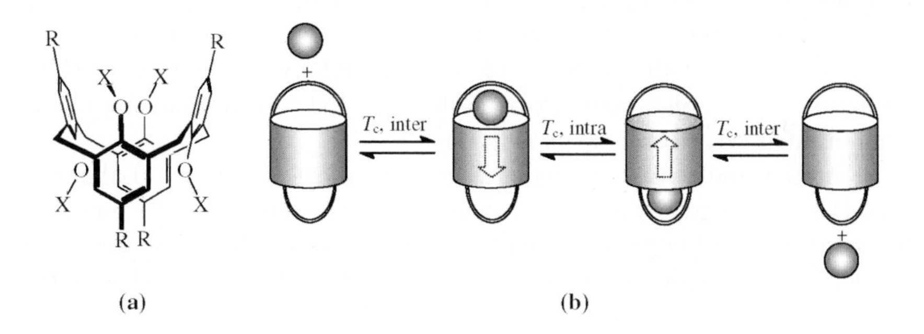

Figure 7-2. Representation of (a) the two metal-binding sites of a 1,3-alternate calix[4]arene and (b) cation oscillation through the π-base tunnel connecting these sites, as detected by NMR spectroscopy.

Substantial support for this notion was provided by detailed studies of the complexation of Ag(I) by 1,3-alternate calix[4]arenes.[15] ^{1}H VTNMR spectroscopy provided evidence that exchange of the metal ion between the two binding sites can be *intramolecular* and thus must involve passage through the macrocyclic ring. In the case of the unsymmetrical 1,3-alternate calix[4]arene **9**, this passage is involved in the establishment of an equilibrium where, at 188 K, 8.1% of Ag(I) resides in the cavity associated with the propyloxy substituents and 91.9% in the cavity with ethoxyethyloxy substituents (Fig. 3).

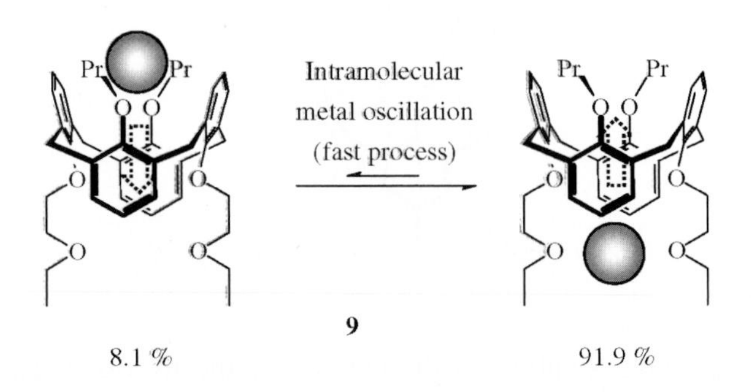

Figure 7-3. Different proportions of Ag^+ in the two different cavities of **9**.

Similar observations were made with Na^+ and K^+. Much more sophisticated exploitation of this behaviour has been applied in the development of a photocontrolled "molecular syringe" **10**.[18] Photoactivated cycloaddition reactions of the pendent anthracenyl lead to the conversion of a cavity where Ag(I) binding is preferred to one where this is less favoured than in the opposing cavity with propoxy of ethoxyethyloxy substituents. Thus, irradiation of the Ag(I) complex of **10** leads to the forced tunneling of the metal through to the other cavity (Fig. 4).

A similar molecular syringe **12** has been based on a 1,3-alternate calix[4]arene bridged by an azacrown unit.[19] Ag(I) is preferentially bound to the azacrown unit but protonation of the nitrogen centre leads to ejection of the metal ion through the calixarene ring and into the ethoxyethyloxy site.

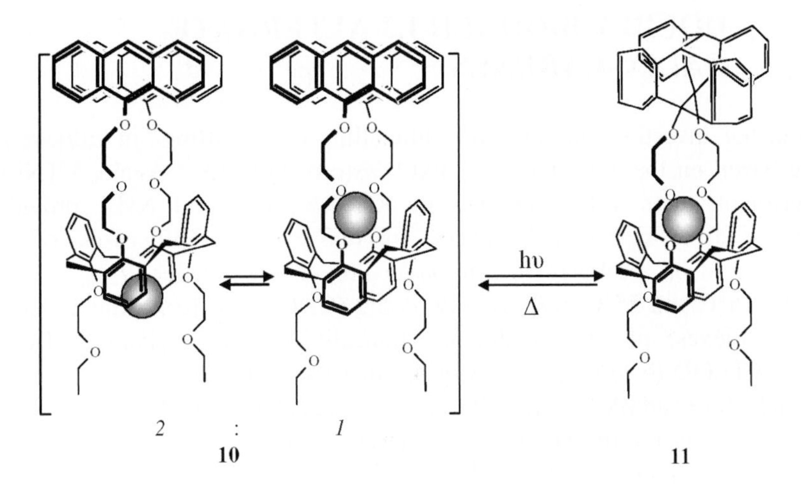

Figure 7-4. The cycloaddition reaction of 1,3-alternate calix[4]arene dianthracene-based light-switch molecular syringes **10**.

Thus, the nitrogen atom can be considered as a plunger, activated by protonation, which can drive the metal through the "syringe barrel" π-base tube of the calixarene (Fig. 5).

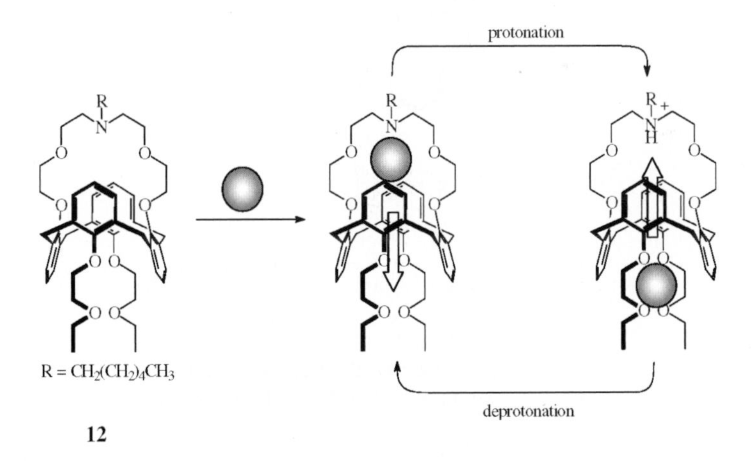

Figure 7-5. Molecular syringe **12** derived from a 1,3-alternate calix[4]arene azacrown.

4. DOUBLY-BRIDGED 1,3-ALTERNATE CALIX[4]ARENES

Further investigations of cation tunnelling or "shuttling" in calixarenes were based on the 1,3-alternate calix[4]*bis*(crown-5) **13**.[20] Again, VTNMR experiments with 1:1 complexes of K^+, Rb^+, Cs^+ and NH_4^+ provided evidence, from signal coalescences, for two exchange processes, one of which could be attributed to *intramolecular* cation transfer ($T_{c,intra}$ 125°C, 105°C, 45°C and 55°C, respectively, in 5:2 $(CDCl_2)_2$:DMF). Similar studies of complexes of the analogous thiacalix[4]arene derivative **14** in $CDCl_3$:CD_3OD (4:1) provided evidence that shuttling here was more rapid. For K(I), for example, $T_{c,intra} = 281$ K, $k_{c,intra} = 26.7\ s^{-1}$ and $\Delta G^{\ddagger}_{c,intra} = 61.5$ kJ mol^{-1}, whereas for the simple calix[4]arene complex in the same solvent, shuttling was not detected. Similarly, Cs^+ was observed to shuttle from one side to the other of 1,3-alternate *thia*calix[4]biscrown-6 (**15**). Interestingly, metal shuttling was observed in the 1:1 complex of K^+ with non-symmetrical 1,3-alternate *thia*calix[4]crown-5,crown-6 (**16**) while it was not in the 1:1 Cs^+ complex.

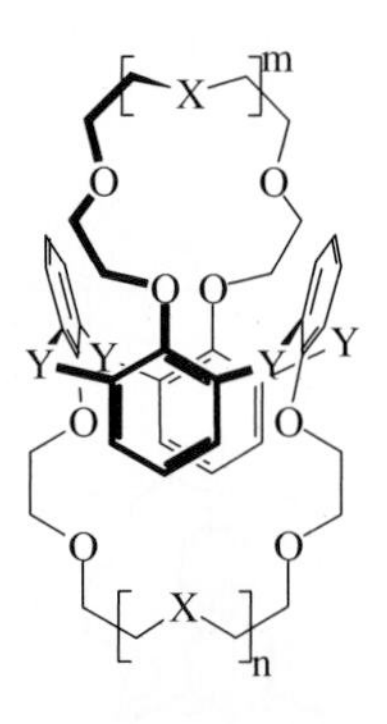

13	X = X' = O, Y = CH_2, m = n = 1
14	X = X' = O, Y = S, m = n = 1
15	X = X' = O, Y = S, m = n = 2
16	X = X' = O, Y = S, m = 2, n = 1
17	X = X' = NH, Y = CH_2, m = n = 1
18	X = O, X' = NH, Y = CH_2, m = n = 1

Given that the *thia*calix[4] macroring is significantly larger[22,23] than that of the classical calix[4]arene, it is clearly probable that a given cation would experience less repulsion in passing through the ring and thus that the rate of passage would be greater. More facile access to the π-donor electrons may also mean weaker binding to the donor atoms of the crown loop and indeed other observations support the notion that the strength of interactions with these donors does affect the rate of shuttling. Thus, with the 1,3-alternate calix[4]azacrowns **17** and **18**[24,25] it appears that strong H-bonding of ammonium ion to the NH centers completely inhibits shuttling, while, as in the molecular syringe described above, passage of Ag(I) into the simple crown cavity of **18** is promoted by protonation of the aza-crown nitrogen atom which weakens the interaction of the metal with this entity. There is also evidence[26,27] that photoexcitation of the coumarin units of the 1, 3-alternate calix[4]biscrown-6 species **19** and **20** modifies the interaction of this loop with Cs(I) and so induces shuttling. In contrast to its behaviour when bound to the azacrown species **17** and **18**, ammonium does appear to traverse the π-base tunnel in its 1:1 complex with the 1,3-alternate calix[4]arene-cryptand-crown-6 (**21b**), the exchange velocity (k_c) being 169 s^{-1} with an activation free energy ($\Delta G_c^{\#}$) of 12.0 kcal mol^{-1}.[28]

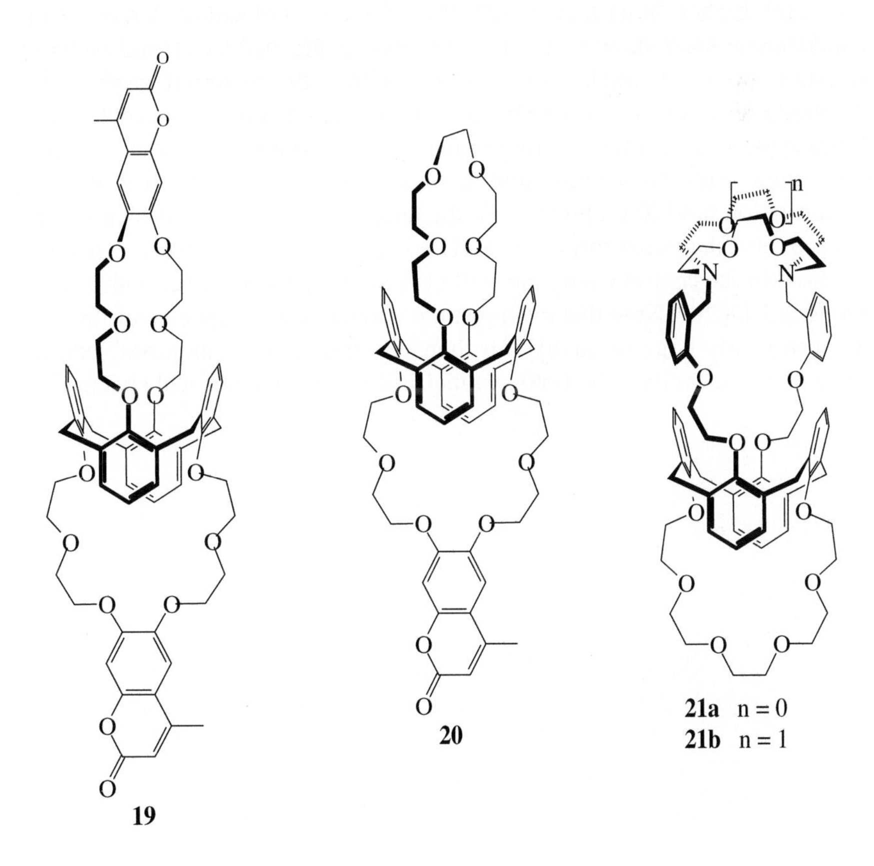

5. MULTI-1,3-ALTERNATE CALIX[4]ARENES : AN APPROACH TO CALIX[4]ARENE NANOTUBES

Early studies[29] of 1,3-alternate calix[4]arene*bis*(crown) species demonstrated that the presence of t-butyl substituents in the phenyl-group para positions inhibited cation binding to the crown loops. Unlike calix tubes derived from cone calix[4]arenes, therefore, those derived from the 1,3-alternate conformers must lack such *para* substituents for any terminal crown loops to be occupied by cations. Thus, the first 1,3-alternate calix tubes, "double calices" obtained by the reaction of *p-tert*-butyl calix[4]arene with an excess of tetraethylene glycol ditosylate in the presence of K_2CO_3, were found to bind K(I) and Rb(I) in the large central cavity only and no exchange involving the terminal cavities was detected.[29] In contrast, analogues and higher homologues (**22-25**) of these molecules derived from calix[4]arene itself show preferential binding, confirmed by crystal structure determinations, of alkali metal cations within the terminal cavities.[30,31] No metal shuttling could be detected. However, while 1,3-alternate calix-thiacalix[4]crown trimers bearing crown-5 and crown-6 bridges at the extremities and with the central unit a *thia*calix[4]arene in the 1,3-alternate conformation bind K(I) and Cs(I) in the terminal crown-5 and crown-6 sites, respectively, they also form 1:1 Ag(I) complexes in which the metal ion is localised in the central cavity but will undergo cavity exchange, with $T_{c,intra}$ = 300 K in $CDCl_3$.[32] Note that even just the "double calyx" species of this type are sufficiently large to qualify for their description as "nanotubes" and the range now known includes both symmetrical and unsymmetrical species.[33,34]

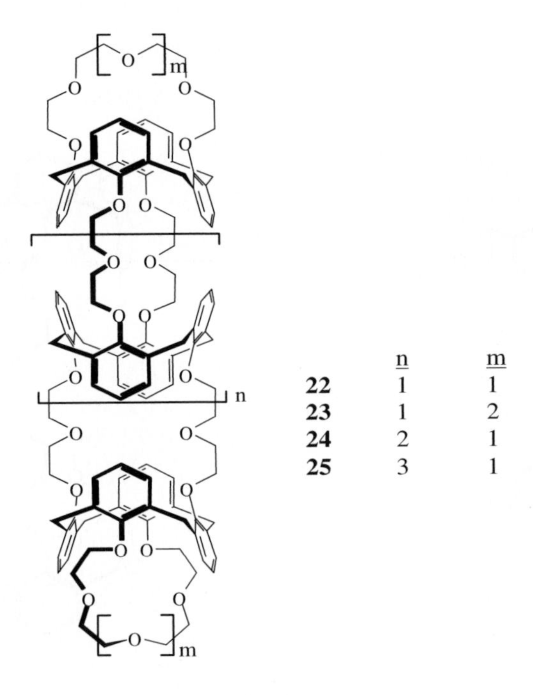

	n	m
22	1	1
23	1	2
24	2	1
25	3	1

The terminal sites of calix tubes may be readily varied, as may be the means of linking the 1,3-alternate calix[4]arene units together, as illustrated by the synthesis of the tricalix tube **26**.[35] While no shuttling could be detected in the Ag(I) complex of **26**,[35] the tubular structure of the related double calix ligand **27** was confirmed by a crystal structure determination.[36] This nanotube possesses a cross-section of 12 Å and a length of 28 Å. Its inside diameter varies between 4.1-4.5 Å and the two terephthaloyl units are parallel and separated by 3.3-3.9 Å.[36]

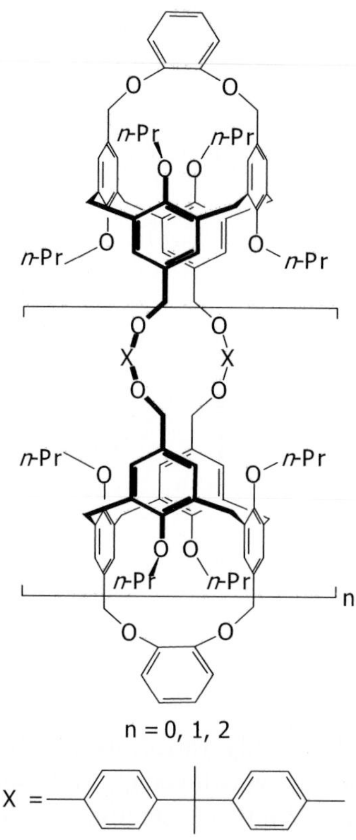

n = 0, 1, 2

X =

26

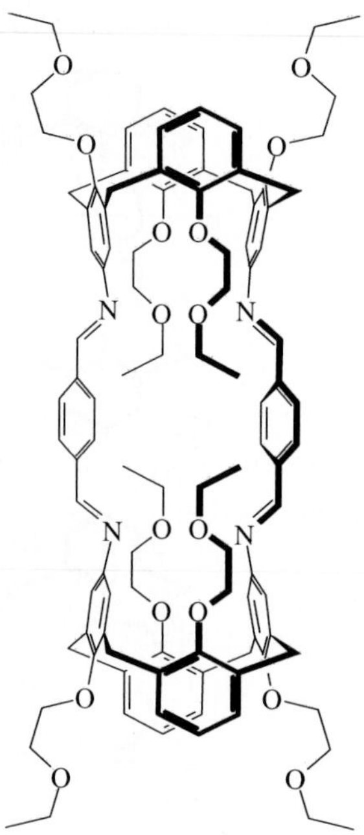

27

6. NANOCONTAINERS FOR NO_X

Small cylindrical molecules such as NO, CO and CO_2 are environmentally and biologically important reactive species for which the development of synthetic receptors is of considerable interest (see Chapter 8). The interaction of NO_x with calix[4]arenes (in all conformations), giving species which can be regarded as complexes of the NO^+ cation, is particularly strong and has been widely explored.[37] Simple tetra-O-alkylated calix[4]arenes in the cone or in the 1,3-alternate conformation, for example, react with NO_2/N_2O_4 ($\equiv NO^+\ NO_3^-$)and entrap reactive NO^+ in complexes of high stability[37,38] which have found application in synthesis as nitrosation reagents.[38] The intense colour developed on formation of NO^+/calixarene complexes has also been the basis of the development of NO_x sensors,[39,40] and calixtubes based on 1,3-alternate calix[4]arenes have been found to act as multiple NO^+ receptors, obviously of interest both as nitrosation reagents and in relation to NO_x pollution problems.[41-43]

7. ANION TRANSPORT BY CALIX TUBES

Calix[4]arene tetrabutyl amide **28** in the 1,3-alternate conformation facilitates the H^+/Cl^- transport in liposomes, forms ion channels in planar lipid bilayers, and supports electric currents in HEK cells.[44] The crystal structure determination of the HCl complex of the related calix[4]arene tetramethyl amide **29** provided a rationale for how the Cl^- is transported across the membrane. Individual calixarenes are bridged by amide NH...Cl^- and NH...OH_2 hydrogen bonds to give a lattice with H_2O-filled and Cl^--filled pores. However, it appears that the macrocyclic structure of the calixarene is not essential to its function, since the acyclic phenoxy acetamide analogs **30** and **31** are equally effective.[45]

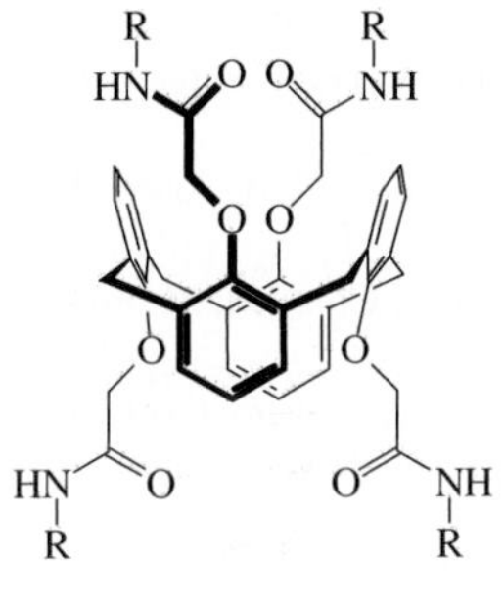

28 R = *n*Bu
29 R = Me

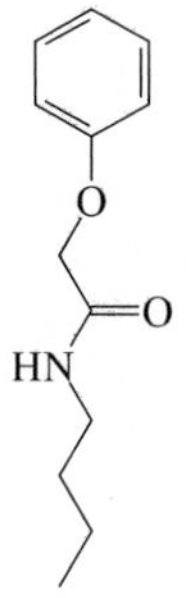

30

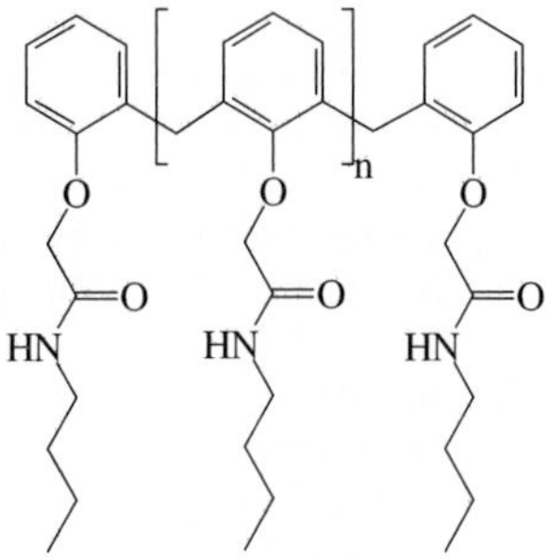

31a, n = 0
b, n = 1
c, n = 2
d, n = 3

8. CONCLUSIONS

In this chapter, the construction of several molecular architectures based on the 1,3-alternate conformation of calix[4]arenes has been described. Examples have been presented on their use as molecular tubes for transporting cations and anions and containers for fixation of gases. There is no doubt that calixarenes in the 1,3-alternate conformation will find applications in the future in various fields of nanotechnology and biotechnology.

9. ACKNOWLEDGEMENTS

This work was partly supported by ARIEL-KOSEF fellowships and by a Grant from the International Collaboration Project of Korea-France (Grant No. F01-2004-000-10023-0).

10. REFERENCES

1. E. O. Wilson *P. Natl. Acad. Sci. USA.* **102**, 6520-6521 (2005).
2. G. M. Whitesides, B. Grzybowski *Science* **295**, 2418-2421 (2002).
3. C. Leduc, O. Campas, K.B. Zeldovich, A. Roux, A. Jolimaitre, L. Bourel-Bonnet, B. Goud, J-F Joanny, P. Bassereau, J. Prost *PNAS* **101**, 17096-17101 (2004).
4. U. Koert, L. Al-Momani, J. R. Pfeifer *Synthesis* 1129-1146 (2004).
5. I. Tabushi, Y. Kuroda, K. Yokota *Tetrahedron Lett.* **23**, 4601-4604 (1982).
6. G. W. Gokel, A. Mukhopadhyay *Chem. Soc. Rev.*, **30**, 274-286 (2001).
7. O. Murillo, S. Watanabe, A. Nakano, G. W. Gokel *J. Am. Chem. Soc.* **117**, 7665-7679 (1995).
8. G. W. Gokel *Chem. Commun.* 1-9 (2000).
9. W. M. Leevy, J. E. Huettner, R. Pajewski, P.H. Schlesinger, G. W. Gokel *J. Am. Chem. Soc.* **126**, 15747-15753 (2004).
10. J. de Mendoza, F. Cuevas, P. Prados, E. S. Meadows, G. W. Gokel *Angew. Chem., Int. Ed.* **37**, 1534-1537 (1998).
11. A. D. Peculis, R. J. Thompson, J. P. Fojtik, H. M. Schwartz, C. A. Lisek, L. L. Frye *Bioorg. Med. Chem.* **5**, 1893-1901 (1997).
12. N. Maulucci, F. De Riccardis, C. B. Botta, A. Casapullo, E. Cressina, M. Fregonese, P. Tecilla, I. Izzo *Chem. Commun.* 1354-1356 (2005).
13. B. Pulpoka, J. Vicens *Collect. Czech. Chem. Commun.* **69**, 1251-1281 (2004).
14. W. Verboom, S. Datta, Z. Asfari, S. Harkema, D. N. Reinhoudt *J. Org. Chem.* **57**, 5394-5398 (1992).
15. A. Ikeda, S. Shinkai *J. Am. Chem. Soc.* **116**, 3102-3110 (1994).
16. A. Ikeda, S. Shinkai *Tetrahedron Lett.* **33**, 7385-7388 (1992).
17. A. Ikeda, H. Tsuzuki, S. Shinkai *Tetrahedron Lett.* **35**, 8417-8420 (1994).
18. T. Tsudera, A Ikeda, S. Shinkai *Tetrahedron*, **53**, 13609-13620 (1997).
19. A. Ikeda, T. Tsudera, S. Shinkai *J. Org. Chem.* **62**, 3568-3574 (1997).
20. K. N. Koh, K. Araki, S. Shinkai, Z. Asfari, J. Vicens *Tetrahedron Lett.* **36**, 6095-6098 (1995).
21. J. K. Lee, S. K. Kim, R. A. Bartsch, J. Vicens, S. Miyano, J. S. Kim *J. Org. Chem.* **68**, 6720-6725 (2003).
22. V. Lamare, J-F. Dozol, P. Thuéry, M. Nierlich, Z. Asfari, J. Vicens *J. Chem. Soc., Perkin Trans. 2* 1920-1926 (2001).
23. A. Bilyk, A. K. Hall, J. M. Harrowfield, M. W. Hosseini, B. W. Skelton, A. H. White *Inorg. Chem.* **40**, 672-686 (2001) and references therein.
24. J. S. Kim, W. K. Lee, K. No, Z. Asfari, J. Vicens *Tetrahedron Lett.* **41**, 3345-3348 (2000).
25. J. S. Kim, S. H. Yang, J. A. Rim, J. Y. Kim, J. Vicens, S. Shinkai *Tetrahedron Lett.* **42**, 8047-8050 (2001).

26. I. Leray, Z. Asfari, J. Vicens, B. Valeur *J. Chem. Soc., Perkin Trans.* **2.** 1429-1434 (2002).
27. I. Leray, Z. Asfari, J. Vicens, B. Valeur *J. Fluorescence* **14**, 451-458 (2004).
28. B. Pulpoka: *Ph.D. Thesis*, Université Louis Pasteur, Strasbourg, France (1997).
29. Z. Asfari, R. Abidi, F. Arnaud, J. Vicens *J. Inclusion Phenom. Mol. Recogn. Chem.* **13**, 163-169 (1992).
30. S. K. Kim, W. Sim, J. Vicens, J. S. Kim *Tetrahedron Lett.* **44**, 805-809 (2003).
31. S. K. Kim, J. Vicens, K-M. Park, S. S. Lee, J. S. Kim *Tetrahedron Lett.* **44**, 993-997 (2003).
32. S. K. Kim, J. K. Lee, S. H. Lee M. S. Lim S. W. Lee, W. Sim, J. S. Kim *J. Org. Chem.* **69**, 2877-2880 (2004).
33. S. K. Kim S. H. Kim, S-G. Kwon, E-H. Lee, S. H. Lee, J. S. Kim *Bull. Korean Chem. Soc.* **25**, 1244-1246 (2004).
34. V. Csokai, B. Balázs, G. Tóth, G. Horváth, I. Bitter *Tetrahedron* **60**, 12059-12066 (2004).
35. A. Ikeda, S. Shinkai *J. Chem. Soc., Chem. Commun.* 2375-2376 (1994).
36. J.-A. Pérez-Adelmar, H. Abraham, C. Sanchez, K. Rissanen, P. Prados, J. de Mendoza: *Angew. Chem. Int. Ed. Engl.* **35**, 1009-1011 (1996).
37. (a) R. Rathore, S. V. Lindeman, K. S. P. Rao, D. Sun, J. Kochi *Angew. Chem. Int. Edn* **39**, 2123-2127 (2000) and references therein; (b) G. V. Zyryanov, Y. Kang, D. M. Rudkevich *J. Am. Chem. Soc.* **125**, 2997-3007 (2003).
38. G. V. Zyryanov, D. M. Rudkevich *Org. Lett.* **5**, 1253-1256 (2003).
39. R. Rathore, S. H. Abdelwahed, I. A. Guzei *J. Am. Chem. Soc.* **126**, 13582-13583 (2004).
40. Y. Kang, D. M. Rudkevich *Tetrahedron* **60**, 11219-11225 (2004).
41. Y. Kang, G. V. Zyryanov, D. M. Rudkevich *Chem. Eur. J.* **11**, 1924-1932 (2005).
42. G. V. Zyryanov, D. M. Rudkevich *J. Am. Chem. Soc.* **126**, 4264-4270 (2004).
43. V. G. Organo, A. V. Leontiev, V. Sgarlata, H. V. Rasika Dias, D. M. Rudkevich *Angew. Chem. Int. Ed.* **44**, 3043-3047 (2005).
44. V. Sidorov, F. W. Kotch, G. Abdrakhmanova, R. Mizani, J. C. Fettinger, J. T. Davis *J. Am. Chem. Soc.* **124**, 2267-2278 (2002).
45. V. Sidorov, F. W. Kotch, J. L. Kuebler, Y-F. Lam, J. T. Davis *J. Am. Chem. Soc.* **125**, 2840-2841 (2003).

Chapter 8

CALIXARENE-BASED NANOMATERIALS

Sensing and fixation of gases

Dmitry M. Rudkevich
Department of Chemistry & Biochemistry, The University of Texas at Arlington, Arlington, TX 76019-0065, USA

Abstract: The role of calixarenes in the design, synthesis and application of supramolecular materials for sensing, storage and fixation of gases is discussed.

Key words: Calixarenes, encapsulation, gases, molecular containers, supramolecular polymers, synthetic nanotubes.

1. INTRODUCTION

Nanoscience and nanotechnology are extensions of materials science. They concern the control of matter at dimensions of 1 to 100 nanometers; and involve the study of how to create and manipulate matter at this scale.[1] One of the most important problems in nanotechnology is how to assemble atoms and molecules into smart materials and working devices. In this case, supramolecular chemistry emerges as a very important tool. Supramolecular chemistry is chemistry beyond the molecule; it is the chemistry of intermolecular interactions and reversible assemblies which lead to larger structures, or nanostructures, and materials.[2] Properties of such nanostructures and materials differ from the properties of individual molecules and yet can be programmed on a molecular level. In such cases, structure and properties of the monomeric molecular units are of crucial importance. In this chapter, we introduce a relatively new area of research, which is supramolecular chemistry of gases, and show how to use concepts and techniques of molecular recognition for chemical utilization of gases and also build materials for sensing, fixation and storage of gases (Fig. 1).[3]

J. Vicens and J. Harrowfield (eds.), Calixarenes in the Nanoworld, 151–172.

Considering the scope of this book, calixarene-based structures and approaches will be emphasized.

Gases compose the atmosphere and also occupy central positions in biology, medicine, science, technology and agriculture. The use of hydrogen (H_2) is extremely promising in the design of energy-rich fuel-cells. Nitrogen (N_2) is utilized in space technology and in ammonia (NH_3) production. Chemical and medical industries use oxygen (O_2), carbon dioxide (CO_2), N_2, NH_3, chlorine (Cl_2), and ethylene. CO_2 and nitrous oxide (N_2O) are major greenhouse gases, and they are also blood gases. N_2O is heavily used in anesthesia. Another crucial group of gases is NO_X - the sum of nitric oxide (NO), nitrogen dioxide (NO_2), N_2O_3, dinitrogen tetroxide (N_2O_4), and N_2O_5. NO serves as an important messenger in signal transduction processes. Other NO_X gases are toxic pollutants and participate in the formation of ground-level ozone and in global warming. Sulfur dioxide (SO_2) and dihydrogen sulfide (H_2S) produce acid rain.

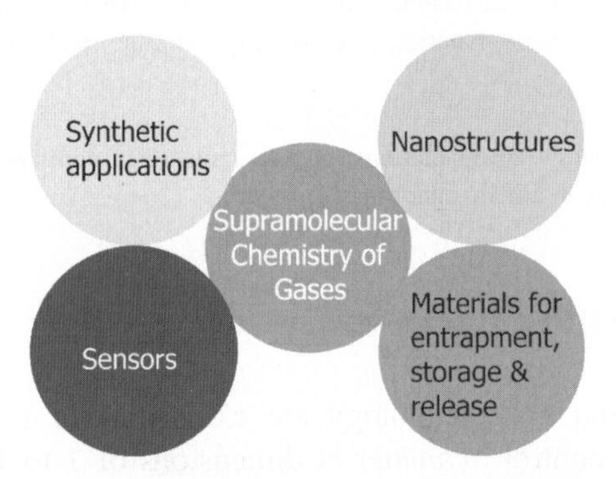

Figure 8-1. Supramolecular chemistry of gases can be used for organic synthesis, in sensing technology, materials science and nanotechnology.

We will briefly overview existing approaches towards encapsulation of gases in calixarene-based molecular containers and in solid materials based on them. Potential applications of encapsulated gas complexes will be discussed. We will further introduce dynamic, reversible chemistry between calixarenes and gases and show how to use it in the design of novel chambers for gas fixation, as well as supramolecular gas sensors and gas storing, releasing and separating materials. Calixarene-based nanotubes and self-assembling nanostructures and their applications will also be overviewed.

2. EARLY GAS-CALIXARENE COMPLEXES

The first generation of host-guest complexes between calixarenes and gases involved carcerands and their relatives. Carcerands are closed-surface host-molecules with enforced inner cavities, which incarcerate smaller organic molecules.[4] Their portals are too narrow to allow the guest to escape without breaking covalent bonds. Carcerands hold their guests permanently. In fact, guests are trapped during the synthesis, upon covalent shell-closure. This was exactly the case with gases.

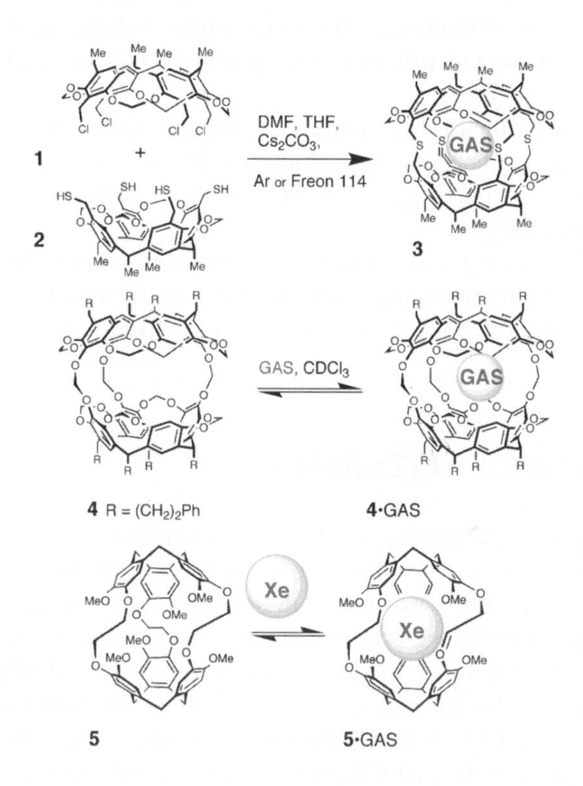

Figure 8-2. Early complexes of carcerands, hemicarcerands and cryptophanes with gases.

In their first syntheses of carcerands, Cram and co-workers coupled concave cavitand derivatives **1** and **2** in the presence of Cs_2CO_3 in DMF-THF solution under Ar atmosphere.[5] The carcerand **3** thus formed contained encapsulated Ar (Fig. 2). The gas presence was confirmed by elemental analysis and FAB mass spectrometry, although the analytical data indicated that only 1 among 150 shell-closures encapsulates the gas. The diameter of

Ar is ~3.1 Å, which is greater than the portal diameter in **3** (~2.6 Å), and the gas simply cannot escape. When the reaction between **1** and **2** was performed under an atmosphere of $(CClF_2)_2$ (Freon 114), this gas was also trapped.

Hemicarcerands are carcerands with larger portals, which allow for the entrapped guest to escape at high temperature, but to remain incarcerated at room temperature.[4] Cram found that hemicarcerand **4** reversibly encapsulated O_2, N_2, CO_2, and Xe (Fig. 2).[6] The exchange between the free and complexed **4** was slow on the NMR time-scale. The K_{assoc} values of 180 M^{-1} (N_2), 44 M^{-1} (O_2) and 200 M^{-1} (Xe) were obtained in $CDCl_3$ at 22°C, assuming a 1:1 stoichiometry. The inner cavity volume in **4** is relatively large (~100 $Å^3$) and the volumes of the gas molecules are within the 40 $Å^3$ range.

Collet's cryptophane-A **5** represents another class of hemicarcerands (Fig. 2). With an inner cavity volume of 95 $Å^3$, cryptophane **5** readily encapsulated CH_4 with K_{assoc} = 130 M^{-1} and Xe with much higher K_{assoc} of ~3000 M^{-1} in $(CDCl_2)_2$.[7]

For molecular containers **3-5**, the gas molecules were obviously better guests than solvent molecules, which were simply too big to enter and occupy the interior. This provided a strong driving force for the gas entrapment.

3. POROUS MATERIALS

With recent developments in materials chemistry and nanotechnology, extensive studies have been undertaken in the search for stable nano- or microporous networks. Most of these networks utilize coordination polymers, the porosity of which can be programmed depending on the applications. A variety of metal organic frameworks (MOFs) have been designed for gas storage and transport.[8] MOFs were found to effectively absorb N_2, O_2, Ar, CO_2, N_2O, H_2 and CH_4. The remarkable ability of certain MOFs in sorption of H_2 and CH_4 makes them very attractive candidates for vehicular gas storage.

Much less effort has been devoted to assessment of pure organic solids as gas sorbents since organic molecules typically adhere to close-packing principles and do not afford porous structures. Calixarenes offer a remarkable exception. Atwood and co-workers showed that CH_4, CF_4, C_2F_6, CF_3Br and other low-boiling halogenated alkanes could be reversibly entrapped and retained within the lattice voids of a crystalline calix[4]arene framework.[9] Such gas-storing crystals appeared to be extremely stable and release their guests only at elevated temperatures, several hundreds of °C above their boiling points. Ripmeester discovered that the calix[4]arene cavities in such crystals are directly involved in the gas complexation.[10] The Atwood team further demonstrated that *p-tert*-butylcalix[4]arene dimerizes in a crystalline phase into a hourglass-shaped cavity, capable of gas entrapment.[11] These crystals soak up gases when stored in air. Absorption of CO_2 was particularly rapid, but CO, N_2, and O_2 were also trapped. Of special importance, the calixarene crystals selectively absorbed CO_2 from a CO_2 - H_2 mixture, leaving the H_2 behind. This phenomenon can be used for purification of H_2. More recently, Atwood showed that calix[4]arene crystals can also absorb H_2 at higher pressures.[12]

Cavity-containing solid materials for gas entrapment and storage have thus emerged. While polymers with intrinsic calixarene cavities have not yet been constructed,[13] He, H_2, N_2, N_2O, and CO_2 were encapsulated in the solid state by hemicarcerand **4** (R = $(CH_2)_{11}CH_3$).[14] These gases were shown to replace each other in the solid **4**. For example, upon flushing the powder containing **4** and **4**•N_2 with CO_2 or N_2O, hemicarceplexes **4**•CO_2 or **4**•N_2O were obtained. The scope of gas encapsulation was thus expanded from solution to the gas-solid interface.

In summary, calixarenes can be employed in the design of cavity-containing solid materials for gas entrapment, storage and release. Polymers with intrinsic calixarene cavities are still not known, but this is just a matter of time. The major drawback of reversible encapsulation complexes with gases is in their low thermodynamic stability. Even well preorganized cavities of calixarenes and hemicarcerands derived from them cannot complex strongly, due to the lack of binding interactions. An alternative approach is based on reversible chemical transformation of gases upon complexation. In this case, they produce reactive intermediates with higher affinities for the receptor molecules. In the following subchapters we will demonstrate this approach for NO_X gases and CO_2.

4. CALIXARENE-BASED MATERIALS FOR NO_X

Kochi and co-workers showed that when converted to the cation-radical, calix[4]arene **6** was able to strongly complex NO gas with the formation of cationic calix-nitrosonium species **7**.[15] In these, the NO molecule is transformed into nitrosonium cation (NO^+) (Fig. 3). Strong charge-transfer interactions between NO^+ and the π-surface of **6** places the guest molecule between the cofacial aromatic rings at a distance 2.4 Å, which is shorter than the typical van der Waals contacts. A value for the association constant $K_{assoc} > 5 \times 10^8\ M^{-1}$ was determined (in CH_2Cl_2). Charge-transfer complex **7** is deeply colored and can be used for colorimetric sensing of NO gas.

Similar complexes were also obtained for calix[4]arenes in their *cone* and *partial cone* conformations.

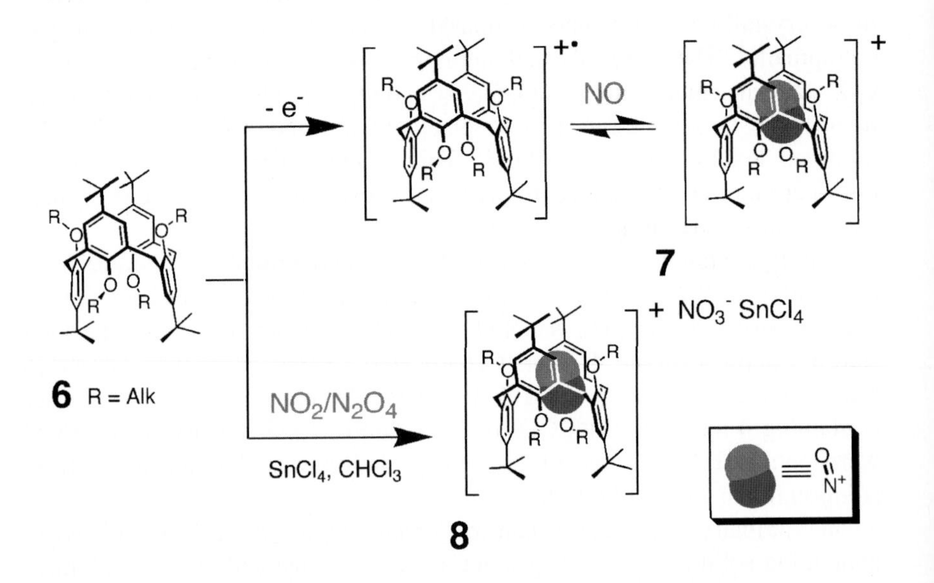

Figure 8-3. Complexes of calix[4]arenes with NO_X gases.

We have studied host-guest complexes formed upon reversible interaction between NO_2 and simple calix[4]arenes.[16] NO_2 is a paramagnetic gas of an intense brown-orange color. It exists in equilibrium with its dimer N_2O_4, which is colorless. The dynamic interconversion between NO_2 and N_2O_4 makes it impossible to study either of these species alone. N_2O_4 may disproportionate to ionic $NO^+NO_3^-$ while interacting with simple aromatic derivatives. We showed that tetrakis-*O*-alkylated calix[4]arenes in their *cone* and *1,3-alternate* conformations react with NO_2/N_2O_4 to form stable nitrosonium complexes, for example **8** (Fig. 3).[17] These complexes are deeply colored. They are strong but dissociate upon addition of water or alcohols. More stable calixarene-nitrosonium complexes were isolated upon addition of Lewis acids such as $SnCl_4$ and BF_3-Et_2O.

The visible spectrum of complex **8** showed broad charge-transfer band at λ_{max}~560 nm ($\varepsilon = 8 \times 10^3\ M^{-1}\ cm^{-1}$). While neither calixarenes nor NO_2 absorb in this region, addition of as little as ~1 eq NO_2 to the solution of **6** in chlorinated solvent results in appearance of the charge-transfer band. Its absorbance grows upon addition of larger quantities of NO_2 and reaches saturation when ~10 eq NO_2 is added.[18] Accordingly, calix[4]arenes can detect NO_2 even at micromolar/ppm concentrations. Interestingly, wider calix[5]-, calix[6]- and calix[8]arenes do not form encapsulation complexes with NO^+.

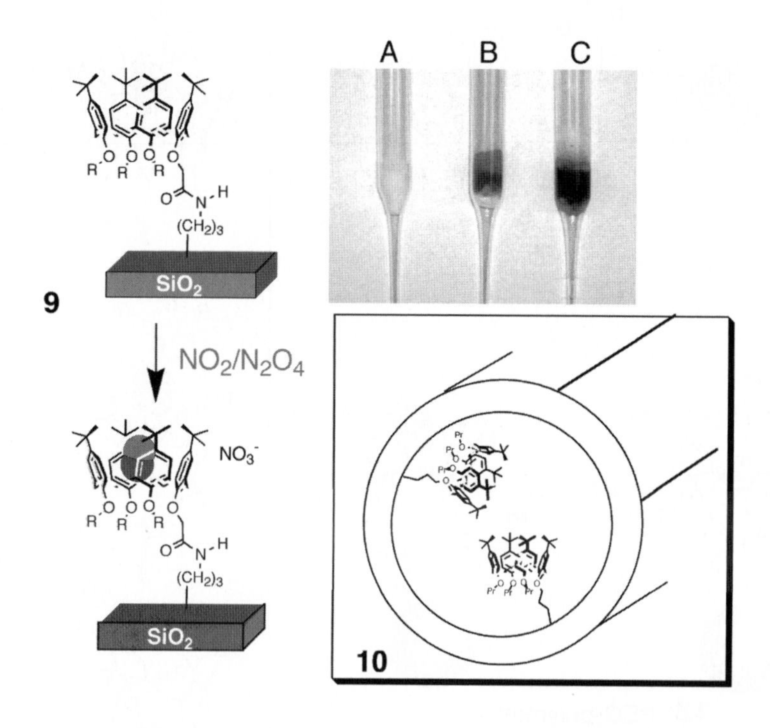

Figure 8-4. Calixarene materials **9** and **10** for NO_2 entrapment. In the upper right corner: NO_2 entrapment "chromatography" experiments. The columns were prepared as follows: A) loaded with commercial aminopropyl silica gel; B) loaded with dry silica gel **9**; C) loaded with **9** and flashed with $CHCl_3$. NO_2 was then passed through all three columns for 30 seconds, and the pictures were made after 2-3 minutes.

Based on these findings, calixarene materials have been prepared for entrapment of NO_X gases. In the first experiments, calixarenes were immobilized on a silica platform (Fig. 4).[17] Calix-silica gel **9** was obtained from commercial 3-aminopropylated silica gel and the corresponding calix[4]arene carboxylic acid, a 17% loading being achieved. Expanding our approach, Economy and co-workers prepared calix[4]arene-based periodic mesoporous silica **10** (Fig. 4).[19] According to adsorption/desorption and TEM studies, the pore dimension in **10** is as large as 2.9 nm.

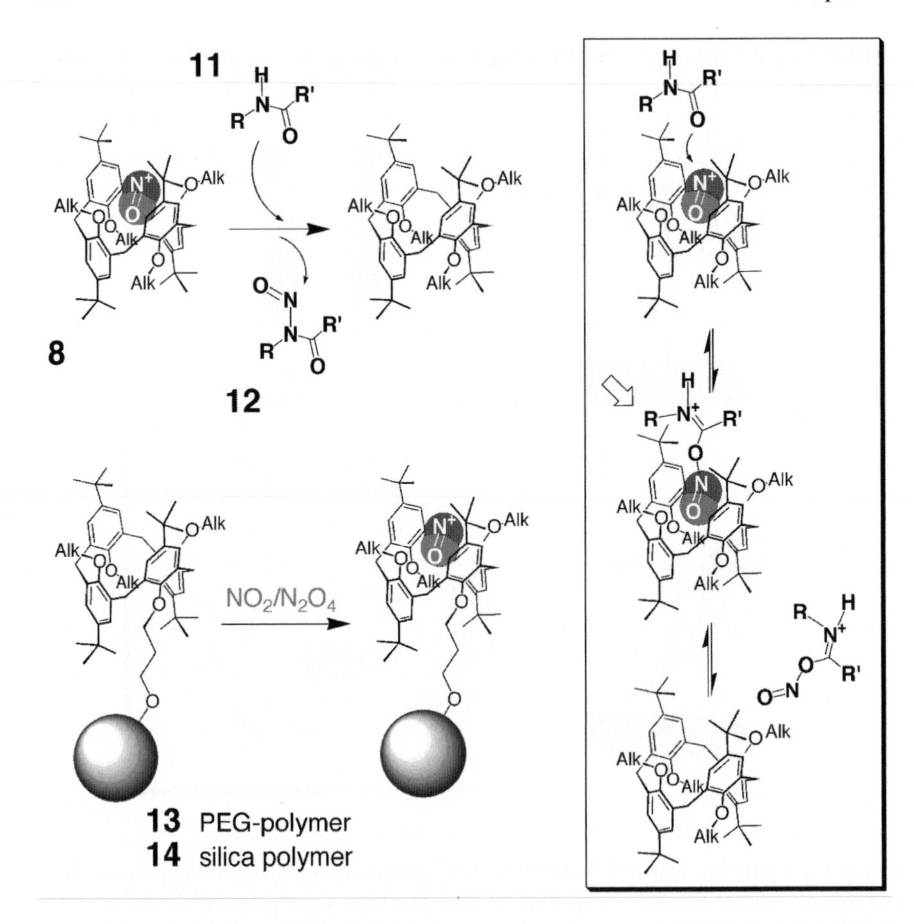

Figure 8-5. Nitrosating reagents and nitrosating materials based on calixarenes. Right box: the proposed mechanism of nitrosation of secondary amides **11** by encapsulated calixarene-based reagent **8**. The dimensions of R-groups of substrate **11** are critical and control the formation of the corresponding *O*-nitroso intermediate.

In the NO_2 entrapment experiments, a stream of the gas was passed through pipettes loaded with the materials **9** and **10**, instantly producing a dark purple color indicative of NO^+ complexation (Fig. 4). The complexation was also confirmed by IR spectroscopy, where the characteristic calixarene-NO^+ stretch was observed at $\nu \sim 1920\ cm^{-1}$. The solid-supported complexes were stable for hours, especially for mesoporous silica material **10**. In this case, the NO^+ entrapment was also detected by elemental analysis. The materials can be regenerated by simply washing with alcohol and reused.

We also found that calixarene-nitrosonium complex **8** and its analogs can release NO^+ and act as nitrosating agents.[20] Nitrosating reagents are important, as in synthetic organic chemistry, nitrosation holds a special place. Alkyl nitrites, nitrosamines, nitrosamides, and nitrosothiols are used in medicine as NO-releasing drugs. In total synthesis and methodology, -N=O is an important activating group, allowing easy transformations of amides to carboxylic acids and their derivatives. In addition, nitrosation mimics interactions between biological tissues and NO_X gases. In preliminary experiments, we discovered that secondary amides **11** reacted with complex **8** with remarkable selectivity. Chemical properties of the encapsulated NO^+ are different from those in bulk solution and controlled by the cavity. The cavity in **8** protects highly reactive NO^+ species from the bulk environment. The complex is quite stable towards moisture and oxygen, and can be handled, for at least several days, without a dry box and/or an N_2 atmosphere. On the other hand, it can be decomposed within few minutes by addition of larger quantities of H_2O or alcohols, regenerating the free calixarene. Accordingly, complex **8** represents an encapsulated reagent.

When mixed with the equimolar solution of amides R'C(O)NHR **11** in $CHCl_3$, complex **8** reacted quickly at room temperature, yielding up to 95% of *N*-nitrosamides **12** (Fig. 5).[20] Dark-blue solutions of **8** lost their color upon addition of the substrates, which is a reasonable visual test for the reaction. Among the variety of amides **11**, only those possessing *N*-CH_3 substituents were transformed to the corresponding *N*-nitrosamides **12**. No reaction occurred for substrates with bulkier groups. As a consequence, no color discharge was observed. The calixarene cavity was obviously responsible for such unique size-shape selectivity. Larger substrates simply cannot reach the encapsulated NO^+ (see mechanism in Fig. 5, right box).

Polymer-supported nitrosating reagents have been subsequently prepared.[18,20] Among the advantages of such reagents are the ease of their separation from the reaction mixture, their recyclability, and the simplification of handling toxic and odorous NO_X gases. Particularly useful are soluble polymers, as they overcome problems associated with the heterogeneous nature of the reaction conditions.

Although a wide variety of polymers is commercially available, NO_2/N_2O_4 (and also other NO_X) react with many of them, causing destruction and aging. As a free radical, NO_2 agressively attacks double bonds in polybutadienes, polyisoprenes and their copolymers, ester groups in poly(methyl)methacrylate, and also amide fragments in polyamides and polyurethanes. Furthermore, NO^+, generated from various NO_X, reacts with alkenes and other double bond containing structures. Considering this, we used polyethylene glycol (PEG) and silica gels as solid supports as these are robust and stable.

Initially, we synthesized PEG-supported polymer **13**, which is soluble in organic solvents (Fig. 5).[18] Deep purple NO^+-storing material **13**•$(NO^+)_n$ was then obtained upon simply bubbling NO_2/N_2O_4 through the solution of **13** in

CH_2Cl_2 for 2-3 min, followed by brief flushing with N_2 to remove the remaining NO_2/N_2O_4 gases. Material **13**•$(NO^+)_n$ effectively nitrosated amides **11** in CH_2Cl_2 and preference for the less bulky *N*-Me amide was observed as well.

In the preparation of insoluble nitrosating materials, commercial silica gel of relatively high porosity (150 Å) was activated in 18% HCl at reflux and then reacted with the corresponding calixarene siloxane in CH_2Cl_2 to give material **14** (Fig. 5).[20] The presence of calix[4]arene units in **14** was confirmed by the appearance of characteristic absorption bands in the IR spectrum. From the TGA and elemental analyses, a calixarene loading of ~10% was estimated. This rather modest level appeared to be reproducible, even when larger quantities of **14** were employed, and may be due to the steric bulkiness of the calixarene fragment.

The dark-blue nitrosonium-storing silica gel **14**•$(NO^+)_n$ was prepared by bubbling NO_2/N_2O_4 through the suspension of **14** in CH_2Cl_2 for 5-10 sec, followed by filtration and washing with CH_2Cl_2. Material **14**•$(NO^+)_n$ is quite robust and does not change the color for several days. For nitrosation, it was suspended in dry CH_2Cl_2, equimolar amount of amides **11** was then added, and the reaction mixture was stirred at room temperature for 24 h. The reactant's color disappeared, thus visually indicating the reaction progress. Material **14** was separated by simple filtration. Yields of nitrosamides **12** were determined by ^1H NMR spectroscopy, integrating signals of the product *vs* the starting compounds. The size-shape selectivity trend, observed for the solution experiments with complexes **8**, was clearly seen in this case as well. After at least three independent runs, the averaged yields of *N*-Me nitrosoamides **12** with less hindered *N*-Me fragments were established as up to 30%, while bulkier *N*-Et and *N*-Pr derivatives **12** formed in much smaller quantities ($\leq$ 8%). In control experiments, involving starting silica gel, no visible amounts of **12** were seen, again emphasizing the role of calixarene cavity in the described reactions.

Tight but reversible encapsulation of NO^+ species by calixarene monomers and materials based on them thus offers size-shape selectivities, previously unknown for existing, more aggressive nitrosating agents. NO^+ species, generated from NO^+-salts, N_2O_3, NO_2/N_2O_4, NO/O_2, NO/air, and $NaNO_2/H_2SO_4$, are typically not selective.

Concluding this part, novel calixarene-based materials are now available for sensing and fixation of NO_X gases. The supramolecular chemistry of calix[4]arenes and NO_X possesses a number of unique features. It is reversible, results in dramatic color changes, and specific for nitrosonium generating gases. This can be used for the detection in the presence of other gases/vapors. Calixarene materials may be employed as traps for NO_2/N_2O_4 and also NO/O_2, NO/air, and N_2O_3. Such procedures are often desirable in purification of commercial gases, especially NO for medical purposes. Nitrosonium complexes of calixarenes and polymers derived from them can be used as encapsulated nitrosating reagents/materials for synthetic methodology and medicinal chemistry.

5. CALIXARENE NANOTUBES

Filling single-walled carbon nanotubes (SWNTs) with foreign guest species is a rapidly emerging research area.[21] The major goal here is to enforce filling materials to adopt one-dimensional morphology for nanowiring and information flow. As far as gases are concerned, potential applications include using SWNTs as gas storing cylinders and, further, as reaction vessels.[22] Among the problems, however, are far from trivial protocols for chemical opening of SWNTs, still not fully understood mechanisms of their filling, as well as identification of the encapsulated material. Synthetic analogs of SWNTs have recently been introduced,[23] some of them based on calixarenes.[24] Organic synthesis offers a variety of sizes and shapes. However, most of these synthetic nanotubes are formed via self-assembly, and thus stable only under specific conditions.[23] Furthermore, the stability of the encapsulation complexes is generally weak,[24] which diminishes their capabilities as storing materials.

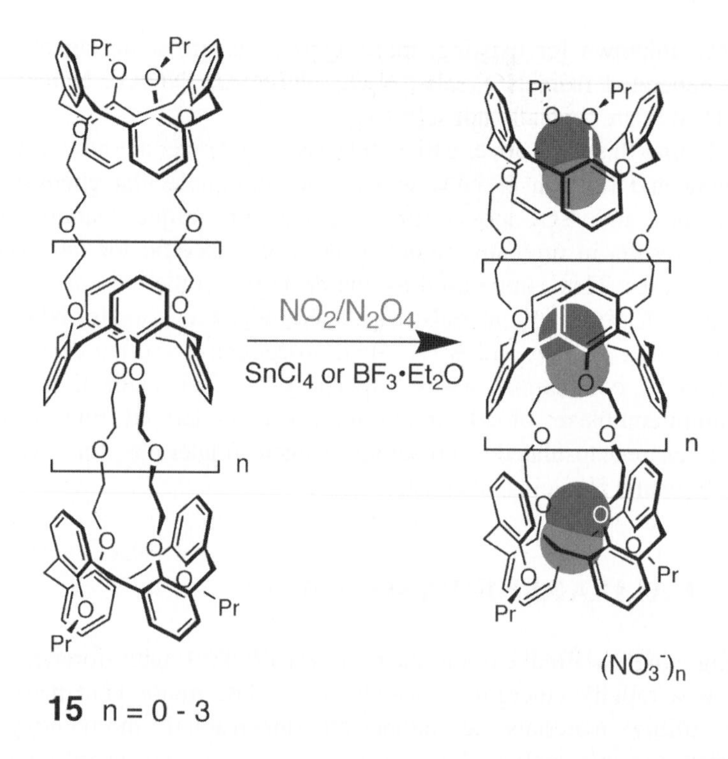

Figure 8-6. Synthetic nanotubes based on calixarenes and their reactions with NO_2/N_2O_4 gases.

As we discussed earlier, calix[4]arenes reversibly interact with NO_2/N_2O_4 and entrap reactive NO^+ cations within their π-electron rich interiors, one per cavity. Very high $K_{assoc} >> 10^6\ M^{-1}$ values ($\Delta G^{295} >> 8$ kcal/mol) for these processes were determined, and the complexes were also kinetically stable. These features were used in the design of calixarene-based nanotubes **15** (Fig. 6).[25-27] In the nanotube, *1,3-alternate* calix[4]arenes are rigidly connected from both sides of their rims with pairs of diethylene glycol linkers. In this calixarene conformation, two pairs of phenolic oxygens are oriented in opposite directions, providing diverse means to modularly enhance the tube length. The bridge length is critical since it not only provides relatively high conformational rigidity of the tubular structure, but also seals the walls, minimizing the gaps between the calixarene modules. The calixarene tubes possess defined inner tunnels of 6 Å diameter and may entrap multiple NO^+, one per each cavity. Finally, they can be emptied at will in a nondestructive manner.

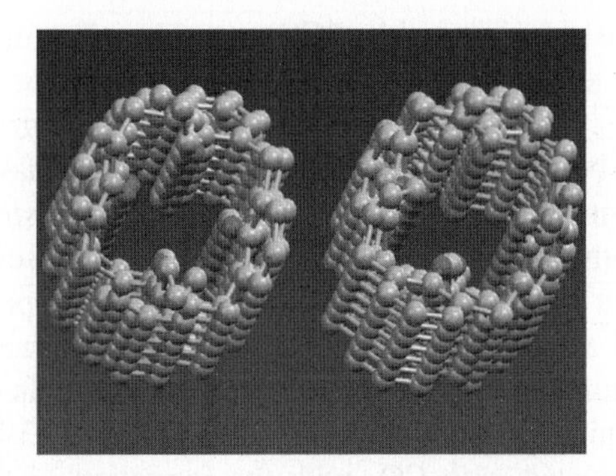

Figure 8-7. Solid state packing of nanotube **15** (n = 1).

Exposure of **15** to NO_2/N_2O_4 in chlorinated solvents results in the rapid encapsulation of NO^+ cations within its interior (Fig. 6). The complexes were characterized by UV-vis, FTIR and ^1H NMR spectroscopies. The NO^+ entrapment process is reversible, and addition of water quickly regenerated starting tube **15**.

The nanotube units **15** (n = 1, Fig. 7) pack head-to-tail, in straight rows, resulting in infinitely long cylinders. The neighboring nanocylinders aligned parallel to each other. In each nanocylinder, molecules **15** are twisted by 90° relative to each other, and the Ar-O-Pr propyl groups effectively occupy the voids between the adjacent molecules. In such an arrangement, the intermolecular distance between two neighboring tubes in the nanocylinder is ~6 Å. The nanocylinders are separated from each other by ~9 Å. This supramolecular order comes with the tube length and is without precedent for conventional, shorter calixarenes. The unique linear nanostructures maximize their intermolecular van der Waals interactions in the crystal through the overall shape simplification.

Thus, in addition to SWNTs, synthetic nanotubes are now available, which pack in tubular bundles and can reversibly react with NO_X gases. At this stage only certain guests, such as NO^+, can fill the interiors, but they are charged, which is important for the design of nanowires. Given the ability of calixarenes to react with NO_X gases even in the solid state,[17,18] it should be interesting to look at the nitrosonium flow along the infinite nanocylinders in the solid-state bundles of **15**.

6. SUPRAMOLECULAR MATERIALS FROM CO_2

It has been known for decades that CO_2 smoothly reacts with amines at ordinary conditions to yield carbamates.[28,29] Carbamates are thermally unstable and release CO_2 upon heating. Polymer-bound amines have been employed in industry as reusable "CO_2 scrubbers," removing CO_2 from industrial exhaust streams.[30] Imprinted polymers have been introduced, in which a template can be attached and then removed through a carbamate linker.[31] Reactions between CO_2 and immobilized amines have been employed for the gas sensing.[32,33] Reversible carbamate chemistry may thus be considered as another case of dynamic, covalent self-assembly of building blocks. It has been shown, that exposure of long-chain primary RNH_2 and secondary RR'NH alkyl amines to CO_2 results in the formation of thermally responsive alkylammonium alkylcarbamate organogels.[34] In another example, the amine containing ionic liquid 1-(3'-aminopropyl)-3-butylimidazolium tetrafluoroborate reacted with CO_2 with the formation of the corresponding carbamate dimers.[35] This process can be used to capture CO_2 in purification of industrial gas mixtures.

Dynamic covalent chemistry (DCC) is quickly emerging as a promising alternative to noncovalent self-assembly.[36] It simply offers an elegant opportunity of performing supramolecular chemistry with covalent bonds. One of the most important advantages here is the robustness of covalently organized structures, which on the other hand can be reversibly broken, at will. We suggested that carbamate bonds could be employed for wider variety of DCC experiments.[32] Of particular interest here are supramolecular polymers and supramolecular materials.

Reversibly formed polymers are usually called supramolecular polymers.[37] They represent a novel class of macromolecules, in which monomeric units are held together by reversible bonds/forces. These are self-assembling polymers, and thus far hydrogen bonds, metal-ligand interactions, and van der Waals forces have been employed to construct them. Supramolecular polymers combine features of conventional polymers with properties resulting from the bonding reversibility. Structural parameters of supramolecular polymeric materials, in particular their two- and three-dimensional architectures, can be switched "on-off" through the main chain association-dissociation processes. On the other hand, their strength and degree of polymerization depend on how tightly the monomeric units are aggregated. We introduced a strategy to build supramolecular polymeric chains, which takes advantage of dynamic chemistry between CO_2 and amines and also utilize hydrogen bonds.[29]

Monomeric units were designed, which *a)* strongly aggregate/dimerize in apolar solution and *b)* possess "CO_2-philic" primary amino groups on the periphery. Calixarenes were employed as self-assembling units. Calix[4]arene tetraurea dimers are probably the most studied class of strong hydrogen bonding aggregates.[38] Discovered ten years ago by Rebek[39] and Böhmer,[40] these dimers form in apolar solution with $K_D \geq 10^6$ M^{-1} and are held together by a seam of sixteen intermolecular C=O—H-N hydrogen bonds. This results in a rigid inner cavity of ~200 Å^3 which reversibly encapsulates a solvent molecule or a benzene-sized guest

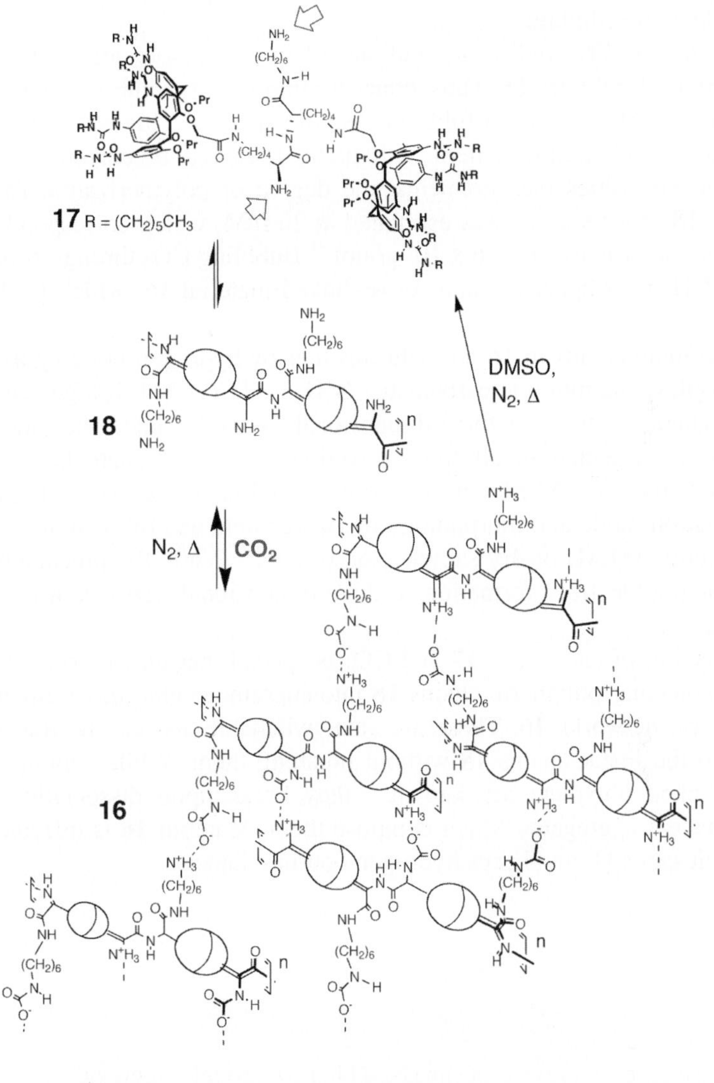

Figure 8-8. Supramolecular calixarene-based polymers can be reversibly cross-linked by CO_2.

Supramolecular polymer **16** is based on carbamate chemistry (Fig. 8). This is a 3D molecular network, which employs CO_2 as a cross-linking agent.[41] In monomer **17**, two calixarene tetraurea moieties are linked with a dipeptide, di-*l*-lysine chain. Calixarenes were attached to the ε-NH_2 ends, so the dilysine module orients them away from each other, in roughly opposite directions. Such arrangement also prevents intramolecular assembly. A hexamethyleneamine chain was then attached to the carboxylic side of the dipeptide. Its amino group and the α-NH_2 group of **17** react with CO_2, providing cross-linking.

First, viscosity studies of solutions **17** in apolar solvents confirmed the formation of polymer **18**. Thus, concentrating $CHCl_3$ solutions of **17** from 5 to just 40 mM leads to ≥5-fold increase in viscosity. This was not observed for the model, non-polymeric calixarene tetraurea precursors. From the specific viscosities measurements, the degree of polymerization for linear chains **18** of ~2.8×10^2 was estimated at 20 mM, which corresponds to the average molar mass of ~7.6×10^5 g/mol.[41] Bubbling CO_2 through solution of **18** in $CHCl_3$ or benzene yields cross-linked material **16**, which is clearly a gel.

The main chains in **16** are held together by hydrogen-bonding assembly of capsules, and multiple carbamate $-N^+H_3$—$^-$OC(O)NH- bridges cross-link these chains. This is a three-dimensional network, since the side amine groups are oriented in all three directions. The carbamate bridges were detected by ^{13}C NMR spectroscopy. Finally, visual insight into the aggregation mode and morphology in **16** was obtained by scanning electron microscopy (SEM) of dry samples, or xerogels. While the precursors show only negligible fiber formation, a three-dimensional network was obvious for **16**.

Reaction of calixarene **17** and CO_2 is special, because it converts linear supramolecular polymeric chains **18** into supramolecular, three-dimensional polymeric networks **16**. These are also switchable and can be transformed back to the linear chains **18** without breaking them. While supramolecular cross-linked polymers are known,[37] they break upon dissociation of the noncovalent aggregates which compose them. Material **16** is different, as it only releases CO_2 and keeps hydrogen bonding intact.

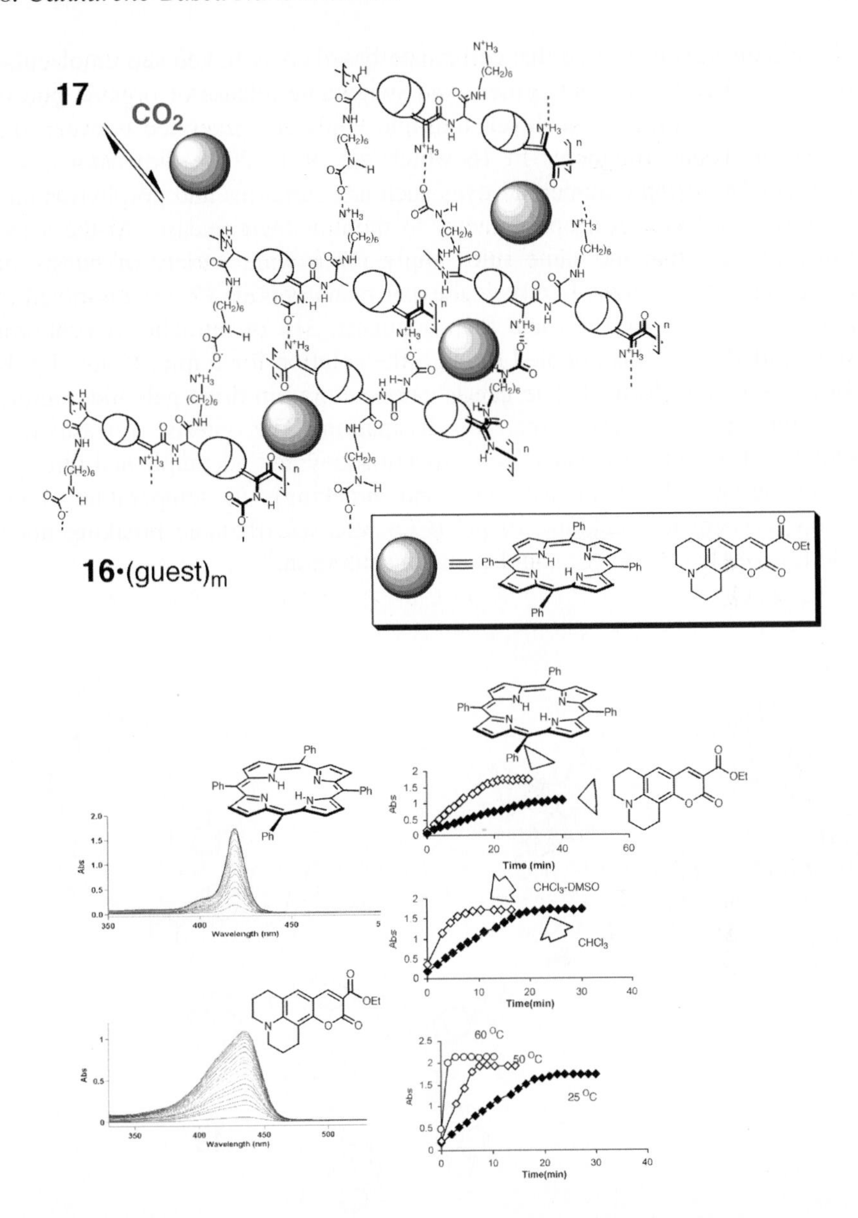

Figure 8-9. Guest entrapment experiments using supramolecular gel **16**. Below: guest release from material **16** can be followed by absorption spectroscopy. The release rates are dependent on temperature, solvent polarity and other factors.

We further discovered that carbamate-based cross-linked supramolecular polymer **16** serves for entrapment and switchable release of organic guests (Fig. 9).[42] On a molecular level, multiple voids are generated between the carbamate-lysine fragments in **16** which are of 15-20 Å dimensions. We used gel **16** to trap commercial dyes such as coumarins and porphyrins and employed UV-vis spectrophotometry to monitor their release. At the same time, we feel that the same rules apply for a wide variety of guests of comparable dimensions. In a typical experiment, peptide **17** was dissolved in a small volume of $CHCl_3$ and then coumarin 314 or tetraphenylporphyrin were added. CO_2 was bubbled through the solution for 5 min. Colored gels **16•**(guest)$_n$ were formed. The guests can be stored in dried gels indefinitely and released only upon the gel dissipation. The guests' release was monitored by conventional UV-vis spectroscopy and accomplished through *a)* changing solvent polarity (H-bond breaking), *b)* temperature (CO_2 release/carbamate breaking), *c)* pH (CO_2 release/carbamate breaking upon addition of HCl, TFA, etc.), and even concentration.[42]

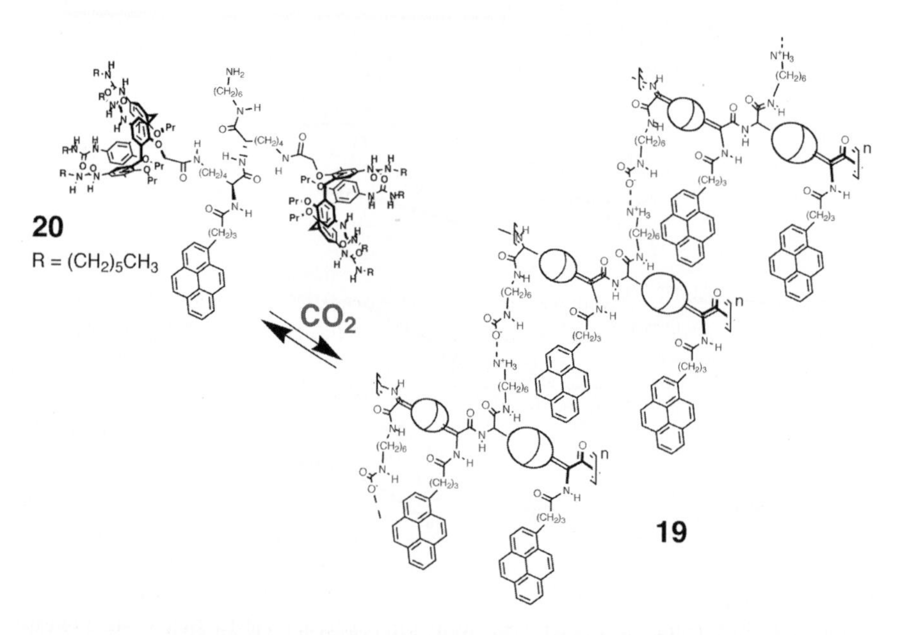

Figure 8-10. Fluorescent supramolecular material **19** is made using CO_2 as a cross-linking agent.

Gel **19** was obtained from benzene and benzene-$CHCl_3$ solutions of biscalixarene **20** and CO_2 (Fig. 10).[43] The polymeric chains here possess multiple fluorophore units - pyrene moieties - brought together through hydrogen bonding and carbamate bridges. Accordingly, material **19** is fluorescent and may act as a vehicle for energy migration. The aggregation degree and therefore the fluorophore local concentrations can be controlled and switched on-off, as described earlier. Formation of the carbamate bridges in **19** was routinely confirmed by ^{13}C NMR spectroscopy. As previously described, they can be broken after heating solution **19** for a few minutes at ~100°C and bubbling N_2 through it. The SEM pictures of the corresponding xerogels revealed, in particular, well-defined pores of ~1-3 μm diameter, which can be used for guest/solvent entrapment.

In preliminary photophysical experiments, we noticed a striking contrast in fluorescent behavior of xerogels **19** obtained from benzene and from 95:5 benzene-nitrobenzene solutions of derivative **20** (Fig. 10).[43] The former is strongly fluorescent (λ_{ex} = 347 nm), but the latter is not. Nitrobenzene is known to quench fluorescence of pyrene. Incorporated within the gel's pores, molecules of nitrobenzene appear to be in close proximity to the multiple pyrene donors, and energy transfer is effective. In another experiment, dropwise addition of nitrobenzene (up to 10% v/v) to the benzene suspension of fluorescent xerogel **19**, initially obtained from benzene, resulted in the fluorescence disappearing within seconds. These observations could be useful in the design of switchable light harvesting materials.

7. CONCLUSIONS

Calixarenes have made an enormous contribution to the field of molecular recognition, sensing and self-assembly, serving as unique objects for studying intermolecular interactions and generating new supramolecular functions.[44] Calixarenes are now entering nanoscience and nanotechnology. In this chapter, we showed that materials for sensing, fixation and storage of environmentally, industrially and biomedically important gases can be prepared from calixarenes. Simple encapsulation complexes with gases led to porous solids for gas separation and storage. Further functionalization of calixarenes brought novel chambers and nanotubes for gas fixation, as well as supramolecular gas sensors and gas storing and releasing materials. Calixarene-based self-assembled and reversibly formed nanostructures and gels have also been prepared. In synthetic organic chemistry, calixarenes help to convert gases into active reagents, which may safely be stored inside the capsule or cage for a prolonged time and released under facile control. Approaches towards polymeric caged reagents have also been introduced. Given the high diversity of the chemistry of calixarenes, the capabilities here are really beyond limits.

ACKNOWLEDGEMENTS

I acknowledge the members of my research group for their most valuable contribution to this research. Financial support was generously provided by the American Chemical Society Petroleum Research Fund, the US National Science Foundation, the Texas Higher Education Coordinating Board – Advanced Technology Program, and the Alfred P. Sloan Foundation.

8. REFERENCES

1. http://www.e-drexler.com/
2. (a) J.-M. Lehn, *Supramolecular Chemistry*, VCH, Weinheim, (1995). (b) J. W. Steed, J. L. Atwood, *Supramolecular Chemistry*, John Wiley & Sons, Ltd., Chichester-New York-Weinheim-Brisbane-Singapore-Toronto, (2000). (c) P. D. Beer, P. A. Gale, D. K. Smith, *Supramolecular Chemistry*, Oxford University Press, (1999). (d) H. Dodziuk, *Introduction to Supramolecular Chemistry*, Kluwer Academic Publishers, Dordrecht-Boston-London, (2002).
3 Reviews on supramolecular chemistry of gases: (a) D. M. Rudkevich, *Angew. Chem. Int. Ed.* **43**, 558-571 (2004). (b) D. M. Rudkevich, A. V. Leontiev, *Aust. J. Chem.* **57**, 713-722 (2004).
4. (a) D. J. Cram, J. M. Cram, *Container Molecules and their Guests,* Royal Society of Chemistry, Cambridge, 1994. (b) A. Jasat, J. C. Sherman, *Chem. Rev.* **99**, 931-967 (1999). (c) R. Warmuth, J. Yoon, *Acc. Chem. Res.* **34**, 95-105 (2001).
5. (a) D. J. Cram, S. Karbach, Y. H. Kim, L. Baczynskyj, G. W. Kalleymeyn, *J. Am. Chem. Soc.* **107**, 2575-2576 (1985). (b) D. J. Cram, S. Karbach, Y. H. Kim, L. Baczynskyj, K. Marti, R. M. Sampson, G. W. Kalleymeyn, *J. Am. Chem. Soc.* **110**, 2554-2560 (1988).
6. D. J. Cram, M. E. Tanner, C. B. Knobler, *J. Am. Chem. Soc.* **113**, 7717-7727 (1991).
7. (a) K. Bartik, M. Luhmer, J.-P. Dutasta, A. Collet, Reisse, *J. Am. Chem. Soc.* **120**, 784-791 (1998). (b) T. Brotin, A. Lesage, L. Emsley, A. Collet, *J. Am. Chem. Soc.* **122**, 1171-1174 (2000). (c) T. Brotin, J.-P. Dutasta, *Eur. J. Org. Chem.* 973-984 (2003).
8. (a) S. L. James, *Chem. Soc. Rev.* **32**, 276-288 (2003). (b) C. Janiak, *J. Chem. Soc., Dalton Trans.* 2781-2804 (2003). (c) N. L. Rosi, J. Eckert, M. Eddaoudi, D. T. Vodak, J. Kim, M. O'Keeffe, O. M. Yaghi, *Science* **300**, 1127-1130 (2003). (d) L. Pan, M. B. Sander, X. Huang, J. Li, M. Smith, E. Bittner, B. Bockrath, J. K. Johnson, *J. Am. Chem. Soc.* **126**, 1308-1309 (2004). (e) T. Düren, L. Sarkisov, O. M. Yaghi, R. Q. Snurr, *Langmuir* **20**, 2683-2689 (2004). (f) M. Eddaoudi, J. Kim, N. Rosi, D. Vodak, J. Wachter, M. O'Keeffe, O. M. Yaghi, *Science* **295**, 469-472 (2002). (g) S. Kitagawa, R. Kitaura, S. Noro, *Angew. Chem. Int. Ed.* **43**, 2334-2375 (2004). (h) A. C. Sharma, A. S. Borovik, *J. Am. Chem. Soc.* **122**, 8946-8955 (2000). (i) K. M. Padden, J. F. Krebs, C. E. MacBeth, R. C. Scarrow, A. S. Borovik, *J. Am. Chem. Soc.* **123**, 1072-1079 (2001). (j) J. L. C. Rowsell, O. M. Yaghi, *Angew. Chem. Int. Ed.* **44**, 4670-4679 (2005).
9. J. L. Atwood, L. J. Barbour, A. Jerga, *Science* **296**, 2367-2369 (2002).
10. G. D. Enright, K. A. Udachin, I. L. Moudrakovski, J. A. Ripmeester, *J. Am. Chem. Soc.* **125**, 9896-9897 (2003).
11. J. L. Atwood, L. J. Barbour, A. Jerga, *Angew. Chem. Int. Ed.* **43**, 2948-2950 (2004). See also: B. F. Graham, J. M. Harrowfield, R. D. Tengrove, A. F. Lagalante, T. J. Bruno, *J. Inclus. Phenom., Macrocycl. Chem.*, **43**, 179-182 (2002).

12. P. K. Thallapally, G. O. Lloyd, T. B. Wirsig, M. W. Bredenkamp, J. L. Atwood, L. J. Barbour, *Chem. Commun.* 5272-5274 (2005).
13. For the preliminary results, see: N. B. McKeown, P. M. Budd, K. J. Msayib, B. S. Ghanem, H. J. Kingston, C. E. Tattershall, S. Makhseed, K. J. Reynolds, D. Fritsch, *Chem. Eur. J.* **11**, 2610-2620 (2005).
14. A. V. Leontiev, D. M. Rudkevich, *Chem. Commun.* 1468-1469 (2004).
15. (a) R. Rathore, S. V. Lindeman, K. S. S. Rao, D. Sun, J. K. Kochi, *Angew. Chem. Int. Ed.* **39**, 2123-2127 (2000). (b) S. V. Rosokha, J. K. Kochi, *J. Am. Chem. Soc.* **124**, 5620-5621 (2002). (c) S. V. Rosokha, S. V. Lindeman, R. Rathore, J. K. Kochi, *J. Org. Chem.* **68**, 3947-3957 (2003).
16. Review from this laboratory: D. M. Rudkevich, Y. Kang, A. V. Leontiev, V. G. Organo, G. V. Zyryanov, *Supramolecular Chemistry,* **17**, 93-99 (2005).
17. (a) G. V. Zyryanov, Y. Kang, S. P. Stampp, D. M. Rudkevich, *Chem. Commun.*, 2792-2793 (2002). (b) G. V. Zyryanov, Y. Kang, D. M. Rudkevich, *J. Am. Chem. Soc.* **125**, 2997-3007 (2003).
18. Y. Kang, D. M. Rudkevich, *Tetrahedron*, **60**, 11219-11225 (2004).
19. C. Liu, L. Fu, J. Economy, *Macromol. Rapid Commun.* **25**, 804-807 (2004).
20. (a) Y. Kang, G. V. Zyryanov, D. M. Rudkevich, *Chem. Eur. J.* **11**, 1924-1932 (2005). (b) G. V. Zyryanov, D. M. Rudkevich, *Org. Lett.* **5**, 1253-1256 (2003).
21. (a) M. Monthioux, *Carbon*, **40**, 1809-1823 (2002). (b) O. Vostrowsky, A. Hirsch, *Angew. Chem. Int. Ed.*, **43**, 2326-2329 (2004). (c) K. Koga, G. T. Gao, H. Tanaka, X. C. Zeng, *Nature*, **412**, 802-805 (2001). (d) D. A. Britz, A. N. Khlobystov, K. Porfyrakis, A. Ardavan, G. A. D. Briggs, *Chem. Commun.*, 37-39 (2005).
22. (a) A. I. Kolesnikov, J.-M. Zanotti, C.-K. Loong, P. Thiyagarajan, A. P. Moravsky, R. O. Loutfy, C. J. Burnham, *Phys. Rev. Lett.*, **93**, 035503-1 – 035503-4 (2004). (b) C. Matranga, B. Bockrath, *J. Phys. Chem. B*, **108**, 6170-6174 (2004). (c) O. Byl, P. Kondratyuk, J. T. Yates, Jr., *J. Phys. Chem. B*, **107**, 4277-4279 (2003). (d) O. Byl, P. Kondratyuk, S. T. Forth, S. A. FitzGerald, L. Chen, J. K. Johnson, J. T. Yates, Jr., *J. Am. Chem. Soc.*, **125**, 5889-5896 (2003). (e) A. Fujiwara, K. Ishii, H. Suematsu, H. Kataura, Y. Maniwa, S. Suzuki, Y. Achiba, *Chem. Phys. Lett.*, **336**, 205-211 (2001).
23. Self-assembling nanotubes: (a) D. T. Bong, T. D. Clark, J. R. Granja, M. R. Ghadiri, *Angew. Chem. Int. Ed.*, **40**, 988-1011 (2001). (b) S. Matile, A. Som, N. Sorde, *Tetrahedron*, **60**, 6405-6435 (2004). (c) T. Yamaguchi, S. Tashiro, M. Tominaga, M. Kawano, T. Ozeki, M. Fujita, *J. Am. Chem. Soc.*, **126**, 10818-10819 (2004). (d) S. Tashiro, M. Tominaga, T. Kusukawa, M. Kawano, S. Sakamoto, K. Yamaguchi, M. Fujita, *Angew. Chem. Int. Ed.*, **42**, 3267-3270 (2003). (e) M. Tominaga, S. Tashiro, M. Aoyagi, M. Fujita, *Chem. Commun.*, 2038-2039 (2002). (f) V. Sidorov, F. W. Kotch, G. Abdrakhmanova, R. Mizani, J. C. Fettinger, J. T. Davis, *J. Am. Chem. Soc.*, **124**, 2267-2278 (2002). (g) L. Baldini, F. Sansone, A. Casnati, F. Ugozzoli, R. Ungaro, *J. Supramol. Chem.*, 219-226 (2002). Covalently linked nanotubes: (h) A. Harada, J. Li, M. Kamachi, *Nature*, **364**, 516-518 (1993). (i) Y. Kim, M. F. Mayer, S. C. Zimmerman, *Angew. Chem. Int. Ed.*, **42**, 1121-1126 (2003). See also ref 24.
24. (a) A. Ikeda, S. Shinkai, *J. Chem. Soc., Chem. Commun.*, 2375-2376 (1994). (b) A. Ikeda, M. Kawaguchi, S. Shinkai, *Anal. Quim. Int. Ed.*, **93**, 408-414 (1997). (c) J.-A. Perez-Adelmar, H. Abraham, C. Sanchez, K. Rissanen, P. Prados, J. de Mendoza, *Angew. Chem. Int. Ed. Engl.*, **35**, 1009-1011 (1996). (d) S. K. Kim, W. Sim, J. Vicens, J. S. Kim, *Tetrahedron Lett.*, **44**, 805-809 (2003). (e) S. K. Kim, J. Vicens, K.-M. Park, S. S. Lee, J. S. Kim, *Tetrahedron Lett.*, **44**, 993-997 (2003).
25. G. V. Zyryanov, D. M. Rudkevich, *J. Am. Chem. Soc.* **126**, 4264-4270 (2004).

26. V. G. Organo, A. V. Leontiev, V. Sgarlata, H. V. R. Dias, D. M. Rudkevich, *Angew. Chem. Int. Ed.*, **44**, 3043-3047 (2005).
27. V. Sgarlata, V. G. Organo, D. M. Rudkevich, *Chem. Commun.*, 5630-5632 (2005).
28. D. B. Dell'Amico, F. Calderazzo, L. Labella, F. Marchetti, G. Pampaloni, *Chem. Rev.*, **103**, 3857-3898 (2003).
29. Review from this laboratory: D. M. Rudkevich, H. Xu, *Chem. Commun.*, 2651-2659 (2005).
30. (a) T. Yamaguchi, C. A. Koval, R. D. Nobel, C. Bowman, *Chem. Eng. Sci.*, **51**, 4781-4789 (1996). (b) T. Yamaguchi, L. M. Boetje, C. A. Koval, R. D. Noble, C. N. Bowman, *Ind. Eng. Chem. Res.*, **34**, 4071-4077 (1995). (c) P. Kosaraju, A. S. Kovvali, A. Korikov, K. K. Sirkar, *Ind. Eng. Chem. Res.*, **44**, 1250-1258 (2005).
31. (a) C. D. Ki, C. Oh, S.-G. Oh, J. Y. Chang, *J. Am. Chem. Soc.*, **124**, 14838-14839 (2002). (b) J. Alauzun, A. Mehdi, C. Reye, R. J. P. Corriu, *J. Am. Chem. Soc.*, **127**, 11204-11205 (2005).
32. (a) E. M. Hampe, D. M. Rudkevich, *Chem. Commun.*, 1450-1451 (2002). (b) E. M. Hampe, D. M. Rudkevich, *Tetrahedron*, **59**, 9619-9625 (2003).
33. (a) L. C. Brousseau, III, D. J. Aurentz, A. J. Benesi, T. E. Mallouk, *Anal. Chem.*, **69**, 688-694 (1997). (b) P. Herman, Z. Murtaza, J. Lakowicz, *Anal. Biochem.*, **272**, 87-93 (1999).
34. (a) M. George, R. G. Weiss, *J. Am. Chem. Soc.*, **123**, 10393-10394 (2001). (b) M. George, R. G. Weiss, *Langmuir*, **18**, 7124-7135 (2002). (c) M. George, R. G. Weiss, *Langmuir*, **19**, 1017-1025 (2003). (d) M. George, R. G. Weiss, *Langmuir*, **19**, 8168-8176 (2003).
35. E. D. Bates, R. D. Mayton, I. Ntai, J. H. Davis, Jr., *J. Am. Chem. Soc.*, **124**, 926-927 (2002).
36. (a) J.-M. Lehn, *Chem. Eur. J.*, **5**, 2455-2463 (1999). (b) S. J. Rowan, S. J. Cantrill, G. R. L. Cousins, J. K. M. Sanders, J. F. Stoddart, *Angew. Chem. Int. Ed.*, **41**, 898-952 (2002).
37. (a) A. W. Bosman, L. Brunsveld, B. J. B. Folmer, B. J. B.; Sijbesma, R. P.; Meijer, E. W. *Macromol. Symp.*, **201**, 143-154 (2003). (b) J.-M. Lehn, *Polym. Int.*, **51**, 825-839 (2002). (c) U. S. Schubert, C. Eschbaumer, C. *Angew. Chem. Int. Ed.*, **41**, 2892-2926 (2002). (d) A. T. ten Cate, R. P. Sijbesma, *Macromol. Rapid Commun.*, **23**, 1094-1112 (2002). (e) L. Brunsveld, B. J. B. Folmer, E. W. Meijer, R. P. Sijbesma, *Chem. Rev.*, **101**, 4071-4097 (2001). (f) R. F. M. Lange, M. van Gurp, E. W. Meijer, *J. Polym. Sci. A*, **37**, 3657-3670 (1999). (h) R. K. Castellano, D. M. Rudkevich, J. Rebek, Jr., *Proc. Natl. Acad. Sci. USA*, **94**, 7132-7137 (1997).
38. (a) J. Rebek, Jr., *Chem. Commun.*, 637-643 (2000). (b) V. Böhmer, M. O. Vysotsky, *Austr. J. Chem.*, **54**, 671-677 (2001). (c) F. Hof, S. L. Craig, C. Nuckolls, J. Rebek, Jr., *Angew. Chem. Int. Ed.*, **41**, 1488-1508 (2002).
39. K. D. Shimizu, J. Rebek, Jr., *Proc. Natl. Acad. Sci. USA*, **92**, 12403-12407 (1995).
40. (a) O. Mogck, V. Böhmer, W. Vogt, *Tetrahedron*, **52**, 8489-8496 (1996). (b) O. Mogck, E. F. Paulus, V. Böhmer, I. Thondorf, W. Vogt, *Chem. Commun.*, 2533-2534 (1996).
41. H. Xu, D. M. Rudkevich, *Chem. Eur. J.*, **10**, 5432-5442 (2004).
42. H. Xu, D. M. Rudkevich, *Org. Lett.*, **7**, 3223-3226 (2005).
43. H. Xu, D. M. Rudkevich, *J. Org. Chem.*, **69**, 8609-8617 (2004).
44. (a) *Calixarenes 2001*, Z. Asfari, V. Böhmer, J. Harrowfield, J. Vicens Eds., Kluwer Academic Publishers: Dordrecht, Netherlands, (2001). (b) C. D. Gutsche: In *Calixarenes Revisited*, J. F. Stoddart, Ed., Monographs in Supramolecular Chemsitry, Royal Society of Chemistry: London, (1998).

Chapter 9

FULLERENES AND CALIXARENES

Toward molecular devices

Juan Luis Delgado de la Cruz and Jean-François Nierengarten
Groupe de Chimie des Fullerènes et des Systèmes Conjugués, Laboratoire de Chimie de Coordination du CNRS, 205 route de Narbonne, 31077 Toulouse Cedex 4, France; E-mail: jfnierengarten@lcc-toulouse.fr

Abstract: This chapter provides an overview on the various aspects of fullerene chemistry involving calix[*n*]arenes and related macrocycles.

Key words: Fullerenes, calixarenes, inclusion complexes, thin films, photoactive molecular devices, self-inclusion.

1. INTRODUCTION

Since the fullerenes became available in macroscopic quantities in 1990,[1] the chemical properties of these fascinating carbon cages have been intensively investigated. While the covalent functionalisation of the fullerenes has seen a rapid development,[2] supramolecular fullerene chemistry has not yet been explored to the same extent.[3] Whereas early efforts were mainly directed at the molecular complexation of the pure carbon sphere in the solid state and in solution, the construction of novel supramolecular architectures bearing properly functionalized carbon cages as structural motifs developed later,[4] since it required the covalent chemistry of fullerenes to first be vanquished. This chapter provides an overview on the various aspects of fullerene chemistry involving calix[*n*]arenes and related macrocycles. Calix[*n*]arenes are polyphenolic molecules with hydrophilic cavities and have been shown to complex fullerenes. This host-guest phenomenon is driven by π-π interactions and/or solvophobic effects, while the complementarity of the curvature of the interacting species maximizes the number of intermolecular contacts. Inclusion complexes obtained from

J. Vicens and J. Harrowfield (eds.), Calixarenes in the Nanoworld, 173–196.

various calixarenes have found applications for the direct purification of fullerenes from carbon soot.[5,6] In addition, such host-guest systems have been used for the incorporation of fullerenes into well-ordered systems such as Langmuir films or self-assembled monolayers. In light of their unique electronic properties, fullerene derivatives are also suitable building blocks for the preparation of molecular devices displaying photoinduced energy and electron transfer processes. Indeed, a few examples of photoactive calixarene-fullerene non-covalent assemblies have been described so far. Finally, covalently bound fullerene-calixarene conjugates have been prepared for their self-inclusion capabilities or for their ionophoric properties. All these results will be summarized in the present chapter.

2. INCLUSION OF FULLERENES BY CALIXARENES

2.1. Calix[8]arenes

The groups of Atwood[5] and Shinkai[6] discovered independently that toluene solutions of C_{60} and *p-t*-Bu-calix[8]arene form a sparingly soluble brown–yellow precipitate which was identified as the 1:1 complex. In this association, the fullerene most likely resides within the cavity of the macrocycle presenting a structure which resembles that of a "ball and socket". The formation of this complex was anticipated based on the perfect complementarity between host cavity and guest size. However, the stability of the supramolecular assembly was limited to the solid state. Effectively, dissociation into each component was observed in solution. Verhoeven and co-workers[7] carried out solid state ^{13}C NMR spectroscopic investigations to explore the nature of the intermolecular interactions in the complex obtained from *p-t*-Bucalix[8]arene and C_{60}. The elemental analysis of the material they investigated was consistent with a 1:1 stoichiometry. They observed a significant upfield shift of 1.4 ppm of the fullerene ^{13}C NMR resonance upon complexation, indicative of interactions between the carbon sphere and the aromatic rings of the calixarene receptor. This complexation-induced shift was accompanied by a sharpening of all calixarene signals and, in particular, by an increase in the number of aromatic resonances of the calixarene (as compared to free solid *p-t*-Bucalix[8]arene). These data not only support the formation of a supramolecular complex but also suggest that complexation-induced conformational changes of the calixarene take place. Such a conformational change is also indicated by the changes in the OH stretch observed in the IR spectrum of the bound macrocycle. The intricate pattern of the solid state NMR signals corresponding to the aryl rings of the host is in agreement with the existence of two conformationally different phenolic rings in a 2:6 ratio; thus the calix[8]arene was proposed to possess "pinched" (double) cone conformation.

Interestingly, the complexation with *p-t*-Bucalix[8]arene is highly selective for C_{60} over higher fullerenes (C_{70}, C_{76}, C_{78}...), thus providing the basis for a convenient purification of C_{60} from the soluble fullerene soot extract. By a protocol involving precipitation of the complex from toluene, followed by recrystallisation from $CHCl_3$, in which the calix[8]arene remains dissolved, high-purity C_{60} is readily obtained.[5,6]

2.2. Calix[6]arenes

Atwood and co-workers[8] have shown that calix[6]arene forms solid state complexes with C_{60} and C_{70} from toluene solutions. Crystals suitable for X-ray analysis were obtained for both complexes (Fig. 1). The X-ray crystal structures revealed that both complexes possessed a 1:2 stoichiometry. The complexes turned out to be isostructural, despite the anisotropic shape of C_{70}, and they both crystallize in the same space group $P\ 4_12_12$. The macrocyclic hosts show a double-cone conformation and each of the associated shallow cavities is occupied by a fullerene. It is worth noting that the lower rim H-bonding network between the OH-groups, which is present in the uncomplexed calixarene, is fully retained in the complexes.

Shinkai and co-workers[9] reported the preparation of inclusion complexes between C_{60} and the calix[6]arene derivatives **1** and **2** (Fig. 2) bearing electron rich *N,N*-dialkylaniline units known to form charge transfer complex with C_{60} in organic solvents.[10] The formation of host-guest complexes in toluene solutions between C_{60} and both **1** and **2** was shown by the continuous changes observed in the UV/Vis spectra upon successive additions of the host to fullerene solutions. The association constants for the binding of C_{60} to **1** and **2** in toluene at 298 K determined with the Benesi-Hildebrand equation[11] were found to be $K_a = 8\ M^{-1}$ for **1** and $K_a = 110\ M^{-1}$ for **2**. The higher affinity observed for **2** was explained in terms of strong donation ability of the *m*-phenylendiamine fragment and the higher preorganization resulting from the bridging of the three pairs of phenyl units of the calix[6]arene.

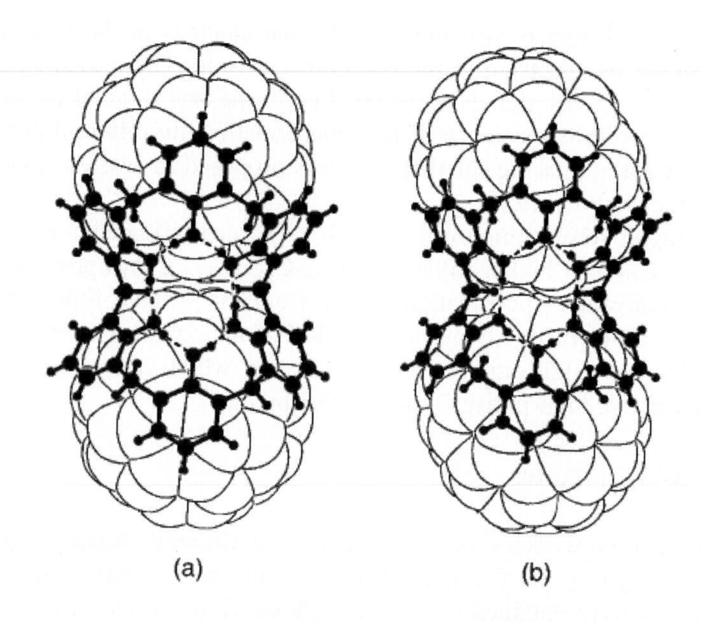

Figure 9-1. X-ray structures of the complexes of [(calix[6]arene)$(C_{60})_2$] (a) and [(calix[6]arene)$(C_{70})_2$] (b) as viewed from below the calix[6]arene double cone. Atoms and bonds of the calixarene are shown in shaded ball-and-stick representation, whereas fullerene molecules are shown as van der Waals surfaces. Hydrogen bonds are indicated by dashed lines (from ref. 8).

2.3. Calix[5]arenes

Comprehensive investigations of the complexation of fullerenes with calix[5]arene receptors **3-5** (Fig. 3) were carried out by Fukuzawa and co-workers.[12] Upon addition of hosts **3-5** to C_{60} in several organic solvents (toluene, CS_2, *o*-dichlorobenzene), changes were observed in the electronic spectra. The stoichiometry of the complexes formed in solution was determined to be 1:1 by Job plot analysis; this stoichiometry was also confirmed by the observation of a clear isosbestic point at 478 nm. The

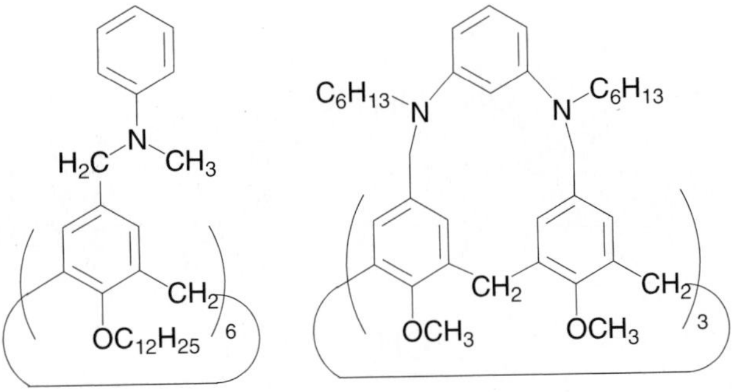

Figure 9-2. Calix[6]arene derivatives **1** and **2**.

binding constants of the complexes were determined by spectrophotometric titration experiments and found to be strongly solvent dependent. With receptor **3**, the measured K_a values increased from 308 M^{-1} in *o*-dichlorobenzene to 660 M^{-1} in CS_2, to 1840 M^{-1} in benzene and to 2120 M^{-1} in toluene. Thus, association strength increases with decreasing solubility of C_{60}: the more weakly solvated guest is more strongly bound by the host. The K_a values were also found to be dependent on the substituents of the host molecule. Receptor **3** with the highly polarizable iodide substituents formed the most stable complexes. Thus, association strength decreased in toluene from K_a = 2120 M^{-1} (**3**) to 1670 M^{-1} (**4**) and to 590 M^{-1} (**5**). This result underlines the importance of van der Waals interactions in the stabilization of the fullerene complexes.

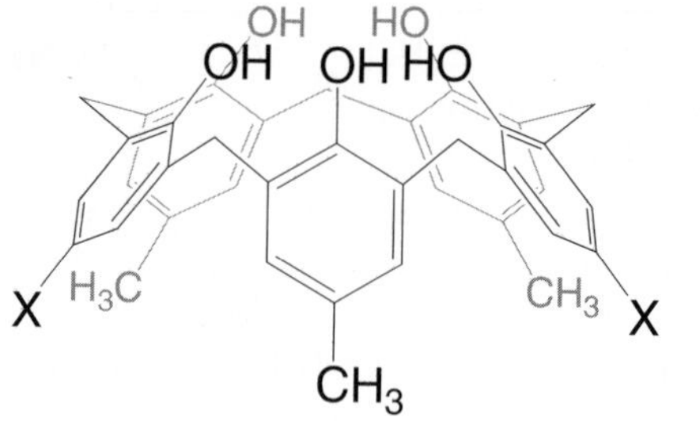

Figure 9-3. Calix[5]arene receptors **3-5**.

Fukuzawa and co-workers were also successful in growing crystals of the complex between C_{60} and receptor **3** which were suitable for X-ray structural analysis.[13] Interestingly, unlike in solution, the stoichiometry of the complex in the solid state was found to be 2:1. The fullerene is effectively encapsulated within a cavity conformed by two calixarene molecules. The apparent contradictory host-guest ratio in solution and in the solid state was also observed in several other examples of fullerene-based supramolecular assemblies[14-16] and it is quite reasonable to have a different (lower) complexation ratio in dilute solution to that in the solid state.

Atwood and co-workers[17] have reported binding studies of C_{60} with *p*-benzylcalix[5]arene showing the formation of 2:1 complexes both in solution and in the solid state. Indeed, the electronic spectra of solutions of C_{60} with the macrocyclic host in toluene show an isosbestic point only for high concentrations of *p*-benzylcalix[5]arene. These findings, along with the binding constants $K_1 = 2800\ M^{-1}$ determined at low concentrations of host and $K_2 = 230\ M^{-1}$ determined at higher concentrations suggest the initial formation of the 1:1 complex, followed by the formation of a 2:1 supermolecule on further addition of *p*-benzylcalix[5]arene. Job plots were also not consistent with a 1:1 stoichiometry in solution, where, at the concentrations used the second equilibrium would have an impact. The formation of a 2:1 complex is this time consistent with the complex obtained in the solid state. When compared to the calix[5]arene derivatives studied by Fukazawa for which a second equilibrium was not detected in solution, it appears that the presence of the benzyl units may play an important role since they can provide additional intramolecular π-π interactions between the two host macrocyles within the 2:1 complex.

2.4. Calix[4]arenes and calix[4]naphthalenes

Whereas inclusion complexes of calix[5,6]arene derivatives with C_{60} and C_{70} can be obtained, the calix[4]arenes are not suited to the formation of analogous inclusion complexes with fullerenes because their cavities are relatively small. Moreover, the well-known C_{4v} cone conformation of these molecules is usually dependent on the presence of hydroxyl groups at the lower rim and substitution at this position gives rise to the so-called pinched-cone C_{2v} conformation with an even smaller cavity. Despite the loss of a cavity allowing the formation of true inclusion compounds, co-crystallization of simple calix[4]arene derivatives with C_{60} was successful.[18] For example, co-crystallization of C_{60} with *p*-iodocalix[4]arene benzyl ether resulted in a remarkably well packed structure in which the C_{60} molecules are intercalated into calixarene bilayers thus preventing appreciably strong fullerene-fullerene interactions.[19]

A class of calix[4]arene-related molecules which possess deeper cavities, are the calix[4]naphthalenes (Fig. 4).[20] Georghiou and co-workers[21] have reported the complexation of the endo-type calix[4]naphthalene **6** and its tetra-*t*-butylated derivative **7** with C_{60} in various organic solvents. Spectrophotometric methods revealed that in dilute solution 1:1 complexes are formed between **6** or **7** and fullerenes. The association constant values were found to be slightly higher than those observed with calix[5]arenes.

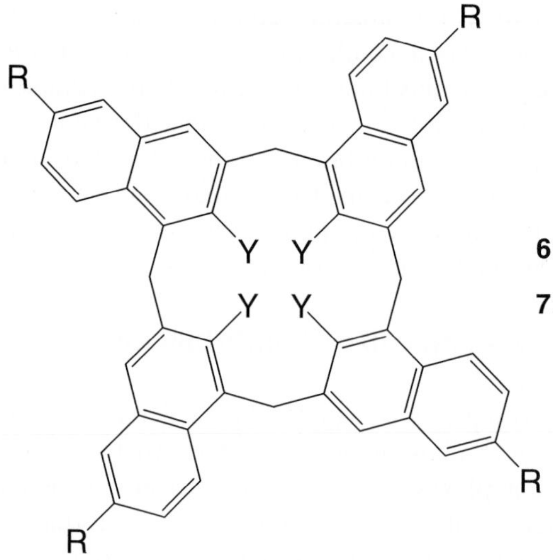

R
R
R
R
Y Y
Y Y
6: R = H, Y = OH
7: R = *t*Bu, Y = OH

Figure 9-4. Calix[4]naphthalenes **6** and **7**.

2.5. Homooxacalix[3]arenes

Shinkai and co-workers[22] have described the formation of an inclusion complex between C_{60} and *t*-butylhexahomooxacalix[3]arene (**8**, Fig. 5) in organic solutions. Supramolecular complexes of C_{60} with various hexahomooxacalix[3]arenes possessing different substituents on their upper rims were also reported by Fuji and co-workers.[23] The interaction between hosts **8-11** (Fig. 5) and C_{60} was examined spectrophotometrically. Upon addition of these hexahomooxacalix[3]arenes to a solution of C_{60} in toluene a slight change in color (from purple to brown) was observed. The stoichiometry of the complexes formed in solution was determined to be 1:1 by Job plot analysis but the K_a values obtained were quite low (9 to 35 M^{-1}). Fuji and co-workers were successful in growing crystals of the 1:1 complex between C_{60} and receptor **11** which were suitable for X-ray structural analysis.[23] The inclusion complex has a C_{3v} symmetric structure in the solid state, in which a six-membered ring of C_{60} is disposed parallel to the mean plane composed of the three phenolic oxygens of **11**. In addition, three six-membered rings around the above-mentioned six-membered ring at the bottom position of C_{60} are approximately parallel to the three phenyl rings of **11**, where the closest distance was found to be 3.615(6) Å.

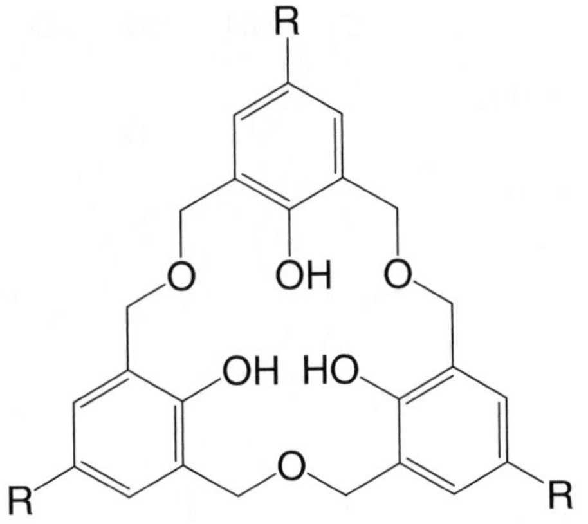

8: R = *t*Bu
9: R = H
10: R = OMe
11: R = Br

Figure 9-5. Hexahomooxacalix[3]arenes **8-11**.

More recently, Shinkai and co-workers[24] reported a water-soluble homooxacalix[3]arene derivative (**12**, Fig. 6), capable of solubilizing C_{60} in aqueous media, which may have important biological applications.[25] Solid C_{60} was extracted into water containing **12** by sonication followed by stirring and centrifugation. Electronic and NMR spectroscopic studies clearly supported the view that C_{60} is solubilized into water by inclusion in **12**. The concentration of solubilized C_{60} was determined by elemental analysis of a dried sample obtained by evaporation of the aqueous solution. The analytical result showed that the ratio between **12** and C_{60} was 2:1 in the original solution. Preliminary results have also revealed that the water-soluble supramolecular host-guest complex obtained from **12** and C_{60} is an efficient DNA photocleavage reagent. The complex was applied to the photocleavage of ColE1 supercoiled plasmid. Under dark conditions, DNA was not cleaved even in the presence of the reagent. Under visible light irradiation, the complex clearly showed DNA-cleaving activity as a result of either the photoinduced electron transfer from guanine units to C_{60} or from the reaction of singlet oxygen photochemically generated by the C_{60}. This study is one of the only successful examples in which pristine C_{60} has been brought into direct contact with DNA and implies that the concept of host-guest chemistry of C_{60} could be fruitfully applied to medicinal chemistry.

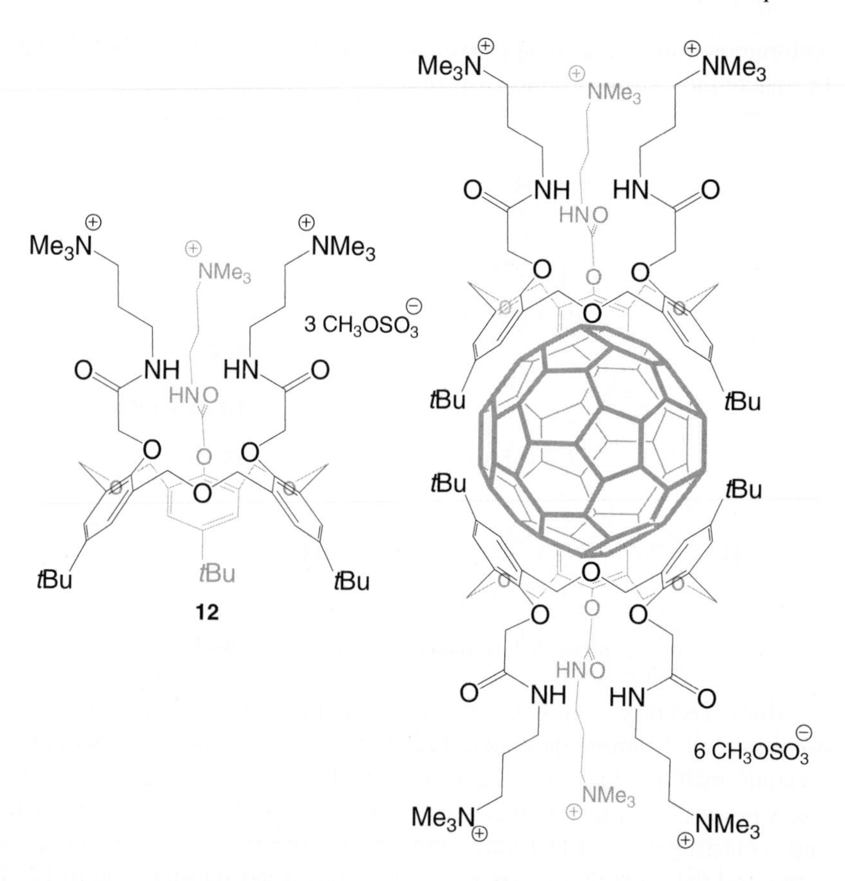

Figure 9-6. Water-soluble homooxacalix[3]arene derivative **12** and schematic representation of the 2:1 complex obtained from **12** and C_{60}.

3. BIS-CALIXARENE RECEPTORS

The discovery of a complex with 2:1 stoichiometry in the solid state led Fukazawa and co-workers[26] to covalently link two calix[5]arene macrocycles **13-15** in order to produce shape-selective receptors with well-defined cavity sizes (Fig. 7). With these bridged calix[5]arene receptors, they achieved a dramatic increase in the association constants for 1:1 complexes with C_{60} in solution. Thus, the K_a value for the complex formed with receptor **13** in toluene is 76000 M^{-1}. The solvent dependency of the association constant followed the same trend as had been observed with **3–5** (Fig. 3). The bridged receptors **13-15** binds C_{70} even better than C_{60}. For example, the binding constant of the 1:1 complex with C_{70} in toluene is K_a = 163000 M^{-1}.

Following these first examples, a few other dimeric calix[*n*]arene receptors have been reported.[27]

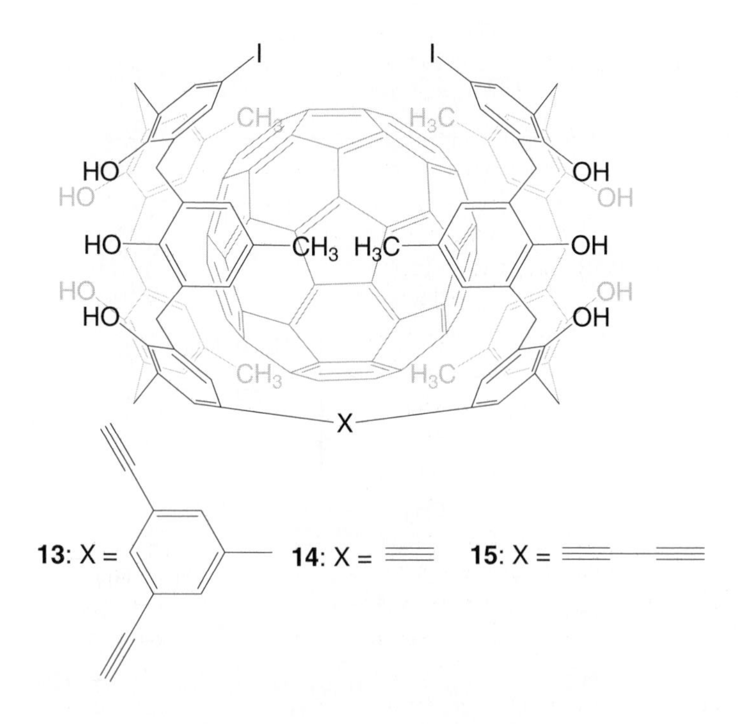

Figure 9-7. Schematic representation of the supramolecular complexes obtained from C_{60} and the bis-calix[5]arene receptors **13-15**.

An alternative way to build well defined cavities with calix[5]arenes to elaborate fullerene receptors was reported by Fukazawa and co-workers.[28] Formation of the *bis*-complex of the bipyridine-appended calix[5]arene **16** with Ag(I) brings the two binding sites to close proximity, thus yielding a well defined cavity allowing the incorporation of fullerenes (Fig. 8). Due to the presence of the Ag cation, it was possible to detect easily the supramolecular complex by mass spectrometry. The formation of the complex was also confirmed by ^{13}C-NMR spectroscopy, in which one upfield shift of 1.8 ppm was observed for the C_{60} fullerene resonance, as a result of its inclusion into the receptor. A similar system using copper(I) coordination for the assembly of the bridged calix[5]arene receptor was also reported by the same authors.[29]

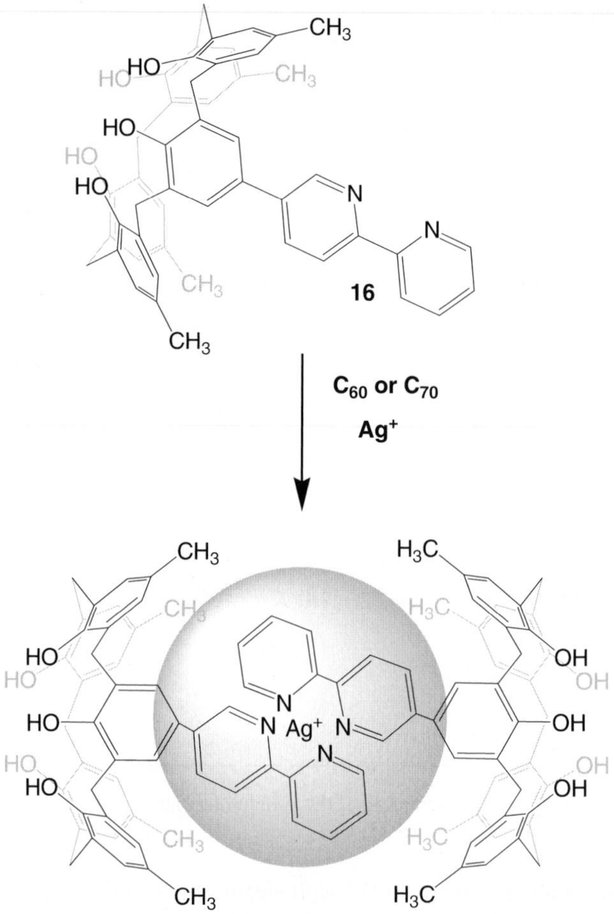

Figure 9-8. Metal-assisted self-assembly of two calix[5]arene derivatives **16** around a fullerene molecule.

The self assembly of a molecular capsule made of two urea-calix[5]arene units (**17**) associated by hydrogen bonds has also been reported (Fig. 9).[30] The binding constants of the hydrogen-bonded dimer were determined by ^{1}H-NMR titration experiments and found to be increased significantly in the presence of C_{60} (K_a = 32 and 110 M^{-1} in the absence and in the presence of C_{60} respectively). The latter observation clearly indicated the formation of the ternary complex. Interestingly, release of the entrapped fullerene can be easily achieved by the addition of trifluoroacetic acid. Due to the protonation of the urea group, the self-association of the host is inhibited, thus promoting the liberation of the encapsulated guest (see chapter 2).

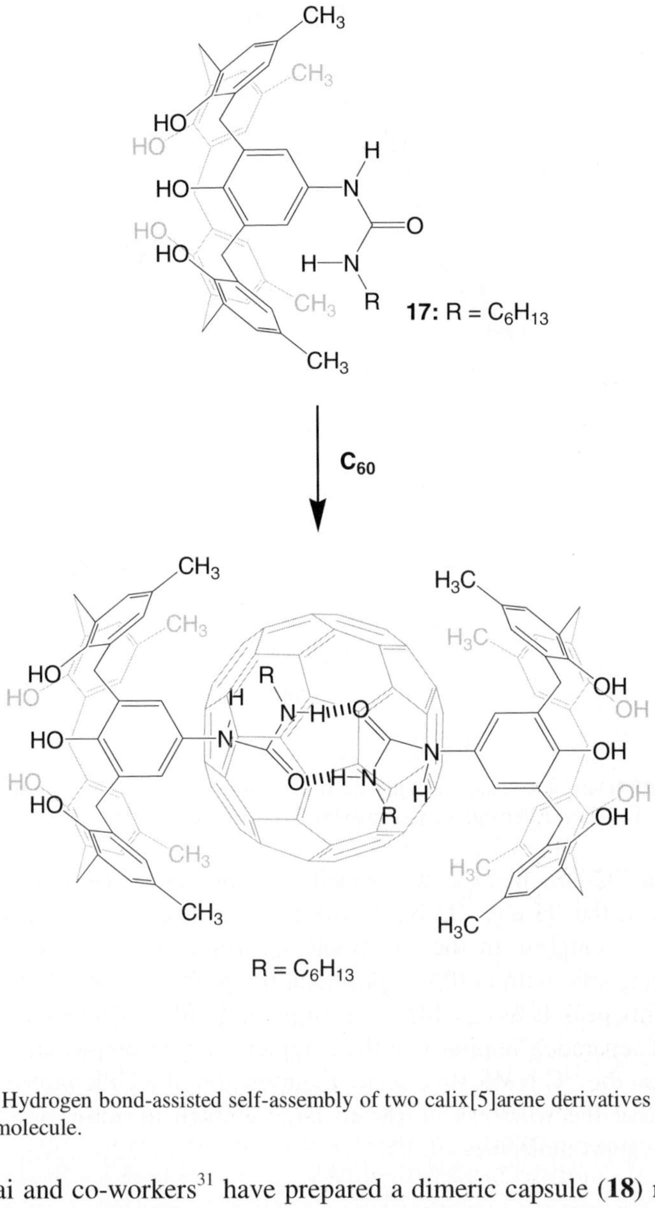

Figure 9-9. Hydrogen bond-assisted self-assembly of two calix[5]arene derivatives **17** around a fullerene molecule.

Shinkai and co-workers[31] have prepared a dimeric capsule (**18**) resulting from the self-assembly of two pyridine-substituted homooxacalix[3]arene derivatives and three Pd(II) centres (Fig. 10). NMR binding studies provided clear evidence for the inclusion of C_{60} in the molecular capsule.

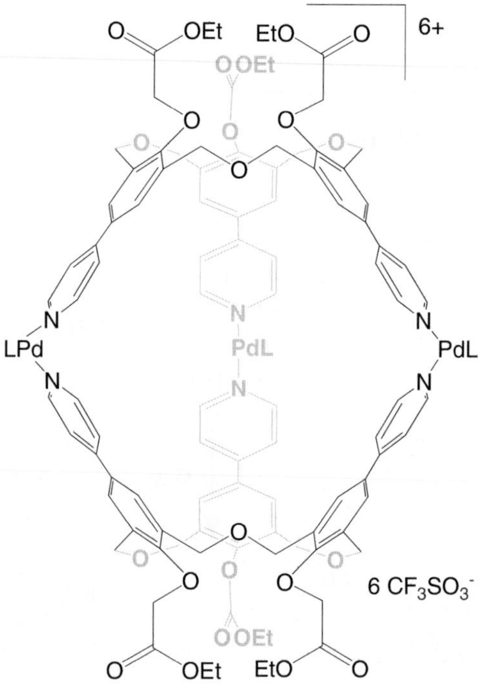

18

Figure 9-10. Dimeric capsule resulting from the self-assembly of two pyridine-substituted homooxacalix[3]arene derivatives and three Pd(II) centres (L = $Ph_2PCH_2CH_2CH_2PPh_2$).

When ^{13}C-enriched C_{60} was added to a solution of **18**, new peaks were detected in the ^1H and ^{13}C NMR spectra, confirming the formation of the host-guest complex. In the ^{13}C NMR spectrum, a new peak appeared at higher magnetic field (140.97 ppm) than the peak for free fullerene (142.87 ppm). This peak is assignable to the supramolecular inclusion complex and the peak separation implies that the complexation-decomplexation exchange is slow on the ^{13}C NMR time scale. Examination of a CPK molecular model reveals that the windows in **18** are large enough to allow passage of the fullerene guest molecule.

4. THIN FILMS

4.1. Langmuir Films

Since the incorporation of fullerenes into thin films is required for the preparation of many optoelectronic devices, the past several years have seen a considerable growth in the use of fullerene-based derivatives at surfaces and interfaces.[32] One possible approach towards structurally ordered fullerene assemblies is the preparation of Langmuir films at the air-water interface and their subsequent transfer onto solid substrates.[33] However, all studies on the spreading behavior of pure fullerenes at the air-water interface revealed the formation of collapsed films due to the non-amphiphilic nature of these compounds and to aggregation phenomena resulting from strong fullerene-fullerene interactions.[34] Furthermore, all attempts to create well defined Langmuir-Blodgett (LB) films have failed. Different approaches have been used to overcome these problems.[33] One of them consists in preventing the fullerene-fullerene interactions by incorporating the fullerenes into a matrix of an amphiphilic compound to produce mixed Langmuir films. Fatty acids or long chain alcohols have been used for this purpose,[35] but the expected protection is not always very effective and fullerene aggregation remains a problem. Amphiphilic molecules containing a cavity able to incorporate the fullerene such as azacrowns[36] or calixarenes[37-39] have been found to be the most suitable matrices for the preparation of fullerene-containing composite Langmuir films of good quality. However, monolayer formation by these calixarene-fullerene systems depends on the initial concentration, is irreversible and therefore cannot be applied to the formation of Langmuir-Blodgett films on solid substrates.

4.2. Self-Assembled Monolayers

Shinkai and co-workers[40] have used their water-soluble capsule-like 2:1 complex obtained from homooxacalix[3]arene derivative **12** (Fig. 6) and C_{60} to form a monolayer on an anion-coated gold surface as schematically depicted in Fig. 11. The surface covered with the hexacationic complex was observed by atomic force microscopy and scanning electron microscopy. In both cases, the surface was smooth without any cluster-like domain. The photoresponsive behavior of a gold electrode covered with the hexacationic complex **12**.C_{60} was also investigated. Photocurrent measurements were carried out with the modified gold electrode as the working electrode, a Pt counter electrode and ascorbic acid as a sacrificial e-donor. Under illumination with visible light, the appearance of a large photocurrent wave was observed. This photoresponse phenomenon could be repeated many times reversibly. In contrast, the working electrode upon which only **12** was deposited showed no photoelectrochemical response at all. Hence, the mechanism of the present photovoltaic response appears to be that upon photoexcitation, ascorbic acid transfers an electron to the excited fullerene and the resulting fullerene anion then injects an electron into the Au electrode.

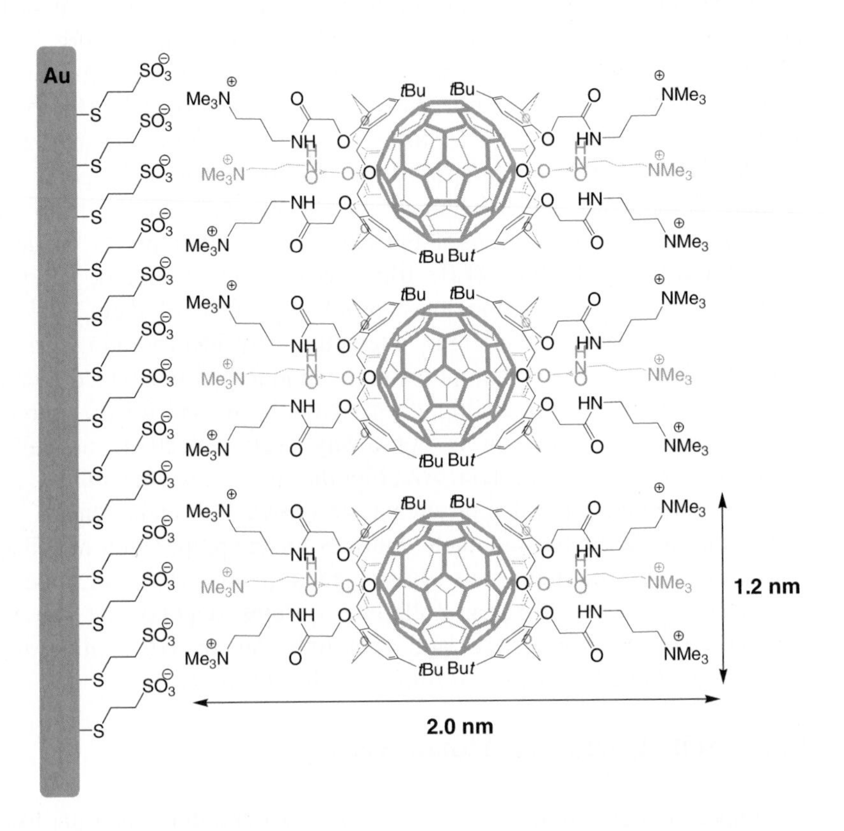

Figure 9-11. Schematic representation of the self-assembled monolayer of [$(\mathbf{12})_2.(C_{60})$] onto a mercaptoethansulfonate-covered gold surface.

A few years later, the same authors showed that the photocurrent generation could be significantly improved by preparing mixed self-assembled multilayers comprised of the hexacationic complex **12**.C_{60} and an anionic porphyrin polymer.[41] The initial photoinduced electron transfer from the excited porphyrinic moieties to the fullerene is very efficient and is followed by charge transfer from the reduced fullerene to the electrode, thus leading to the photocurrent with a quantum yield of 21%.

5. PHOTOACTIVE SUPRAMOLECULAR FULLERENE-CALIXARENE ENSEMBLES

In light of their unique electronic properties, fullerene derivatives are suitable building blocks for the preparation of molecular devices displaying photoinduced energy and electron transfer processes.[42] Whereas research focused on the use of C_{60} as the acceptor in covalently bound donor-acceptor pairs has received considerable attention,[42] only a few related examples of fullerene-containing non-covalent assemblies have been described so far.[43] As part of this research, Fukazawa and co-workers have developed a non-covalent approach for the assembly of the fullerene acceptor with a Re(I)-bipyridine donor by using the inclusion capabilities of a calix[5]arene receptor (Fig. 12).[44]

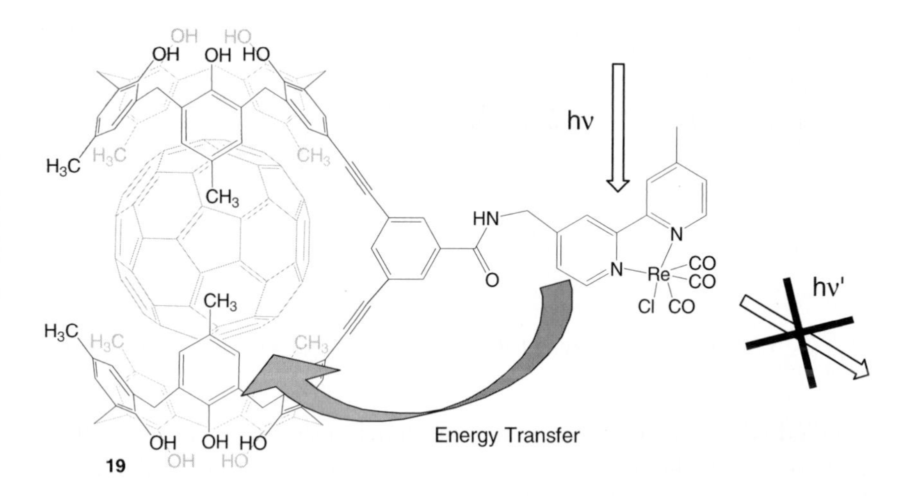

Figure 9-12. Photoactive supramolecular device resulting from the self-assembly of C_{60} with a functionalized calix[5]arene derivative.

The host derivative **19** bearing a Re(I) complex group shows a strong orange luminescence upon photoexcitation at 365 nm. Addition of C_{60} or C_{70} to a solution of **19** caused a dramatic change. The luminescence was immediately extinguished upon the addition of C_{60} or C_{70}. This quenching has been attributed to an intramolecular photoinduced energy transfer from the excited state of the Re(I) complex to the bound fullerene within the supramolecular ensembles.

6. COVALENT FULLERENE-CALIXARENE CONJUGATES

The synthesis of calix[*n*]arene derivatives bearing a fulleropyrrolidine group on the upper-rim was reported by Wang and Gutsche.[45] However, evidence for self-complexation was ambiguous due to the conformational mobility of the macrocyclic rings making NMR studies difficult. The synthesis of a related calix[4]arene-fullerene conjugate (**20**) was reported by Nierengarten and co-workers (Fig. 13).[46] Thanks to the conformationally immobile tetra-*O*-alkylated cone-calix[4]arene skeleton used for the functionalization of the fullerene sphere, full analysis of the system was possible.

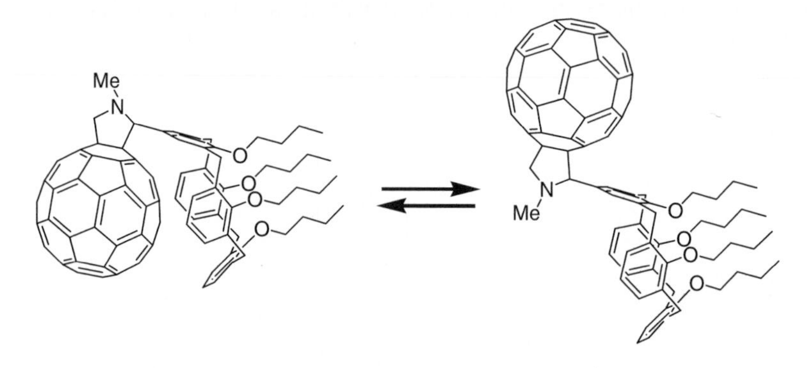

Figure 9-13. Schematic representation of the conformational equilibrium observed for the calix[4]arene-fullerene conjugate **20**.

Detailed NMR studies revealed that the fulleropyrrolidine group in **20** is rotating freely at high temperature but that only the self-complexed conformer is observed at low temperature. Compound **20** can be seen as a covalent assembly of two components able to perform mechanical movements of relatively large amplitudes (rotation of the fulleropyrrolidine group) as a consequence of an external stimulus (temperature). Therefore, calix[4]arene-fullerene conjugate **21** presents characteristic features that makes it an interesting building block for the preparation of new molecular machines.[47] Computational studies were also performed to evaluate the relationship between the potential energy and the relative position of the calix[4]arene macrocycle and the fulleropyrrolidine moiety in **20** (Fig. 14).

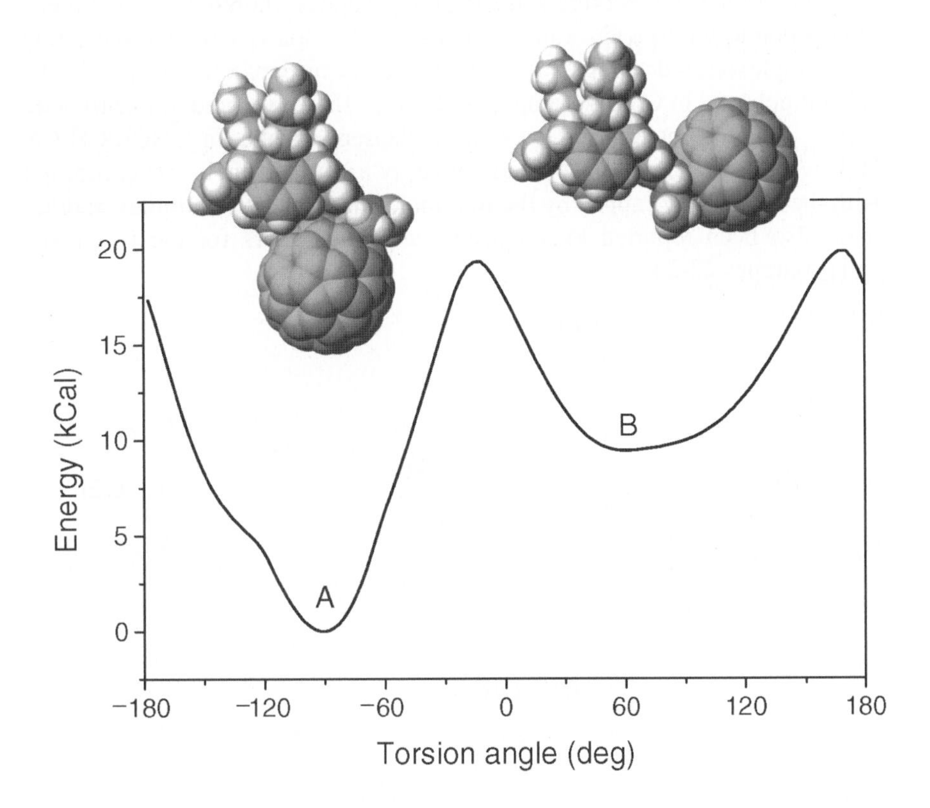

Figure 9-14. Calculated potential energy diagram of compound **20** for rotation about the bond between the pyrrolidine ring and the benzene group attached to it and theoretical structures of the two conformers corresponding to the two minima.

The molecule has been optimized with fixed values of the torsion angle for rotation about the bond between the pyrrolidine ring and the benzene group attached to it. This angle has been increased stepwise from 0 to 360°, leading to the potential energy diagram shown in Fig. 14. Two minima of different energy have been found, the lowest-energy conformation being the one with the fulleropyrrolidine moiety located atop the cavity of the calix[4]arene macrocycle, *i.e.* the self-complexed conformer (A). The second stable conformer (B) in which the two moieties are far apart is located *ca.* 10 kcal/mol higher in energy. The calculations are therefore in good agreement with the ^{1}H-NMR studies, suggesting that compound **20** adopts a self-complexed conformation at low temperature.

Shinkai and co-workers[48] have shown that a homooxacalix[3]arene moiety connected to a C_{60} unit through a flexible spacer exhibits interesting self-complexation-decomplexation properties in response to changes in the solvent polarity. In $CDCl_3$, compound **21** (Fig. 15) exists predominantly with a free fullerene moiety and an open calixarene cavity. In $CHCl_3/CH_3CN$ (1:1, v/v), however, **21** exists predominantly as the self-inclusion conformer with the C_{60} moiety capped by the intramolecular calixarene Similar findings have also been reported by Fukazawa and co-workers for the C_{60}-linked calix[5]arenes **22-24**.[49]

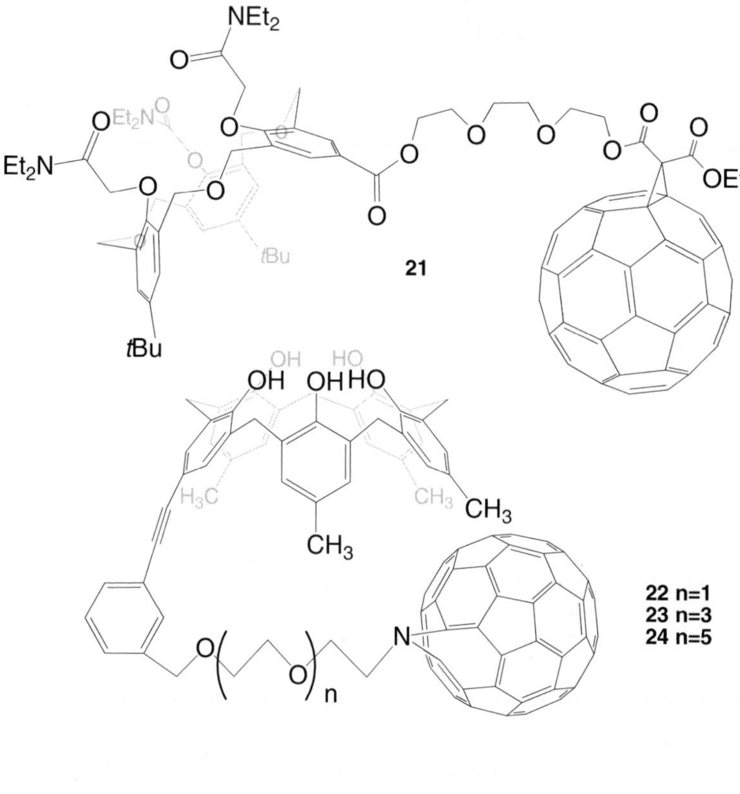

Figure 9-15. Fullerene derivatives **21-24**.

The metal-binding properties of fullerene-linked calix[4]arenes **25** and **26** (Fig. 16) have been studied in detail by Shinkai and co-workers.[50,51] The absorption spectrum of compound **25** in which C_{60} is covalently linked through two ionophoric chains to a 1,3-alternate-calix[4]arene is almost unchanged by addition of metal cations. It is known that 1,3-alternate-calix[4]arenes have two metal binding sites composed of two phenolic oxygens and two benzene π-systems where the cation-π interactions significantly participate in the metal-binding event.

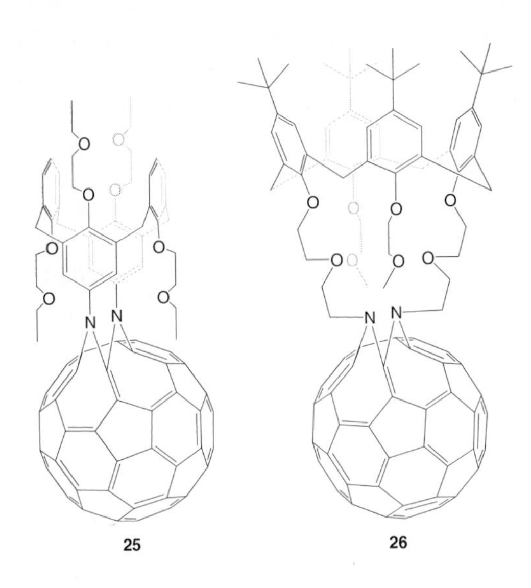

Figure 9-16. Fullerene-substituted calix[4]arenes **25** and **26**.

The distribution ratio between the two metal-binding sites is dependent upon the relative strength of their ionophoricity. In the case of **25**, the binding occurs preferentially on the opposite site relative to the fullerene moiety, thus the bound cation remains relatively far from the C_{60} unit and cannot influence its absorption properties. In contrast, the absorption of the related system functionalized with a cone-calix[4]arene (**26**) is affected by addition of Li^+, Na^+ and Ag^+, indicating the formation of exohedral fullerene-metal complexes. Although the spectral changes in the presence of Ag^+ are much greater than those in the presence of Li^+ and Na^+, the association constant for Ag^+ (K_a = 3000 M^{-1}) is only slightly larger than those for Li^+ (K_a = 2200 M^{-1}) or Na^+ (K_a = 2000 M^{-1}). This supports the idea that Ag^+ is bound mainly by the nitrogen atoms and favorably interacts with the π-basic surface fullerene surface. This view was further confirmed by NMR binding studies.

7. CONCLUSION

Since the pioneering work of Shinkai and Atwood, a diversity of inclusion complexes between C_{60} or C_{70} and calix[*n*]arene derivatives have been prepared in the solid state or in solution. The major driving force in all these molecular recognition events is undoubtedly the van der Waals interactions between the large complementary surfaces of the binding partners. In recent years, research in this field has moved towards the creation of functional systems with increased attention to potential applications. Despite some remarkable recent achievements, it is clear that the examples discussed in this chapter represent only the first steps towards the design of fullerene/calixarene molecular assemblies which can display functionality at the macroscopic level. More research in this area is needed to fully explore the possibilities offered by these materials, for example, in medicinal chemistry or in photovoltaics.

REFERENCES

1. W. Krätschmer, L.D. Lamb, K. Fostiropoulos and D.R. Huffman, *Nature* **347**, 354-357 (1990).
2. A. Hirsch, *The Chemistry of Fullerenes*, Thieme, New York (1994).
3. F. Diederich and M. Gomez-Lopez, *Chem. Soc. Rev.* **28**, 263-277 (1999).
4. D.M. Guldi and N. Martin, *J. Mater. Chem.* **12**, 1978-1992 (2002).
5. J.L. Atwood, G.A. Koutsantonis and C.L. Raston, *Nature* **368**, 229-231 (1994).
6. T. Suzuki, K. Nakashima and S. Shinkai, *Chem. Lett.* **4**, 699-702 (1994).
7. R.M. Williams, J.M. Zwier and J.W. Verhoeven, *J. Am. Chem. Soc.* **116**, 6965-6966 (1994).
8. J.L. Atwood, L.J.Barbour, C.L. Raston and I.B.N. Sudria, *Angew. Chem. Int. Ed.* **37**, 981-983 (1998).
9. K. Araki, K. Akao, A. Ikeda, T. Suzuki and S. Shinkai, *Tetrahedron Lett.* **37**, 73-76 (1996).
10. Y.-P. Sun, C.E. Bunker, B. Ma, *J. Am. Chem. Soc.* **116**, 9692-9699 (1994).
11. H. Benesi and J.H. Hildebrand, *J. Am. Chem. Soc.* **71**, 2703-2707 (1949).
12. T. Haino, M. Yanase and Y. Fukazawa, *Angew. Chem. Int. Ed.* **36**, 259-260 (1997).
13. T. Haino, M. Yanase and Y. Fukazawa, *Tetrahedron Lett.* **38**, 3739-3742 (1997).
14. S. Mizyed, M. Ashram, D. O. Miller and P. E. Georghiou, *J. Chem. Soc., Perkin Trans.* **2,** 1916-1919 (2001).
15. D. Felder, B. Heinrich, D. Guillon, J.-F. Nicoud and J.-F. Nierengarten, *Chem. Eur. J.* **6**, 3501-3507 (2000).
16. Y. Rio and J.-F. Nierengarten, *Tetrahedron Lett.* **43**, 4321-4324 (2002).
17. J.L. Atwood, L.J. Barbour, P.J. Nichols, C.L. Raston and C.A. Sandoval, *Chem. Eur. J.* **5**, 990-996 (1999).
18. L.J. Barbour, G. William Orr and J.L. Atwood, *Chem. Commun.* 1901-1902 (1998); E. Hugues, J.L. Jordan and T. Gullion, *J. Phys. Chem. B* **104**, 691-694 (2000); A. Saha, S. K. Nayak, S. Chottopadhyay and A.K. Mukherjee, *J. Phys. Chem. B* **108**, 7688-7693 (2004).

19. L.J. Barbour, G. William Orr and J.L. Atwood, *Chem. Commun.* 1439-1440 (1997).
20. P.E. Georghiou, Z. Li, M. Ashram, S. Chowdhury, S. Mizyed, A.H. Tran, H. Al-Saraierh and D.O. Miller, *Synlett.* 879-891 (2005).
21. P.E. Georghiou, S. Mizyed and S. Chowdhury, *Tetrahedron Lett.* **40**, 611-614 (1999); S. Mizyed, P.E. Georghiou and M. Ashram, *J. Chem. Soc., Perkin Trans.* **2**, 277-280 (2000).
22. A. Ikeda, M. Yoshimura and S. Shinkai, *Tetrahedron Lett.* **38**, 2107-2110 (1997).
23. K. Tsubaki, K. Tanaka, T. Kinoshita and K. Fuji, *Chem. Commun.* 895-896 (1998).
24. A. Ikeda, T. Hatano, M. Kawawuchi, H. Suenaga and S. Shinkai, *Chem. Commun.*, 1403-1404 (1999).
25. T.D. Ros and M. Prato, *Chem Commun*, 663-669 (1999).
26. T. Haino, M. Yanase and Y. Fukazawa, *Angew. Chem. Int. Ed.* **37**, 997-998 (1998).
27. J. Wang and C.D. Gutsche, *J. Am. Chem. Soc.* **120**, 12226-12231 (1998); J. Wang, S.G. Bodige, W.H. Watson and C.D. Gutsche, *J. Org. Chem.* **65**, 8260-8263 (2000); Y. Van, O. Mitkin, L. Barnhurst, A. Kurchan and A. Katateladze, *Org. Lett.* **2**, 3817-3819 (2000).
28. T. Haino, H. Araki, Y. Yamanaka and Y. Fukazawa, *Tetrahedron Lett.* **42**, 3203-3206 (2001).
29. T. Haino, Y. Yamanaka, H. Araki and Y. Fukazawa, *Chem. Commun.* 402-403 (2002).
30. M. Yanase, T. Haino and Y. Fukazawa, *Tetrahedron Lett.* **40**, 2781-2784 (1999).
31. A. Ikeda, M. Yoshimura, H. Udzu, C. Fukuhara and S. Shinkai, *J. Am. Chem. Soc.* **121**, 4296-4297 (1999); A. Ikeda, H. Udzu, M. Yoshimura and S. Shinkai, *Tetrahedron* **56**, 1825-1832 (2000).
32. D. Felder, M. Gutiérrez Nava, M. del Pilar Carreon, J.-F. Eckert, M. Luccisano, C. Schall, P. Masson, J.-L. Gallani, B. Heinrich, D. Guillon and J.-F. Nierengarten, *Helv. Chim. Acta* **85**, 288-319 (2002).
33. J.-F. Nierengarten, *New J. Chem.* **28**, 1177-1191 (2004).
34. C. A. Mirkin, and W. B. Caldwell, *Tetrahedron* **52**, 5113-5130 (1996).
35. C. Ewins, and B. Steward, *J. Chem. Soc., Faraday Trans.* **90**, 969-972 (1994).
36. J. Effing, U. Jonas, L. Jullien, T. Plesnivy, H. Ringsdorf, F. Diederich, C. Thilgen, and D. Weinstein, *Angew. Chem. Int. Ed.* **31**, 1599-1602 (1992).
37. R. Castillo, S. Ramos, R. Cruz, M. Martinez, F. Lara and J. Ruiz-Garcia, *J. Phys Chem.* **100**, 709-713 (1996).
38. L. Dei, P. L. Nostro, G. Capuzzi, and P. Baglioni, *Langmuir* **14**, 4143-4147 (1998).
39. Z.I. Zakansteva, N.V. Lavrik, A.V. Nabok, O.P. Dimitriev, B.A. Nesterenko, V.I. Kalchenko, S.V. Vysotsky, L.N. Markovskiy and A.A. Marchenko, *Supramolecular Science* **4**, 341-347 (1997).
40. T. Hatano, A. Ikeda, T. Akiyama, S. Yamada, M. Sano, Y. Kanekiyo and S. Shinkai, *J. Chem. Soc. Perkin Trans.* **2**, 909-912 (2000).
41. A. Ikeda, T. Hatano, S. Shinkai, T. Akiyama and S. Yamada, *J. Am. Chem Soc.* **123**, 4855-4856 (2001).
42. N. Martin, L. Sanchez, B. Illescas and I. Perez, *Chem. Rev.* **98**, 2527-2547 (1998).
43. J.-F. Nierengarten, U. Hahn, T. M. Figueira Duarte, F. Cardinali, N. Solladié, M. E. Walther, A. Van Dorsselaer, H. Herschbach, E. Leize, A.-M. Albrecht-Gary, A. Trabolsi and M. Elhabiri, *Comptes Rendus Chimie*, in press.
44. T. Haino, H. Araki, Y. Fujiwara, Y. Tanimoto and Y. Fukazawa, *Chem. Commum.*, 2148-2149 (2002).
45. J. Wang and D. Gutsche, *J. Org. Chem.* **65**, 6273-6275 (2000).
46. T. Gu, C. Bourgogne, J.-F. Nierengarten, *Tetrahedron Lett.* **42**, 7249-7252 (2001).
47. V. Balzani, A. Credi, F. M. Raymo and J. F. Stoddart, *Angew. Chem. Int. Ed. Engl.* **39**, 3348-3391 (2000).

48. A. Ikeda, S. Nobukuni, H. Udzu, Z. Zhong and S. Shinkai, *Eur. J. Org. Chem.* 3287-3293 (2000).
49. T. Haino, M. Yanase and Y. Fukazawa, *Tetrahedron Lett.* **46**, 1411-1414 (2005).
50. M. Kawaguchi, A. Ikeda and S. Shinkai, *J. Chem. Soc. Perkin Trans 1* 179-184 (1998).
51. For related systems, see: C. Luo, D.M. Guldi, A. Soi and A. Hirsch, *J. Phys. Chem. A* **109**, 2755-2759 (2005); C.-F. Chen, J.-S. Li, Q.-Y. Zheng, G.-J. Ji and Z.H. Huang, *J. Chem. Res.* 808-809 (1998).

Chapter 10

CALIXARENES AS CLUSTER KEEPERS

The isolation of multinuclear metal complexes

Jack Harrowfield[a] and George Koutsantonis[b]

[a]Institut de Science et d'Ingénierie Supramoléculaires, 8 allée Gaspard Monge, 67083 Strasbourg, France; [b]Chemistry M313, University of Western Australia, 35 Stirling Highway, Crawley WA6009, Australia

Abstract: Phenolic calixarenes in particular exhibit a tendency to form multinuclear metal ion complexes, many of which can be regarded as hydrolytic or anion-bridged clusters enveloped within a lipophilic calixarene sheath, justifying the designation of the calixarenes as "cluster keepers". Of the factors which determine the nuclearity of the clusters, the calixarene ring size and the capacity of phenoxide donors to act as bridges are obviously important but another which offers prospects for considerable development of known cluster chemistry is that of the substitution of S for CH_2 to give the family of thiacalixarenes. Clusters involving up to 10 metal ions are now well-characterised, many of unique geometry and properties.

Key words: Metal clusters, nanoparticles, cluster keepers.

1. INTRODUCTION

An attractive method for the formation of nanoparticles derived from metal compounds is the use of a particular ligand to excise fragments (clusters) from the lattice of a simple species.[1] Indeed, many examples, such as metal sulfide clusters stabilised by phosphines[2] or gold clusters stabilised by thiols,[3] are known of systems which may be considered as seemingly of such origin. In nearly all such cases, however, the "excision" is a formal process, in that while the cluster may be recognisable as a portion of an extended lattice, it is not, in fact, formed by direct fragmentation of that lattice.[4,5] The simple clusters found in many iron/sulfur proteins provide well-known biological examples of this situation,[6] though many details of their actual formation

J. Vicens and J. Harrowfield (eds.), Calixarenes in the Nanoworld, 197–212.

mechanisms remain obscure. There is no reason to presume that true excision would be impossible with a calixarene ligand but all currently known systems[7-9] in which a metal aggregate is bound to calixarene ligands are those in which the aggregate has been built up from mononuclear or, in a few cases, binuclear precursors.

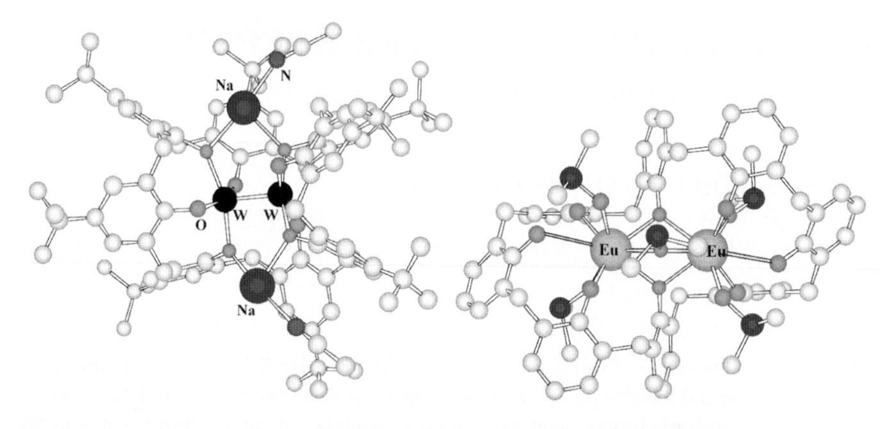

Figure 10-1. (a) The ditungsten (disodium) complex **1** of p-t-butylcalix[8]arene.[13] For clarity, part of the coordination spheres of the Na atoms and one component of disorder in the t-butyl groups are not shown. W...W = 2.298 Å. (b) The dieuropium complex **2** of p-t-butylcalix[8]arene; Eu...Eu = 3.700 Å.[14]

Aside from the fact that the calixarenes are a family of ligands within which an extraordinary variety of donor atoms and donor atom arrays is known,[8] their utility as metal-ion complexants lies principally in the lipophilicity of their complexes. This has been exploited in numerous processes based on solvent extraction and related procedures,[10-12] including the treatment of nuclear wastes. Another possible application, which has received far less attention to date, is the use of lipophilic calixarene/metal-clusters to incorporate the clusters themselves into porous solids suitable for catalysis. It is our purpose in this chapter to discuss the prospects for such an application. We will use the term "cluster" in a general sense to include metal-ion aggregates which may be held together simply by bridging donor atoms and not necessarily metal-metal bonds, and will include consideration of systems in which multiple metal ion binding is due to the presence of multiple but essentially independent binding sites on a calixarene framework, since obviously such systems might ultimately be decomposed to give clusters. It is worthy of note, however, that even in bridged binuclear systems (Fig. 1) situations are known in which metal-metal interactions may be extremely strong, as in **1**,[13] or exceedingly weak, as in **2**.[14] Obviously, such differences may be of some consequence if the calixarene complexes are to be used as precursors to clusters produced by thermal decomposition within a porous lattice. To consider systems in which the maximum metal-metal separations might genuinely at least approach 1 nm (10 Å), we will largely ignore complexes with only 3 or fewer metal atoms present. Small systems and some of their alkali- and alkaline-earth-metal derivatives have been reviewed elsewhere.[15]

2. CALIXARENE AND HOMO-OXACALIXARENE COMPLEXES OF URANYL CLUSTERS

The smallest methylene-bridged calixarene, calix[4]arene, appears to be too small to envelop even a single uranyl ion, though the slight expansion of the macrocycle resulting from the substitution of S for the methylene bridges does allow this to occur,[16] as do other means of macrocycle expansion such as in dihomo-oxacalix[4]arene and calix[5]arene.[17] On passing to calix[6]arene ligands, an intriguing situation is reached where a single U(VI) may be transformed between aryloxo-uranyl and (hexakis) aryloxo-uranium forms,[18] and where the species isolated depends markedly on the conditions of synthesis, both di-[19,20] and tri-nuclear[21] species thus being known, though only the latter case can it be said that the U(VI) unit is a cluster, and it could be argued that in no case is the metal truly encapsulated by the ligands. (Functionalised calix[6]arene ligands have, of course, been designed and used as "uranophiles", the specific intention being to envelop a single uranyl species,[22] though structural evidence[23] indicates that the calix[6]arene skeleton may in fact be particularly unsuited to such a purpose.) As well, even the trimeric species has dimensions only barely comparable to 1 nm, the U_3O core and its attached O-donors forming a strongly oblate sphere approximately 0.9 nm in (the major) diameter and 0.5 nm thick (Fig. 2). The chemistry of uranium(VI) with calix[6]arenes, however, presages more important results with larger calixarenes.

In the case of *p*-*t*-butylcalix[7]arene, a mononuclear uranyl ion complex is readily isolated[24] but substitution of benzyl for the t-butyl substituents and achievement of a higher degree of deprotonation of the calixarene lead to the isolation of a hexauranate species, $[(UO_2)_3$(*p*-benzylcalix[7]arene – 7H)(O)$(CH_3CN)]_2$.[25] Here, somewhat like the di-uranyl derivative of p-t-butylcalix[6]arene,[20] the metal unit is trapped between two calixarene molecules which are too small to simply encircle the cluster. Encirclement is seen with p-t-butylcalix[9]arene[21] but only of a (carbonate-bridged) diuranyl unit and even p-t-butylcalix[12]arene is just sufficiently large enough to encircle four uranyl units (though as two dimeric units not directly linked other than through the calixarene framework).[26]

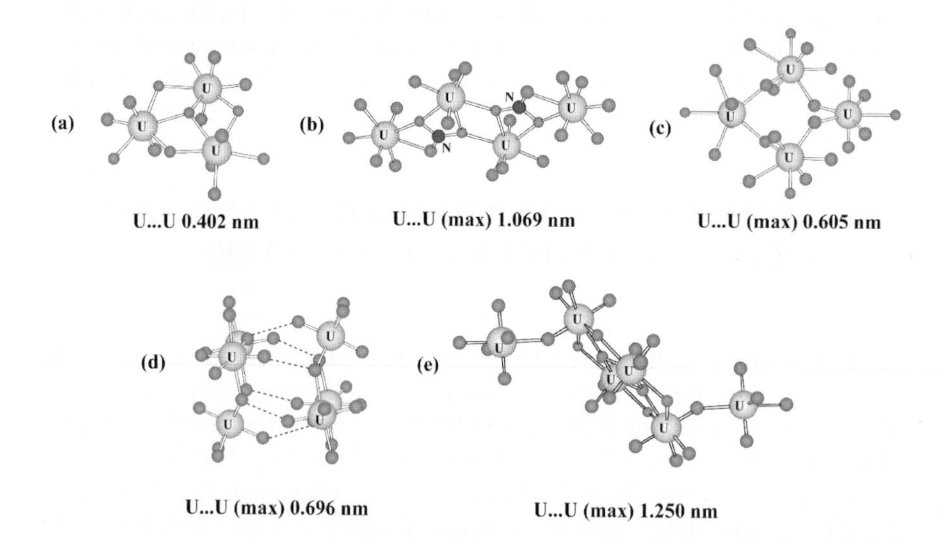

Figure 10-2. The U(VI)-oxo entities found in: (a) the hydroxo-centred trinuclear complex of p-t-butylcalix[6]arene[21]; (b) the nitrato-bridged tetranuclear complex of p-t-butyltetrahomodioxacalix[10]arene[23]; (c) one of the oxo/hydroxo-bridged tetranuclear complexes of p-t-butyloctahomotetraoxacalix[8]arene[29]; (d) the hydrogen-bonded assembly (H-bonding interactions indicated by dashed lines) of two hydroxo-bridged trinuclear units found in the complex of hexahomotrioxacalix[6]arene[30] and; (e) the oxo/hydroxo-bridged hexanuclear complex of p-benzylcalix[7]arene.[25]

The family of homo-oxacalixarenes,[27] one to which addition of new members seems to be relatively facile compared to that to the calixarenes themselves,[23] provides some very large rings suited to envelopment of uranate clusters and, in the case of p-t-butyltetrahomodioxacalix[10]arene, a macrocycle formally just slightly smaller than p-t-butylcalix[12]arene, the decaanion forms a tetranuclear uranyl ion complex involving nitrate bridges between the U centres.[23] With the in turn slightly smaller p-t-butyloctahomotetraoxacalix[8]arene, a tetranuclear uranate is again enveloped but it is one involving smaller bridging entities (oxide/hydroxide/water) and for which two isomeric forms have been obtained.[28,29] A hexanuclear uranyl species is found in the complex of hexahomotrioxacalix[6]arene but, as in the case of the p-benzylcalix[7]arene complex, the metal cluster is sandwiched between two calixarene entities and as well can be considered to consist of two trimeric entities linked by hydrogen bonds.[30]

Consideration of both the exceptionally extensive studies of uranyl complexes of homo-oxacalixarenes[23,28-36] and related work on the complexes of parent calixarenes[21,37] has led to the conclusions[23] that the nature of the polyuranate entrapped by a calixarene may depend upon: (i) the size, symmetry (shape) and conformation of the macrocycle; (ii) the extent of deprotonation of the phenol units; (iii) the presence of small anions (hydroxide, nitrate, carbonate, carbamate) able to bridge uranium(VI) centres; (iv) the presence of other cations, alkali metals in particular, able to compete with U(VI) for phenoxide donor atoms, and (v) the presence of coordinating solvents, such as pyridine. It has been argued[23] that p-t-butylcalix[12]arene, for example, might well form complexes of nuclearity higher than four in the absence of pyridine. Of course, it must be noted that it is only by inclusion of the calixarene sheath that the size of any of the known uranyl ion complexes approaches even just 2 nm and very much higher nuclearities[4] would be required to achieve true nanoparticle dimensions for the metal clusters. None of the clusters yet defined can be seen as significant fractions of the UO_3 lattice.[38]

3. METAL AGGREGATES HIGHER THAN TRIMETALLIC IN CALIXARENE COMPLEXES

The uranyl ion complexes described above all involve coordination through phenoxide groups of the calixarenes, with these but rarely involved in bridging the metal centres. In lanthanide ion complexes of phenolic calixarenes,[37] it is much more common to observe such bridging and this may explain the generally higher nuclearity of clusters bound to particular calixarenes. This cannot be the only factor of importance, though, since, as noted above, the multimetallic units bound to calixarenes frequently incorporate small anions, oxide and hydroxide being prominent. This may indicate that the relatively high affinity of aryloxide-*O* for H^+ facilitates hydrolytic reactions, especially in the case of metals conventionally regarded as "oxophilic". The nature of the macrocyclic ligand is important not simply because the size of any cavity or "hole" defined by the ligand might limit the number of metals to be found in it. It is a remarkable fact[8] that all structurally characterised complexes of phenolic calix[4]arenes contain the ligand in a cone conformation which places all four O-donors to one side of the macrocyclic ring mean plane, albeit that the inclination of the cone may vary widely. Such an array is not incompatible with all four donors binding to a single metal ion, though in general this seems only to arise where the interactions involve long bonds and are presumably weak (*e.g.* where Na^+ is involved), where they are assisted by chelation involving donor atoms of O-substituents, or where they are especially strong (highly charged cations). This last case may indicate that strong interactions can overcome unfavourable conformational changes associated with bringing the O-donor array from a situation favouring divergence to one favouring convergence. In any case, the highest nuclearity yet observed (Fig. 3(a)) in a complex of a simple phenolic calixarene is but 7 in the Eu(III) complex of p-t-butylcalix[9]arene and even in this case the nuclearity is sensitive to the solvent used for synthesis.[39]

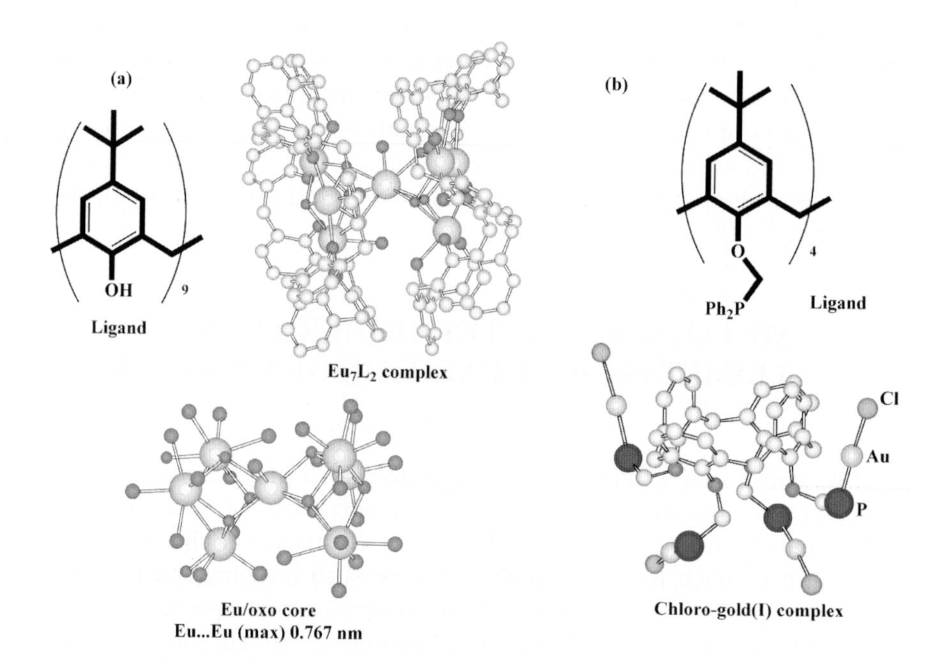

Figure 10-3. Different modes of formation of multimetallic derivatives of calixarenes, illustrated by: (a) envelopment of an hydroxo-bridged hepta-europium(III) species[39] by two molecules of p-t-butylcalix[9]arene (For clarity, the t-butyl substituents and coordinated dmso molecules, both partially disordered, are not shown.); (b) binding of independent Au(I) centres to the four pendent phosphine centres of a calix[4]arene derivative.[44(a)] (Again for clarity, t-butyl groups of the calixarene and phenyl groups on phosphorus are omitted.)

Phenol/Phenoxide-*O* is not the only donor centre in parent calixarenes and the aromatic rings can serve as donors towards metals such as Cr(0),[40] Rh(I), Ir(I) and Ru(II),[41,42] giving, in the case of calix[4]arenes, tetrametallic derivatives. Introduction of new donor centres by means such as substitution of cyano groups for the p-alkyl substituents or alkylation of the phenolic oxygens by phosphorus-containing chains can be used to provide tetrametallic metallocarcerands[43] and tetra-gold(I) (Fig. 3(b)) derivatives of calix[4]arenes.[44] Presumably, higher nuclearity species could be obtained from higher calixarene derivatives. Metal centres can of course be part of the unit used to functionalise a calixarene, ferrocene and cobalocenium derivatives providing the best known examples.[45] Both tetra- and penta-nuclear complexes of phosphinite-substituted resorc[4]arenes are known,[46] halide ion mediated interactions giving rise to the involvement of a fifth metal. The multiple functionality of resorc[4]arenes can in fact be exploited to produce octagold and even, by reaction with a binuclear metal complex, hexadecacobalt derivatives.[47] Further, metallocarcerand analogues can be produced when, for example, pendent tridentate units from separate functionalised resorcarenes complex octahedral transition metal ions.[48] These complexes can have a nuclearity higher than four due to the association with main group metal ions bound to the exterior of the metallocarcerand capsule.

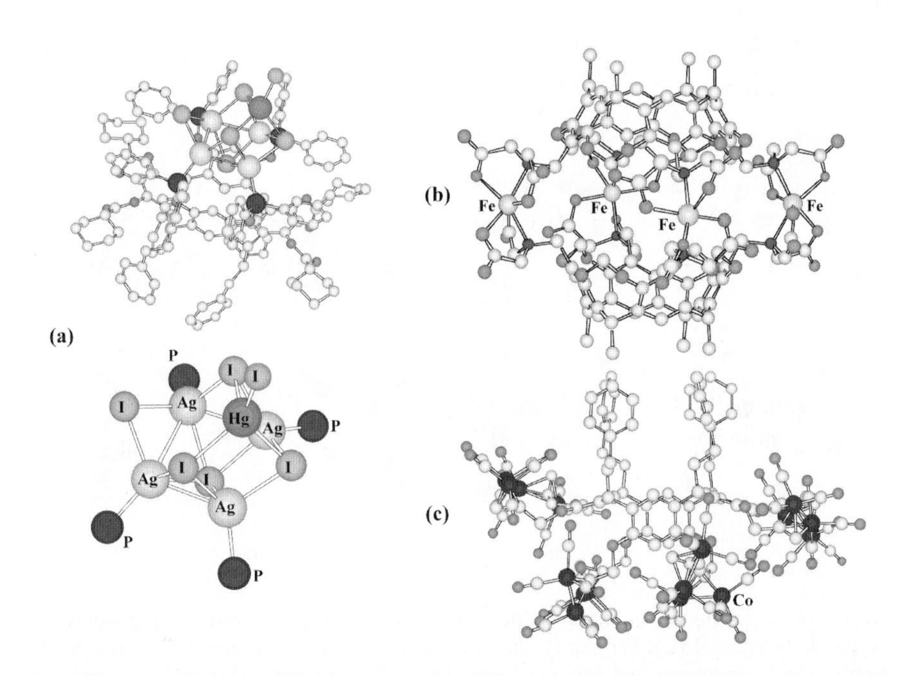

Figure 10-4. Illustrations of the varied exploitations of the resorc[4]arene structure to obtain multinuclear complexes: (a) an Ag_4HgI_6 cluster supported on a tetraphosphonite derivative of resorc[4]arene[46(a)]; (b) four Fe(II) cations sandwiched between two resorc[4]arene units functionalised by four iminodiacetate groups each (associated Sr(II) cations are not shown)[48(a)] and; (c) the hexadecacobalt complex[47] produced by addition of $Co_2(CO)_6$ units to each of the eight triple-bond units of propargyl ether substituents of a resorc[4]arene.

As a means of enhancing the metal ion binding ability of calixarenes, a remarkably effective yet simple structural modification is that of replacing the methylene bridges of the macrocycle by sulphur to give the "thiacalixarenes".[49] Though the thiaether links alone appear to be useful binding sites, they have further value in their ready conversion to sulphoxide and sulphone centres.[49,50] The first thiacalixarene to be widely studied,[38,49,50] p-t-butyltetrathiacalix[4]arene, gives, with certain exceptions,[51,52] metal complexes involving one to four metals (in structurally characterised aggregates),[16,38,49,53-55] as do some of its oxidised/sulphonated derivatives,[56] but its tetrasulphone derivative forms lanthanide ion complexes involving octanuclear clusters,[57] while higher members of the thiacalixarene series[58-63] give rise to even decanuclear clusters with transition metals. A decanuclear species is also one of the exceptional cases found for p-t-butyltetrathiacalix[4]arene.[52] The as yet less studied mercaptothiacalix[4]arene[50,53(a)] is known to give a hexanuclear Hg(II) derivative.

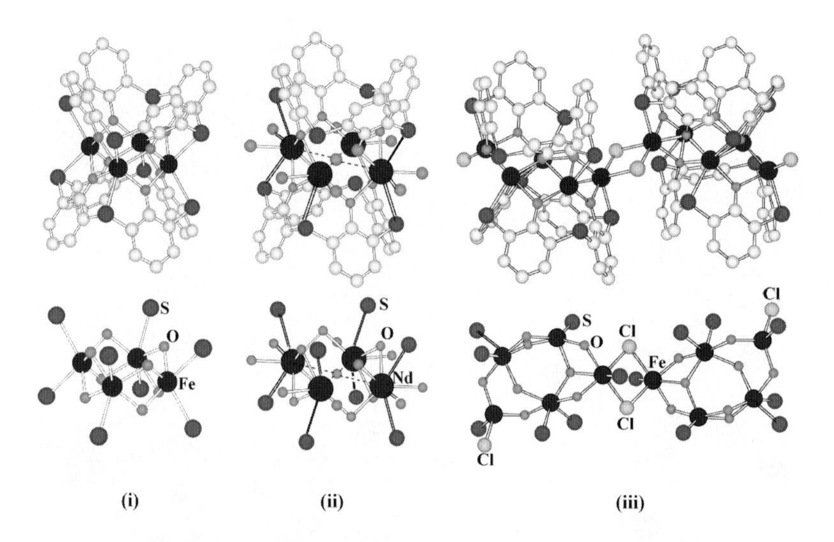

Figure 10-5. Examples of tetrametallic and higher clusters formed with p-t-butyltetrathiacalix[4]arene (LH_4). For clarity, the calixarenes are shown without their (often disordered) t-butyl substituents and only the donor atoms of ligands other than the calixarene are shown. (i) The complex $[Fe_4L_2]\bullet H_2O$ containing essentially trigonal prismatic Fe(II) with an O_4S_2 coordination environment and with each adjacent pair of metals doubly-bridged by phenoxide donors.[52] The metals form a square array of edge 0.319 nm. (ii) The complex $[Nd_4(OH)L_2(dmf)_6(dmso)_2](NO_3)_3$, containing essentially tricapped trigonal prismatic NdO_7S_2 species. The Nd_4 unit is close to square, with a mean edge length of 0.366 nm.[53(c)] (iii) The decanuclear, chloro-bridged species found in $[Fe_{10}L_4Cl_4]\bullet 2H_2O$. Both 5- and 6-coordinate Fe atoms are present, the maximum Fe...Fe separation being 1.574 nm.[52]

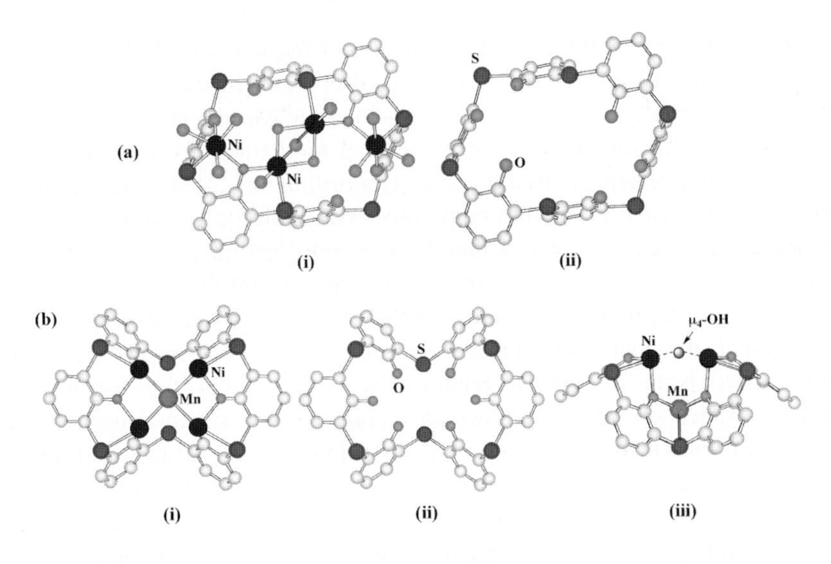

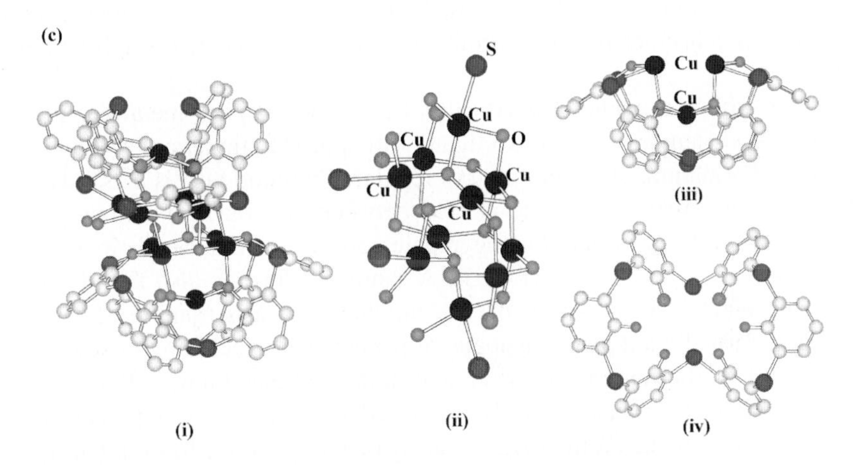

Figure 10-6. Polynuclear metal complexes of p-t-butylhexathiacalix[6]arene.[58,59] In all views showing the calixarene, the t-butyl groups are omitted. Only donor atoms are shown for the non-calix ligands. (a)(i) The tetranuclear Ni(II) complex, viewed perpendicular to the mean Ni_4 plane and (ii) the same view with the metal ions omitted to show the twisted, cone-like conformation of the ligand. (b)(i) The pentanuclear$MnNi_4$ complex, again viewed perpendicular to the mean Ni_4 plane and with the μ_4-OH unit omitted; (ii) the same view without the metal ions, showing the ligand conformation; and (iii) a view orthogonal to (i), showing the μ_4-OH bridge linking the Ni ions. (c) The decanuclear, twofold-symmetric "sandwich" complex of Cu(II): (i) The complex, showing the envelopment of the cluster by the two ligand moieties; (ii) the Cu_{10} cluster, essentially an oxo-bridged dimer, where the maximum Cu…Cu separation is 0.761 nm; (iii), (iv) views of the "monomer" component of the cluster and of the ligand, showing the close similarities to the properties of the pentanuclear species.

It is interesting that the small expansion of the macrocycle that results in passing from p-t-butylcalix[4]arene to p-t-butyltetrathiacalix[4]arene causes the Cs(I) complex to pass from a mononuclear species involving π-coordination, presumably at close to optimal separations from the rings, to a multinuclear species involving oxygen coordination.[53(b)] Calix[4]arenes in the cone conformation seem to be unique in providing a π-coordination environment which can envelop a metal, as π-coordination can certainly occur with higher calixarenes but not in such a way as to inhibit O-donor bridging of the interacting metal to another.[64] While it is perhaps an open question as to whether the methylene bridge hydrogen atoms of simple calixarenes might participate in agostic bonding, it is clear that the S-bridges of thiacalixarenes are effective donor sites and an obvious consequence of this is that no known metal centre could coordinate to all oxygen and sulphur sites (even those of the smallest thiacalix[4]arene) simultaneously, so that full exploitation of the O,S donor sites of a thiacalixarene must necessarily lead to it acting as a bridging ligand, quite possibly in several conformations. Note that a significant advance in recent work on thiacalixarene complexes is the application of solvothermal methods in synthesis,[52,55] this appearing to result in much more readily reproducible stoichiometry of crystalline complexes.

As with methylene-bridged calix[4]arene derivatives, metal complexes of thiacalix[4]arenes as phenoxo ligands all appear to involve the ligand in a cone conformation, albeit not always regular. Common also is a "sandwich" form where a metal aggregate is held between two (and sometimes more) calixarene units and where that aggregate consists of four metals in a square array (*e.g.* [53(a),(e),(c),54(a),(d),55]) (Fig. 5). A sandwich form is also found for the decanuclear Cu(II) complex of p-t-butylhexathiacalix[6]arene (Fig. 6),[59] which is formally a dimeric analogue of pentanuclear, including mixed-metal, complexes of the same ligand with other transition metal ions.[58] Remarkably, these pentanuclear species can be produced from tetranuclear precursors in which the four metals form a zig-zag array bound to a cone form of the ligand and, when the fifth metal is added, the four initially present rearrange to a square array with a form and donor atom environment very similar to that in tetrametallic derivatives of tetrathiacalix[4]arene. However, these transition metal squares (Fig. 6) contain a μ_4-OH bridge at their centre, as in lanthanide metal squares[37,53(c)] but unlike other transition metal squares[53(e),55] bound to tetrathiacalix[4]arene (Fig. 5). The conversion of the tetranuclear to the pentanuclear as well as to the oxo-bridged, twofold symmetric decanuclear species involves a significant conformational change of the ligand to a more symmetrical form, though still one describable as essentially a cone and rather similar to that of the "free" ligand,[62] *i.e.* the proton complex. Although studies of the thiacalix[6]arene are still relatively limited, these results may be indicative of different and perhaps more marked conformational preferences for the thiacalix than appears to be the case for its methylene-bridged analogue, which adopts a cone form as the proton complex but frequently a 1,2,3-alternate conformation in its complexes with both metal cations and quaternary ammonium ions.[64(b),65]

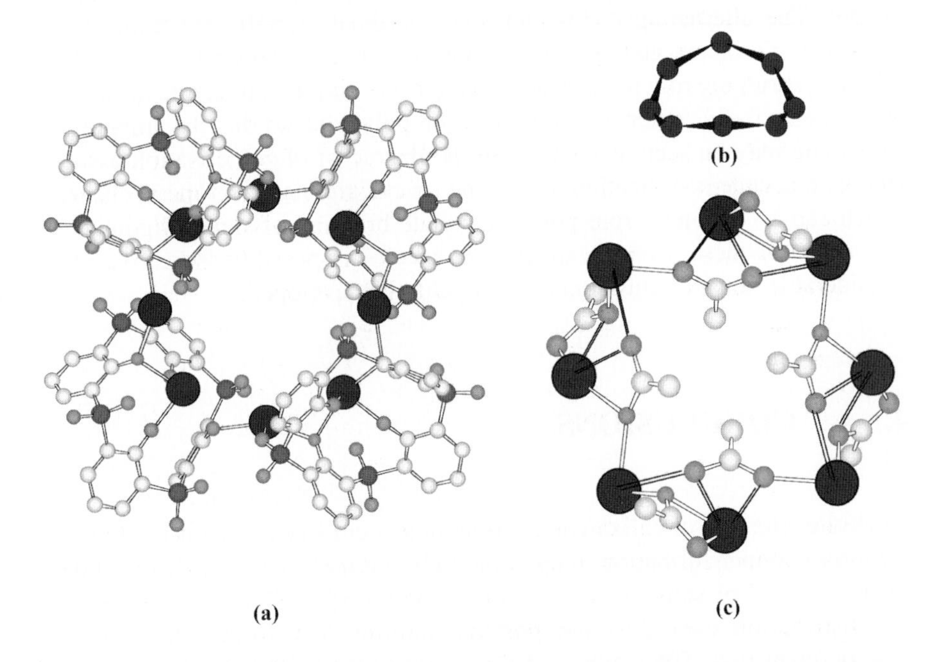

Figure 10-7. Octanuclear lanthanide "wheels" derived from p-t-butyltetrasulphonylcalix[4]arene (L'H$_4$).[59] (a) A simplified view of the complex aggregate present in [Sm$_8$L'$_4$(OAc)$_8$(EtOH)$_4$(OH$_2$)$_4$], showing just the Sm atoms (black) and the atoms of the calixarene units (excluding t-butyl groups) and depicting the alternation of the calixarene cones with respect to the mean plane of the metal array. (b) A perspective view of the Sm$_8$ "wheel" showing the puckering of the ring. (c) A view of the metal ion and acetate units only, showing the two bridging-chelation modes of acetate coordination.

As a ligand to generate metal cluster complexes, p-t-butylhexathiacalix[6]arene seems particularly effective[58,59] and its mixed-metal derivatives are obviously of interest as possible precursors to specific metal clusters for catalysis.[49] A significant feature of recent work on metal complexes of thiacalixarenes in general has been the association of a wider range of simple anions with the metallocalixarene species. A spectacular example is provided in the octanuclear "lanthanide wheels" (Fig. 7) isolated by reaction of lanthanide acetates with p-t-butyltetrasulphonylcalix[4]arene.[57] These involve Ln$_8$ rings encircled by four calixarene units in their cone form, with four Ln(III) occupying sites such that each is bound to all four phenolic oxygens of a given calixarene and has overall LnO_8 coordination, while the other four Ln(III), having LnO_9 coordination, bridge these units.

Subtle differences are apparent for different Ln in the currently known species but the essential structure is retained for at least the first half of the lanthanide series. The Ln_8 rings are close to planar and have a diameter ~1 nm. The alternating eight- and nine-coordinate Ln(III) are bridged by phenoxide-*O* donors and by acetate units which are also chelating. Such a chelating/bridging role for acetate is seen in the simple lanthanide acetates[66] and leads to their existence as coordination polymers, so that the sulphonylcalixarene may be seen as controlling the degree of oligomerisation associated with acetate coordination. Given the spectacular nature of metal clusters which can be obtained from polycarboxylate bridging of small oligomers,[67] the nature of these acetato-calixarenato species is possibly indicative of a considerable range of cluster chemistry yet to be developed.

4. CONCLUSIONS

While a feature of calixarene coordination chemistry is the rarity of coordination polymer formation, only commonly observed with alkali metal derivatives,[15,68,69] or sulfonated calixarene complexes,[64(b),70] factors controlling the distribution over different possible multinuclear forms are still not readily quantified. The subtlety of these influences is illustrated by marked differences in nuclearity which can result merely from the change of crystallisation solvent.[39] With large calixarenes and homo-oxacalixarenes, able to adopt close to planar conformations, the capacity to encircle a metal or metal cluster with a tendency to form coordinate bonds in one plane, as in the case of U(VI), may be of primary importance, though such instances serve also to illustrate the fact that there may be a complicated interplay between metal ion coordination preferences, calixarene ring size and calixarene conformation. As discussed in Chapters 3 and 4, the conformation of a calix[4]arene can influence the ease of synthesis of polytopic ligands (“calixtubes”) which can be used to form multinuclear complexes but the importance of conformation alone in general is difficult to assess other than in the rather obvious sense that orientation of sets of donor groups in divergent directions enhances the likelihood of obtaining at least binuclear species.[56(c)] An elegant illustration of the significance of donor atom orientation is provided in the octanuclear structures of Ln(III) complexes of p-t-butyltetrasulphonylcalix[4]arene.[57] Four Ln(calixarene) units in which the Ln(III) can be regarded as centred within the phenoxide-*O* square are bridged by four other Ln(III) in part as a result of lateral suphonyl-*O* coordination which must be compatible with the essentially octagonal array of the whole octametal unit. Of course, the striking lattice arrays seen in these compounds serve as reminders of the possible importance of the total lattice interactions in determining the forms of the complexes and of the fact that for calixarene complexes generally only rather limited information is available on the forms they may adopt in solution.

Of the practical variables which appear to influence the nuclearity of metal complexes of calixarenes, perhaps least exploited in a rational manner to date is the use of small anionic co-ligands. Acetate[57,59] and chloride[52] have provided pathways to some of the largest species yet known and, given the huge variety of carboxylate anions which might be employed, their use, in particular, would seem to offer prospects for a large range of further developments which, as noted above, have parallels in established nanochemistry. Since most simple metal acetates are hydrated, they are particularly useful in the synthesis of multinuclear complexes in that not only is acetate a good bridging ligand but also the enhanced basicity of acetate in (commonly used) aprotic solvents leads to the conversion of water to hydroxide and even oxide, both excellent bridging ligands.[71] While most of the complexes presently discussed are soluble in weakly polar solvents and therefore should be useful simply for the loading of mesoporous solids with metal units of defined composition,[72] they also have unique electronic and magnetic properties determined by the particular nature of the calixarene involved and their exploitation should certainly not be limited to that of nanoparticle precursors only. Nearly all paramagnetic multinuclear complexes of calixarenes have been found[52-57] to exhibit antiferromagnetic coupling, varying from extremely weak to moderately strong, between the metal centres, but the decanuclear Cu(II) complex of p-t-butylhexathiacalix[6]arene shows ferromagnetic coupling,[59] and the mixed metal complexes that can be obtained with the same ligand[58] must presumably have complicated magnetic properties. It also appears possible that changing from a calixarene to a thiacalixarene could be used to control the oxidation state (and hence the magnetism) of the bound metal, since Fe(III) complexes are readily prepared for calixarenes[73] but Fe(II) species are those favoured by at least p-t-butyltetrathiacalix[4]-arene.[52]

5. REFERENCES

1. I. G. Dance in G. Wilkinson, R. D. Gillard, J. McCleverty (Eds.), *Comprehensive Coordination Chemistry*, Pergamon Books, Oxford, 1987, Vol. 1, Ch. 4, p. 135.

2. I. G. Dance, K. Fisher, *Progr. Inorg. Chem.* **41**, 637-803 (1994).
3. M.-C. Daniel, D. Astruc, *Chem. Rev.* **104**, 293-346 (2004).
4. D. Fenske, C. E. Anson, A. Eichhöfer, O. Fuhr, A. Ingendoh, C. Persau, C. Richert, *Angew. Chem. Int. Ed.* **44**, 5242-5246 (2005) and references therein.
5. G. H. Woehrle, J. E. Hutchison, *Inorg. Chem.* **44**, 6149-6158 (2005) and references therein.
6. P. V. Rao, R. H. Holm, *Chem. Rev.* **104**, 527-560 (2004).
7. C. Redshaw, *Coord. Chem. Rev.* **244**, 45-70 (2003).
8. Z. Asfari, V. Böhmer, J. M. Harrowfield, J. Vicens (Eds.), *Calixarenes 2001*, Kluwer Academic Publishers, Dordrecht, 2001 (Chs. 28-30 in particular).
9. C. Wieser, C. B. Dieleman, D. Matt, *Coord. Chem. Rev.* **165**, 93-161 (1997).
10. (a) D. M. Roundhill, J. Y. Shen in ref. 8, Ch. 22, pp. 407-420; (b) F. Arnaud, M.-J. Schwing-Weill, J.-F. Dozol in ref. 8, Ch. 35, pp. 642-662.
11. L. Mandolini, R. Ungaro, *Calixarenes in Action*, Imperial College Press, London, 2000.
12. G. Lumetta, R. D. Rogers, A. Gopalan, *Calixarenes for Separation*, ACS Symposium Series 757, American Chemical Society, Washington, 2000.
13. V. C. Gibson, C. Redshaw, M. R. J. Elsegood, *Chem. Commun.* 1200-1201 (2002).
14. J.-C. G. Bünzli, P. Froidevaux, J. M. Harrowfield, *Inorg. Chem.* **32**, 3306-3311 (1993).
15. A. J. Petrella, C. L. Raston, *J. Organomet. Chem.* **689**, 4125-4136 (2004).
16. Z. Asfari, A. Bilyk, J. W. C. Dunlop, A.K. Hall, J.M. Harrowfield, M.W. Hosseini, B.W. Skelton, A.H. White, *Angew. Chem., Int. Edn,* **40**, 721-723 (2001).
17. J. M. Harrowfield, *Gazz. Chim. Ital.* **127**, 663-671 (1997) and references therein.
18. P. C. Leverd, M. Nierlich, *Eur. J. Inorg. Chem.* 1733-1738 (2000).
19. P. C. Leverd, P. Berthault, M. Lance, M. Nierlich, *Eur. J. Inorg. Chem.* 1859-1862 (1998).
20. P. Thuéry, M. Lance, M. Nierlich, *Supramol. Chem.* **7**, 183-185 (1996).
21. X. Delaigue, C. D. Gutsche, J. M. Harrowfield, M. I. Ogden, B. W. Skelton, D. R. Stewart, A. H. White, *Supramol. Chem.* **16**, 603-609 (2004).
22. S. Shinkai, H. Koreishi, K. Ueda, T. Arimura, O. Manabe, *J. Am. Chem. Soc.* **109**, 6371-6376 (1987).
23. B. Masci, P. Thuéry, *New J. Chem.* **29**, 493-498 (2005).
24. P. Thuéry, M. Nierlich, M. I. Ogden, J. M. Harrowfield, *Supramol. Chem.* **9**, 297-303 (1998).
25. P. Thuéry, M. Nierlich, B. Souley, Z. Asfari, J. Vicens, *J. Chem. Soc., Dalton Trans.* 2589-2594 (1999).
26. P. C. Leverd, I. Dumazet-Bonnamour, R. Lamartine, M. Nierlich, *Chem. Commun.* 493-494 (2000).
27. B. Masci in ref. 8., Ch. 12, pp. 235-249.
28. P. Thuéry, M. Nierlich, J. Vicens, B. Masci, *J. Chem. Soc., Dalton Trans.* 867-874 (2001).
29. P. Thuéry, B. Masci, *Polyhedron* **22**, 3499-3505 (2003).
30. P. Thuéry, M. Nierlich, J. Vicens, B. Masci, *J. Chem. Soc., Dalton Trans.* 3410-3412 (2001).
31. P. Thuéry, M. Nierlich, J. Vicens, B. Masci, H. Takemura, *Eur. J. Inorg. Chem.* 637-643 (2001).
32. B. Masci, M. Nierlich, P. Thuéry, *New J. Chem.* **26**, 120-128; 766-774 (2002).
33. B. Masci, M. Gabrielli, S. Levi Mortera, M. Nierlich, P. Thuéry, *Polyhedron*, **21**, 1125-1131 (2002).
34. B. Masci, P. Thuéry, *Supramol. Chem.* **15**, 101-108 (2003).
35. P. Thuéry, B. Masci, *J. Chem. Soc., Dalton Trans*, 2411-2417 (2003).

36. P. Thuéry, B. Masci, *Polyhedron* **23**, 649-655 (2004).
37. P. Thuéry, M. Nierlich, J. M. Harrowfield, M. I. Ogden in ref. 8., Ch. 30, pp. 561-582.
38. M. T. Weller, P. G. Dickens, D. J. Penny, *Polyhedron* **7**, 243-244 (1988) and references therein.
39. S. Fleming, C. D. Gutsche, J. M. Harrowfield, M. I. Ogden, B. W. Skelton, D. F. Stewart, A. H. White, *J. Chem. Soc., Dalton Trans.* 3319-3327 (2003).
40. H. Iki, T. Kikuchi, H. Tsuzuki, S. Shinkai, *Chem. Lett.* 1735-1738 (1993).
41. M. Staffilani, K. S. B. Hancock, J. W. Steed, K. T. Holman, J. L. Atwood, R. K. Juneja, R. S. Burkhalter, *J. Am. Chem. Soc.* **119**, 6324-6335 (1997).
42. Z. Asfari, A. Bilyk, C. Bond, J. M. Harrowfield, G. A. Koutsantonis, N. A. Lengkeek, M. Mocerino, B. W. Skelton, A. N. Sobolev, S. Strano, J. Vicens, A. H. White, *Org. Biomol. Chem.* **2**, 387-396 (2004).
43. D. Zuccaccia, L. Pirondini, R. Pinalli, E. Dalcanale, A. Macchioni, *J. Am. Chem.Soc.* **127**, 7025-7032 (2005).
44. (a) C. B. Dieleman, D. Matt, I. Neda, R. Schmutzler, H. Thönnessen, P. G. Jones, A. Harriman, *J. Chem. Soc., Dalton Trans* 2115-2121 (1998); (b) C. B. Dieleman, D. Matt, A. Harriman, *Eur. J. Inorg. Chem.* 831-834 (2000).
45. S. E. Matthews, P. D. Beer in ref. 8., Ch. 23, pp. 421-439.
46. (a) D. J. Eisler, R. J. Puddephatt, *Inorg. Chem.* **42**, 8192-8202 (2003); (b) D. J. Eisler, R. J. Puddephatt, *Inorg. Chem.* **44**, 4666-4678 (2005).
47. D. J. Eisler, W. Hong, M. C. Jennings, R. J. Puddephatt, *Organometallics* **21**, 3955-3960 (2002).
48. (a) O. D. Fox, N. K. Dalley, R. G. Harrison, *Inorg. Chem.* **38**, 5860-5863 (1999); (b) R. G. Harrison, J. L. Burrows, L. D. Hansen, *Chem. Eur. J.* **11**, 5881-5888 (2005).
49. N. Iki, S. Miyano, *J. Incl. Phenom. Macrocyclic Chem.* **41**, 99-105 (2001).
50. M. W. Hosseini in ref. 8., Ch. 6, pp. 110-129.
51. Ni(II), under certain conditions, forms a Ni_{32} cluster binding six thiacalixarenes – A. Bilyk, A. H. White, University of Western Australia, unpublished.
52. C. Desroches, G. Pilet, P. A. Szilagyi, G. Molnár, S. A. Borshch, A. Boussekou, S. Parola, D. Luneau, *Eur. J. Inorg. Chem.* (2005), in press; DOI 10.1002/ejic.200500640.
53. (a) H. Akdas, E. Graf, M. W. Hosseini, A. De Cian, A. Bilyk, B. W. Skelton, G. A. Koutsantonis, I. W. Murray, J. M. Harrowfield, A. H. White, *Chem. Commun* 1042-1043 (2002); (b) A. Bilyk, A. K. Hall, J. M. Harrowfield, M. W. Hosseini, B. W. Skelton, A. H. White, *Inorg. Chem.* **40**, 672-686 (2001); (c) A. Bilyk, A. K. Hall, J. M. Harrowfield, M. W. Hosseini, B. W. Skelton, A. H. White, *Aust. J. Chem.* **53**, 895-898 (2000); (d) A. Bilyk, A. K. Hall, J. M. Harrowfield, M. W. Hosseini, G. Mislin, B. W. Skelton, C. Taylor, A. H. White, *Eur. J. Inorg. Chem.* 823-826 (2000); (e) G. Mislin, E. Graf, M. W. Hosseini, A. Bilyk, A. K. Hall, J. M. Harrowfield, B. W. Skelton, A. H. White, *Chem. Commun.* 373-374 (1999).
54. (a) N. Kon, N. Iki, T. Kajiwara, T. Ito, S. Miyano, *Chem. Lett.* **33**, 1046-1047 (2004); (b) H. Katagiri, N. Morohashi, N. Iki, C. Kabuto, S. Miyano, *J. Chem. Soc., Dalton Trans.* 723-726 (2003); (c) N. Morohashi, T. Hattori, K. Yokomakura, C. Kabuto, S. Miyano, *Tetrahedron Lett.* **43**, 7769-7772 (2002); (d) N. Iki, N. Morohashi, C. Kabuto, S. Miyano, *Chem. Lett.* 219-220 (1999).
55. C. Desroches, G. Pilet, S. A. Borshch, S. Parola, D. Luneau, *Inorg. Chem.* **44**, 9112-9120 (2005).
56. (a) Q. Guo, W. Zhu, S. Dong, S. Ma, X. Yan, *J. Mol. Struct.* **650**, 159-164 (2003); (b) N. Morohashi, N. Iki, S. Miyano, T. Kajiwara, T. Ito, *Chem. Lett.* 66-67 (2001); (c) T. Kajiwara, S. Yokozawa, T. Ito, N. Iki, N. Morohashi, S. Miyano, *Chem. Lett.* 6-7 (2001).

57. T. Kajiwara, H. Wu, T. Ito, N. Iki, S. Miyano, *Angew. Chem. Int. Ed.* **43**, 1832-1835 (2004).
58. T. Kajiwara, R. Shinagawa, T. Ito, N. Kon, N. Iki, S. Miyano, *Bull. Chem. Soc. Jpn* **76**, 2267-2275 (2003).
59. T. Kajiwara, N. Kon, S. Yokozawa, T. Ito, N. Iki. S. Miyano, *J. Am. Chem. Soc.* **124**, 11274-11275 (2002).
60. N. Morohashi, N. Iki, M. Aono, S. Miyano, *Chem. Lett.* 494-495 (2002).
61. N. Kon, N. Iki. S. Miyano, *Tetrahedron Lett.* **43**, 2231-2234 (2002).
62. N. Iki, N. Morohashi, T. Suzuki, S. Ogawa, M. Aono, C. Kabuko, H. Kumagai, H. Takeya, S. Miyanari, S. Miyano, *Tetrahedron Lett.* **41**, 2587-2590 (2000).
63. Y. Kondo, K. Endo, N. Iki, S. Miyano, F. Hamada, *J. Incl. Phenom. Macrocyclic Chem.* **52**, 45-49 (2005).
64. (a) R. Asmuss, V. Böhmer, J.M. Harrowfield, M.I. Ogden, W.R. Richmond, B.W. Skelton, A. H. White, *J. Chem. Soc., Dalton Trans.* 2427-2433 (1993); (b) Z. Asfari, J. M. Harrowfield, P. Thuéry and J. Vicens, *Supramolecular Chem.* **15**, 69-77 (2003).
65. J. M. Harrowfield, W. R. Richmond, A. N. Sobolev, *J. Incl. Phenom. Mol. Recognit. Chem.* **19**, 257-276 (1995). See also I. Thondorf in ref. 8, Ch. 15, pp. 280-295.
66. P. C. Junk, C. J. Kepert, W.-M. Lu, B. W. Skelton, A. H. White, *Aust. J. Chem.* **52**, 437-457 (1999).
67. G. Férey, C. Mellot-Draznieks, C. Serre, F. Millange, *Acc. Chem. Res.* **38**, 217-225 (2005) and references therein.
68. (a) P. Thuéry, Z. Asfari, J. Vicens, V. Lamare, J.-F. Dozol, *Polyhedron* **21**, 2497-2503 (2002); (b) P. Thuéry, B. Masci, *J. Chem. Soc., Dalton Trans.* 2411-2417 (2003).
69. N. Iki, T. Horiuchi, H. Oka, K. Koyama, N. Morohashi, C. Kabuto, S. Miyano, *J. Chem. Soc. Perkin 2* 2219-2225 (2001).
70. S. J. Dalgarno, M. J. Hardie, J. L. Atwood, C. L. Raston, *Inorg. Chem.* **43**, 6351-6356 (2004) and references therein.
71. J. M. Harrowfield, B. W. Skelton, A. H. White, *C. R. Chimie* **8**, 169-180 (2005).
72. L. M. Bronstein, *Top. Curr. Chem.* **226** (Colloid Chem.), 55-59 (2003).
73. J. Zeller, S. Koenig, U. Radius, *Inorg. Chim. Acta* **357**, 1813-1821 (2003).

Chapter 11

CALIXARENES ON MOLECULAR PRINTBOARDS

Multivalent binding, capsule formation, and surface patterning

Manon J. W. Ludden, Mercedes Crego-Calama, David N. Reinhoudt, and Jurriaan Huskens

Laboratory of Supramolecular Chemistry and Technology, MESA⁺ Institute for Nanotechnology, University of Twente, P.O. Box 217, 7500 AE Enschede, The Netherlands

Abstract: The divalent binding of bis(adamantyl)-calix[4]arene **4** in solution and at β-cyclodextrin (β-CD) self-assembled monolayers (SAMs) (**2**) has been investigated. At β-CD SAMs the binding constant is three orders of magnitude higher than the binding constant for the divalent binding of **4** to the β-CD dimer **3** in solution (1.2×10^7 M^{-1}). A model that treats the sequential binding events as independent, and takes into account an effective concentration term for the second, intramolecular, binding event explains these results. The build-up and subsequent break-down of a non-covalent capsule consisting of tetra(adamantyl)-calix[4]arene **5** and tetrasulfonate-calix[4]arene **6** at the β-CD SAM was studied by surface plasmon resonance (SPR) spectroscopy. The association constant for capsule formation at β-CD SAMs ($(3.5 \pm 1.6) \times 10^6$ M^{-1}) is comparable to the association constant of the capsule in solution ($(7.5 \pm 1.2) \times 10^5$ M^{-1}). Microcontact printing (μCP) and dip-pen nanolithography (DPN) were applied in the patterning of β-CD SAMs. Stable features were obtained upon printing and writing of the bis(adamantyl)-calix[4]arene **4** at the β-CD SAMs. The features could not be removed upon rinsing with water, while rinsing with 10 mM β-CD resulted in partial removal of the patterns. In contrast features printed at OH-terminated SAMs were removed instantly upon rinsing with water.

Key words: Cyclodextrins; self-assembled monolayers; host-guest interactions; multivalency; non-covalent capsules; surface patterning; microcontact printing; dip-pen nanolithography.

J. Vicens and J. Harrowfield (eds.), Calixarenes in the Nanoworld, 213–231.

1. INTRODUCTION

The precise positioning of molecules at a surface is a prerequisite for the build-up of nanosized objects at surfaces. In our group, β-cyclodextrin (β-CD) self-assembled monolayers (SAMs)[1,2] are utilized as a template in this positioning process.[3,4] β–CD is a very well known host in aqueous media for a variety of small hydrophobic organic guest molecules, e.g. ferrocene, aromatic compounds, and adamantyl derivatives,[5] each with its own intrinsic binding affinity.[5] A β-CD SAM can therefore serve as a host surface for these molecules. A β-CD SAM consists of self-assembled β-CDs which are modified at the primary side with seven heptathioether chains.[1,2] On gold, the formed SAM is quasi-hexagonally, densely packed[1,2] and has a well-defined lattice constant (2.1 nm),[6] as determined by atomic force microscopy (AFM). All guest-binding sites in the β-CD SAM are equivalent. Since these host molecules are positioned in a very regular pattern, these surfaces are referred to as "molecular printboards".[3,7] According to surface plasmon resonance (SPR)[2,8] and electrochemical impedance spectroscopy (EIS) measurements,[2] the binding constants (*K*) of small monovalent molecules at the molecular printboard are comparable to the binding constants of these molecules to native β-CD in solution.[6]

Multivalent interactions are widespread in nature and supramolecular chemistry and describe the simultaneous binding of multiple guest entities on one molecule to multiple host entities on another.[9] Multivalent binding processes differ markedly from monovalent binding processes, e.g. they consist of inter- and intramolecular interactions, and dissociation is in general slow and can be influenced by a competitor in solution.[9]

An example of a multivalent system is that of a tris(vancomycin) derivative and a tris(di-peptide) (D-alanine-D-alanine).[10,11] This system has a very high binding constant ($K_a = (4 \pm 1) \times 10^{17}\ M^{-1}$) even higher than that of the binding between (strept)avidin and biotin, which is regarded to be one of the strongest interactions in nature. The dissociation mechanism of the trivalent vancomycin system involves a sequence of successive dissociation events at the vancomycin binding sites, the rate of which can be influenced by a monovalent competitor in solution. Another example is the interaction between the heat-labile enterotoxin from E. Coli and some pentavalent ligands.[12] These pentavalent ligands are excellent inhibitors for the enterotoxin, while the lower valent ligands do not show the same level of inhibition. Also at surfaces multivalent binding processes have been studied, e.g. the aggregation of membrane-bound synthetic receptors[13] and the binding of multivalent ligands at cell surfaces.[14] These studies, however, do not provide many quantitative details, mainly because models that describe multivalent binding at surfaces are still under development.[15]

Multivalent interactions are not only important in nature, but they are also crucial when one wants to develop stable assemblies at surfaces. Multivalent host-guest interactions allow for controllable adsorption and desorption by variation of the type and number of host-guest interactions.[3] By making use of host-guest interactions, it may become possible to build nanosized structures such as capsules in a controlled fashion on a molecular printboard. The same interactions can also be exploited for creating patterns of molecules on surfaces through supramolecular microcontact printing (μCP) or dip-pen nanolithography (DPN).

In this chapter we describe a model which allows the quantitative understanding of multivalent interactions at surfaces.[15,16] In this model, a β-CD SAM serves as a host surface, and a bis(adamantyl)-functionalized calix[4]arene serves as a multivalent guest. Furthermore, applications are shown that require kinetically and/or thermodynamically stable interactions. We show that it is possible to build a non-covalent capsule on a β-CD SAM making use of orthogonal host-guest and ionic interactions,[17] and that it is possible to pattern β-SAMs by microcontact printing (μCP) and dip-pen nanolithography (DPN).[7]

2. MULTIVALENT CYCLODEXTRIN HOST-GUEST INTERACTIONS IN SOLUTION AND AT THE MOLECULAR PRINTBOARD

Multivalent binding assuming independent binding sites, i.e. without cooperativity, can be described quantitatively in several ways. One is to analyze multivalent binding based on intrinsic binding constants and effective concentrations.[12,18,19] The intrinsic binding constant (K_i) describes the interaction of the monovalent recognition motif, while the effective concentration is used to differentiate between inter- and intramolecular binding steps, and represents the probability of interaction between two complementary interacting sites in an intramolecular binding event. Effective concentration[18-21] is conceptually similar to effective molarity,[20,22] which represents the ratio of association constants for intra- and intermolecular processes. The effective concentration symbolizes a "physically real" concentration of one of the interacting sites as experienced by its complementary counterpart in the probing volume determined by the covalent or non-covalent linker between them. The effective concentration is thus dependent on the linker length and the structure of the molecules. Such a model was used by Lees et al.[12] to describe the binding of multivalent inhibitors for Shiga-like toxins. Another way to describe multivalency is in entropy terms as proposed by Whitesides and coworkers.[9,11] When dealing with independent binding sites, the overall binding enthalpy is the sum of the binding enthalpies of the individual interactions. Although the basic assumptions of both treatments are the same, the correct interpretation of the entropy terms in multivalent binding is not trivial. Therefore, the approach based on intrinsic binding constants and effective concentration will be used here, because stability constants of the interactions between guest and host molecules in solution as well as at the molecular printboard can be easily obtained.

As a model system to understand multivalent binding at interfaces in a quantitative sense, we have compared the binding of calix[4]arene **4** to β-CD (**1**) and a β-CD dimer **3** in solution and to the β-CD printboard **2** (see Fig. 1).[16]

Thermodynamic data on the binding of **4** in solution were obtained from isothermal titration calorimetry (ITC). Figure 2 shows enthalpograms obtained for the binding of **4** to native β-CD (**1**) (Fig. 2a) and to the β -CD dimer **3** (Fig. 2b). When β-CD was titrated to **4**, the inflection point of the curve occurred at a [β-CD]/[**4**] ratio of 2, which indicates a 2:1 host-guest binding mode, *i.e.* two β-CDs are bound to **4**. The ITC curve was fitted to a model in which independent binding sites are assumed, and in which the intrinsic association constant, K_i, and the binding enthalpy, ΔH_i, of the monovalent interaction were used as independent fitting parameters. The intrinsic binding constant (4.6×10^4 M^{-1}) and the binding enthalpy (-7.0 kcal/mol) obtained after fitting are typical for a β-CD-adamantyl interaction (see Table 1). The quality of the fit indicates that the binding mode is indeed 2:1, and that the assumption of independent binding sites is valid.

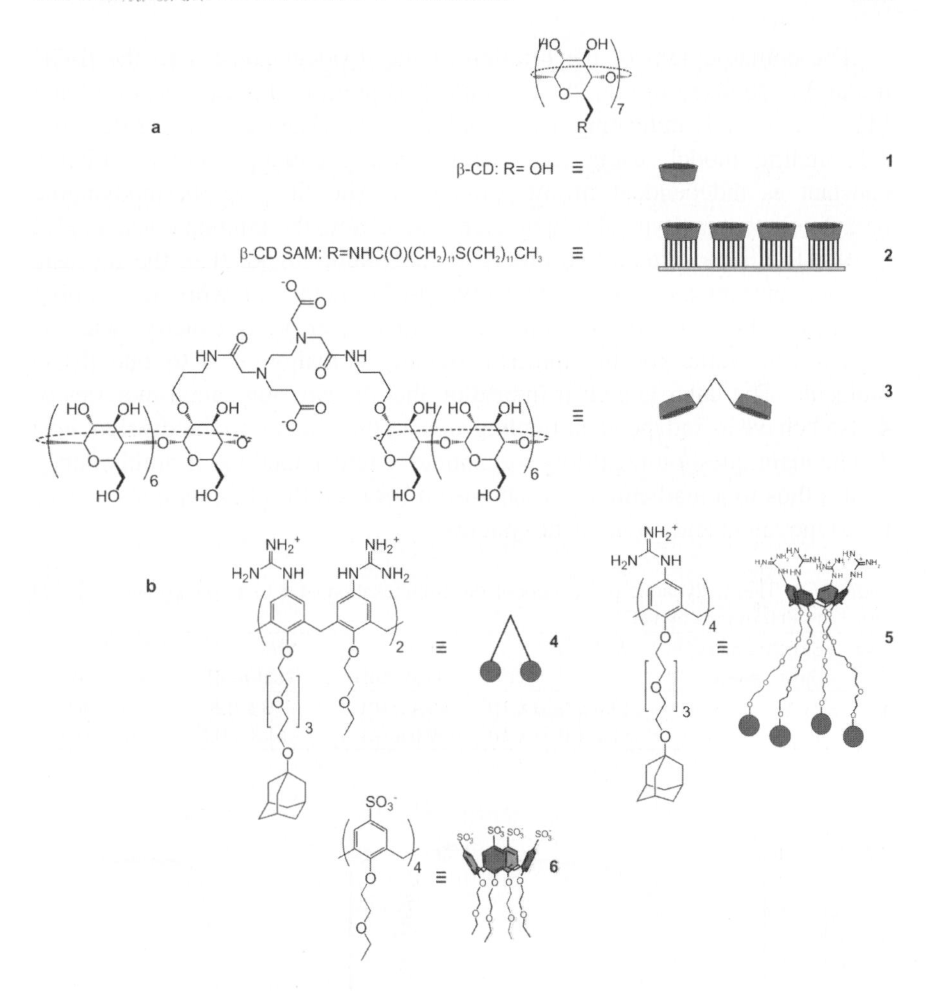

Figure 11-1. a) Host molecules: β-CD (**1**), the β-CD SAM **2**, and an EDTA-linked β-CD dimer **3**. b) Guest molecules employed here: a bis(adamantyl)-calix[4]arene **4**, a tetra-(adamantyl)-calix[4]arene **5**, and a tetrasulfonate-calix[4]arene **6**.

The enthalpogram of the titration of the divalent guest **4** to the β-CD dimer **3** is depicted in Fig. 2b. The inflection point in this case occurred at a [**4**]/[**3**] ratio of 1, indicating a 1:1 binding mode. The curve was fitted to a 1:1 binding model, using the overall binding enthalpy and association constant as independent fitting parameters. The fit gave thermodynamic parameters that are typical of a divalent interaction, the binding constant (1.2×10^7 M^{-1}) is more than two orders of magnitude higher than the intrinsic binding constant for a single adamantyl- β-CD interaction, while the binding enthalpy (–14.8 kcal/mol), within experimental error, is exactly twice as high as the value for the intrinsic binding enthalpy of **4** to one β-CD molecule. The latter is a clear indication that the two adamantyl moieties of **4** also behave as independent binding sites in the binding to the divalent host **3**. The main question regarding a quantitative understanding of multivalency relates thus to a mathematical relationship between the stability constants of the monovalent and multivalent systems.

Table 11-1. Thermodynamic parameters of the complexation of **4** to **1** (β-CD) and **3** (β-CD dimer), as determined by ITC.

host	Stoichiometry (host : guest)	K (M^{-1})	$\Delta G°$ (kcal/mol)	$\Delta H°$ (kcal/mol)	$T\Delta S°$ (kcal/mol)
1	2 : 1	(4.6 ± 0.3) x 10^4	–6.4 ± 0.1	–7.0 ± 0.5	–0.6 ± 0.6
3	1 : 1	(1.2 ± 0.3) x 10^7	–9.6 ± 0.1	–14.8 ± 0.5	–5.1 ± 0.6

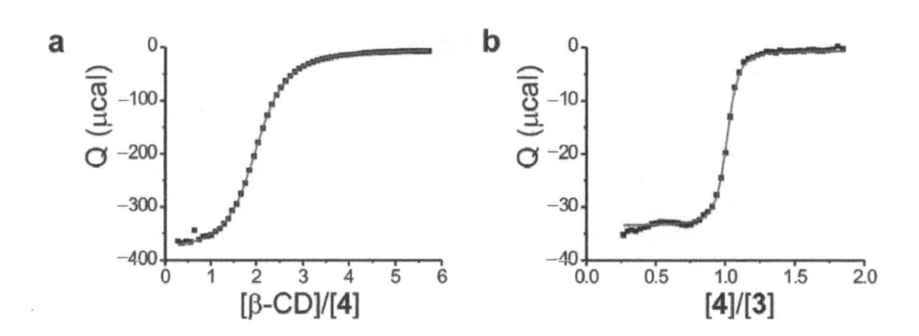

Figure 11-2. Heat evolved per injection plotted against the molar ratio (markers) and fits (solid lines) to 2:1 and 1:1 models, respectively, for the isothermal microcalorimetric titrations (25 °C) of **1** (10 mM) to **4** (0.4 mM) and of **4** (0.4 mM) to **3** (0.05 mM) in water.

Figure 3a schematically depicts the stepwise binding process of **4** to **3**. The first, intermolecular, interaction (K_1) is directly related to the intrinsic binding constant ($K_{i,l}$): $K_1 = 4K_{i,l}$, where the coefficient 4 is statistical. The subsequent intramolecular association step is the product of the intrinsic association rate constant and an effective concentration term, C_{eff}, which reflects the concentration of uncomplexed host (β-CD) that is experienced by the uncomplexed adamantyl moiety that is linked to this free host site via a non-covalent linker incorporating the host-guest pair formed in the first step. Since the second dissociation step is twice as likely as the intrinsic dissociation, the equilibrium constant of the second step is given by: $K_2 = ½K_{i,l}C_{eff}$.

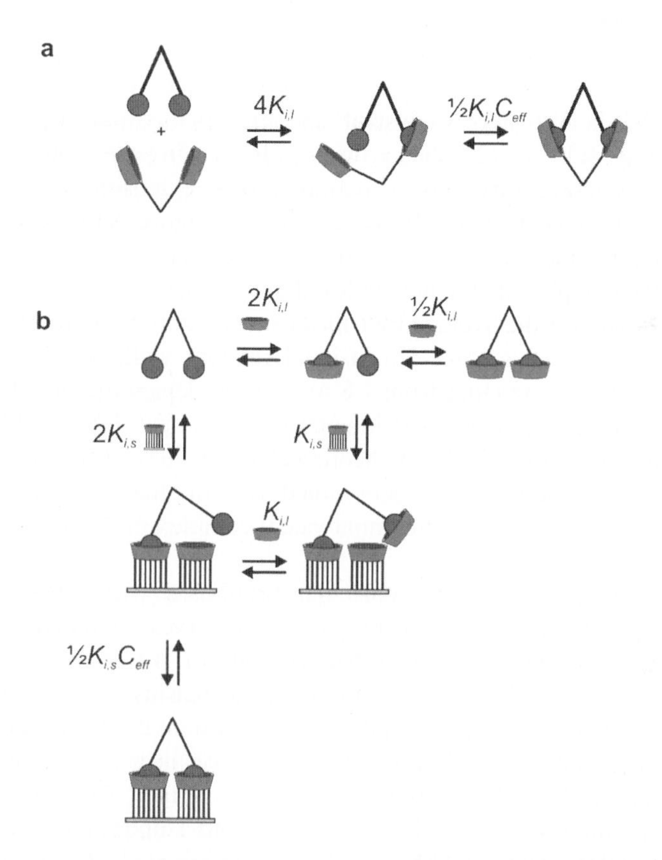

Figure 11-3. Equilibria for the sequential binding of **4** to **3** (a) and for the sequential binding of **4** to the molecular printboard in the presence of competing β-CD in solution (b).

Thus, assuming independent binding sites for the divalent interaction between **4** and **3**, the overall binding process can be described according to Eq. (1):

$$K = K_1 K_2 = 2K_i^2 C_{eff} \tag{1}$$

As described above, the empirical effective molarity, EM, here given by $EM = \frac{1}{2}K/K_i^2$ (Eq. 1), can now be calculated to be 2.8 ± 0.6 mM.

The effective concentration can be estimated theoretically using the formula for cyclization probability.[21,23] It can be conveniently approximated by Eq. 2:

$$C_{eff} = \frac{3}{4\pi < r_0 >^3 N_{AV}} \tag{2}$$

Here, N_{AV} is Avogadro's constant, and $<r_0>$ is the radius of the probing volume (Fig. 4a). Eq. (2), shows that C_{eff} has an inverse cubic relation to $<r_0>$ Furthermore, this approximation allows extension to multivalent binding at the molecular printboard (see below; Figure 9.4b). The radius of the probing volume is in this case defined by the average end-to-end distance between the complexed and uncomplexed guest moieties.[19,20]

An estimate for the average root-mean-square end-to-end distance, $<r_0>$, can be obtained using three-dimensional random walk statistics,[23] which gave values for C_{eff} ranging from 1.8 to 92 mM, depending on which chain stiffness is assumed. The experimentally determined *EM* of 2.8 mM is within this range of calculated theoretical C_{eff} values. The relatively low value for the experimentally determined *EM* implies that the rotational mobility of the linker within the monovalent complex of **4** and **3** (Fig. 4a) is limited.

The binding of **4** to the molecular printboard was studied by SPR spectroscopy. Figure 5 shows five titration curves obtained from SPR titration experiments at different backgrounds of β-CD in solution. The experiments consisted of the addition of an increasing amount of a 1 μM solution of **4** to the molecular printboard at a constant β-CD concentration in solution. The addition of **4** resulted in each case in a change of the SPR angle, which indicated adsorption of **4** to the surface. In strong contrast, similar experiments performed on an 11-mercapto-1-undecanol SAM did not show adsorption. These SAMs are also OH-terminated like the molecular printboards, but lack the β-CD cavities and thus serve as reference layers.

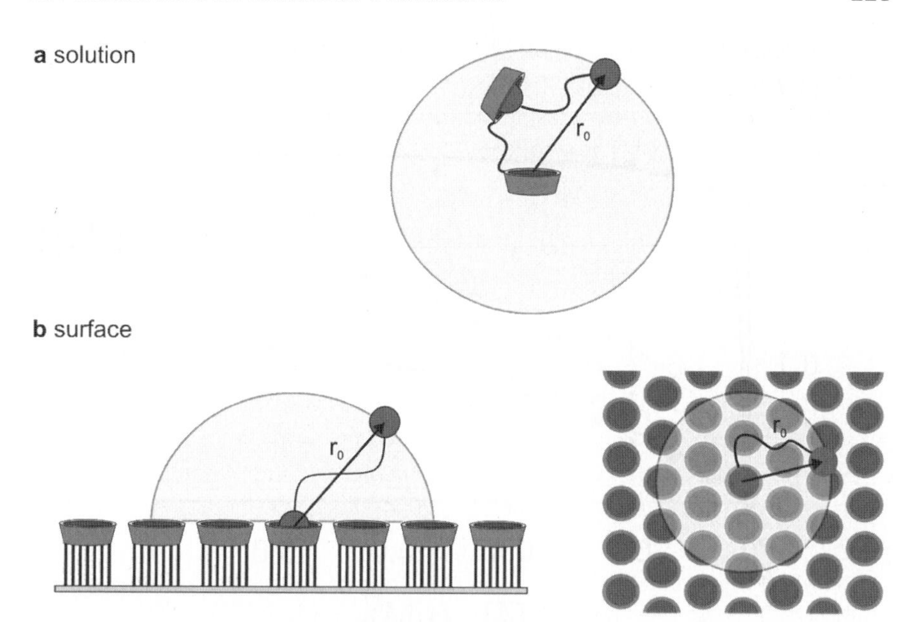

Figure 11-4. Schematic representation of the concept of effective concentration for the interactions between **4** and **3** (a) and of **4** to the molecular printboard (b).

After every addition, repeated rinsing with a competing solution, i.e. containing a high concentration of β-CD, resulted in desorption of **4** from the surface. The binding curve at 0.1 mM β-CD showed near-quantitative binding of **4** to the molecular printboard, prohibiting accurate determination of the binding constant. Reliable binding constants could only be obtained when using β-CD concentrations higher than 0.1 mM, to induce competition between binding of **4** to the molecular printboards vs. binding to β-CD in solution. From fitting the curves of Figure 5 to Langmuir isotherms, all complexation constants were about 10^{10} M^{-1}. This means that the binding constant at the surface is two to three orders of magnitude higher than the binding constant of **4** to **3** in solution.

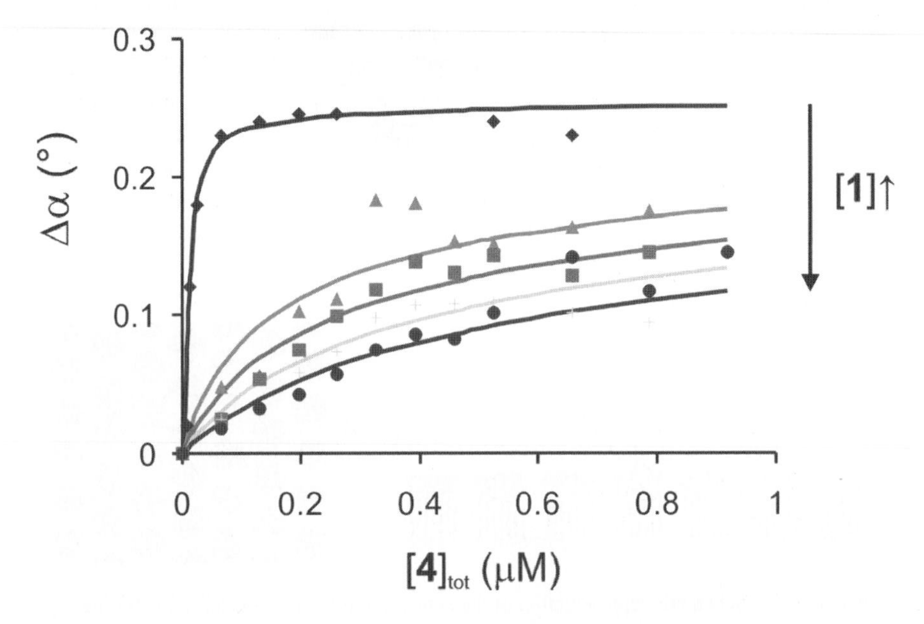

Figure 11-5. SPR angle data (markers) and corresponding fits to the sequential binding model (solid lines) for titrations of **4** to the molecular printboard at five different β-CD (**1**) background concentrations (♦ = 0.1 mM; ▲ = 0.5 mM; ■ = 1 mM; + = 2.5 mM; ● = 5 mM). Errors on the data points are approx. 0.02°. Arrows indicate a decrease of the slope upon an increasing concentration of β-CD (**1**).

This difference in binding affinities between solution and surface can be explained by the higher effective concentration at the surface. Again, the binding process is described by two sequential, independent binding events, which can take place in solution and/or at the surface (Fig. 3b). The solution species that are present are shown in the top row; from top to bottom, the equilibria for binding to the surface are shown. The binding events at the surface are considered to be equal and independent, just as in solution. This means that all binding events can be expressed in terms of intrinsic binding constants. The intrinsic binding constants in solution and at the surface are defined as $K_{i,l}$ (shown above to be 4.6 x 10^4 M^{-1}) and $K_{i,s}$, respectively.

The only intramolecular binding event at the surface, the formation of the bottom species in Fig. 3b, is again associated with an effective concentration term, similar to the second binding event for the sequential divalent binding of **4** in solution. Within this probing volume, several host molecules can be found to which the uncomplexed guest moiety may bind in an intramolecular fashion. Depending on the surface coverage, Θ, of the immobilized host cavities, however, not all hosts may be accessible.

The methodology used for the approximation of the maximal effective concentration $C_{eff,\max}$ *i.e.* the C_{eff} at low coverage, at the molecular printboard is similar to that used for the approximation of C_{eff} in solution. The difference is that at the surface the number of accessible host molecules is larger than 1, and that the linker length, $<r_0>$ results from the guest only, leading to a smaller probing volume. If the effective concentration concept is applied to the molecular printboard, Eq. 3 is obtained, in which A_{CD} is the surface area covered by a single β-CD host on the molecular printboard.

$$C_{eff} = \frac{3}{2A_{CD}N_{AV} < r_0 >}(1-\Theta) \tag{3}$$

In this case, $C_{eff,max}$ scales with $<r_0>^{-1}$ and therefore, the effective concentration at the surface is less dependent on $<r_0>$ than in solution (in solution C_{eff} scales with $<r_0>^{-3}$ see above). Consequently, the approximation of $C_{eff,max}$, (*i.e.* C_{eff} at $\Theta = 0$) based on a range of $<r_0>$ values, gives a relatively narrow range of $C_{eff,max}$ values. Analogous to the solution case, the lower limit of $C_{eff,max}$ of 0.20 M was chosen for fitting of the SPR titration curves.

The SPR curves were fitted to this model using a least-squares optimization routine and treating $K_{i,s}$ and $\Delta\alpha_{max}$ as variables, while using fixed values for $K_{i,l}$ and $C_{eff,max}$. The fitting of the SPR curves of Fig. 5 using this multivalency model gave $K_{i,s}$ values at different β-CD background concentrations of approx. 3×10^5 M^{-1}. This value is comparable to the K_i value of small monovalent adamantyl guest molecules at the molecular printboard. Thus the assumption of independent binding sites at the interface also holds for the divalent binding of **4**. Furthermore, it clearly shows that the multivalency model provides a deeper insight into the true nature of the observed binding enhancements, both relative to monovalent binding and relative to divalent binding in solution. The high effective concentration at the interface is the main cause for these binding enhancements, and no cooperativity needs to be assumed to explain the experimental observations.

3. CAPSULE FORMATION AT THE MOLECULAR PRINTBOARD

Multivalent interactions can provide such high binding constants that molecules can be positioned on molecular printboards in a both thermodynamically and kinetically stable fashion. This phenomenon can be applied in the localized assembly formation leading to patterns of host-guest complexes (see section 4 below) and in the formation of more complicated non-covalent structures at interfaces. Regarding the latter, we chose to build an ionic capsule at the printboard based on host-guest interactions.[17]

Much research has already been performed on the formation of non-covalent capsules in solution, *e.g.* for the encapsulation of drugs and the active transport or delivery of these drugs,[24] or for catalysis.[25] There are, however, only few cases in which capsules are self-assembled at a surface.[26-28] In those cases, the bottom part of the capsule was connected directly at the solid substrate via self-assembly of thiols on gold. Here we use orthogonal host-guest and ionic interactions to allow a stepwise build-up and break-down of the capsules at the molecular printboards (Fig. 6).

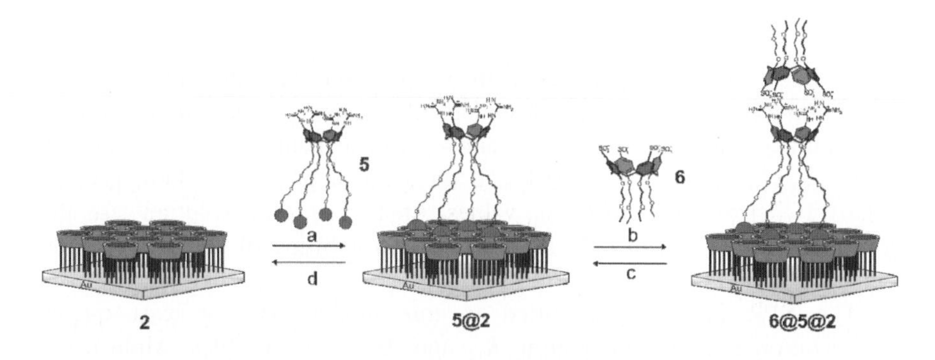

Figure 11-6. Schematic representation of the build-up (a,b) and subsequent break-down (c,d) of the capsule **6@5** at the molecular printboard.

The capsule consists of calix[4]arenes **5** and **6** (Fig. 1). The bottom part of the capsule is calix[4]arene **5**, the lower rim of which is modified with four oligo(ethyleneglycol) chains which each possess an adamantyl functionality, while the upper rim is modified with four guanidinium groups to increase water solubility. The top part of the capsule is the tetrasulfonate calix[4]arene **6**. The resulting capsule is based on the ionic interactions between the two oppositely charged upper rims of these calix[4]arenes.

ITC studies, in the presence of 10 mM β-CD to complex all adamantyl groups of **5**, showed that capsule formation in solution is an endothermic, entropy-driven process, with an association constant $K = (7.5 \pm 1.2) \times 10^5$ M^{-1}. The driving force for the capsule formation is the desolvation of the charged groups upon complex formation. The ITC enthalpogram indicates a 1:1 complex stoichiometry. Also, NMR experiments showed the formation of a well defined assembly.

For studying the surface assembly, the binding of **5** to the molecular printboard was studied first. In an SPR measurement, **5** was adsorbed at the molecular printboard at a β-CD background of 4 mM. The absolute increase of the signal, ~0.2° is very close to the absolute increase of the SPR signal when **4** was adsorbed, which indicates a surface coverage of **5** comparable to **4**. In contrast to the binding of the divalent **4** however, the binding of **5** to the molecular printboard proved to be irreversible, as it appeared impossible to remove **5** by extensive rinsing procedures in which competition was induced by using a high concentration of native β-CD in solution (8 mM). Similarly, rinsing with 1 M KCl did not result in the removal of **5**. This is explained by the multivalency model described in section 2; the association constant of **5** to **2** is expected to be in the order of ~10^{15} M^{-1}. In contrast, subsequent rinsing procedures with methanol, ethanol and 2-propanol did result in the removal of **5** from the surface, by weakening the intrinsic hydrophobic interactions between the β-CD cavities and the adamantyl functionalities. Thus we have shown that the lower halve of the capsule can be strongly immobilized at the molecular printboard in aqueous solutions, but that the application of organic solvents provides a way of removing it from the surface again by lowering of $K_{i,s}$ and thus the stability of the assembly as a whole.

SPR titrations of the addition of **6** to a monolayer of **5** on the molecular printboard were performed, and fitted to a Langmuir isotherm. The association constant was (3.5 ± 1.6) × 10^6 M^{-1}, which is slightly higher than the association constant found in solution (7.5 × 10^5 M^{-1}). This could be due to some form of positive cooperativity, resulting from stronger electrostatic interactions of the many calixarenes **5** at the surface. However, the slightly higher association constant found on the surface compared to solution cannot be due to the formation of a 1:2 complex, because in that case an 8+/4- ion pair is to be expected, which should give rise to an association constant of approx. 10^{12} M^{-1}. It is obvious that the association constant observed here is far lower than this value.

The capsule could be built up in two steps at the molecular printboard, and broken down again in two steps (Fig. 6: a→b→c→d). This assembly and disassembly process can clearly be followed by SPR spectroscopy (Fig. 7). First a monolayer of **5** was formed at the molecular printboard (Figs. 6 and 7, step a). Subsequently, **6** was attached through ionic interactions on top of the monolayer of **5** (step b). At this point, the capsule is present at the molecular printboard. The stepwise assembly of the capsule is followed by the stepwise disassembly of the capsule. First, a rinsing procedure with 1 M KCl was performed (step c), in which the top part of the capsule, **6**, was removed by weakening of the ionic interactions due to charge screening at this high salt concentration. As noted before, this rinsing step does not affect the binding of **5** at the molecular printboard. After restoring the 10^{-2} M KCl background solution, a rinsing procedure with 2-propanol was applied in order to remove the bottom part of the capsule (step d). Hereafter the molecular printboard appeared to be clean, since the whole procedure could be repeated without loss of efficiency (Fig. 7).

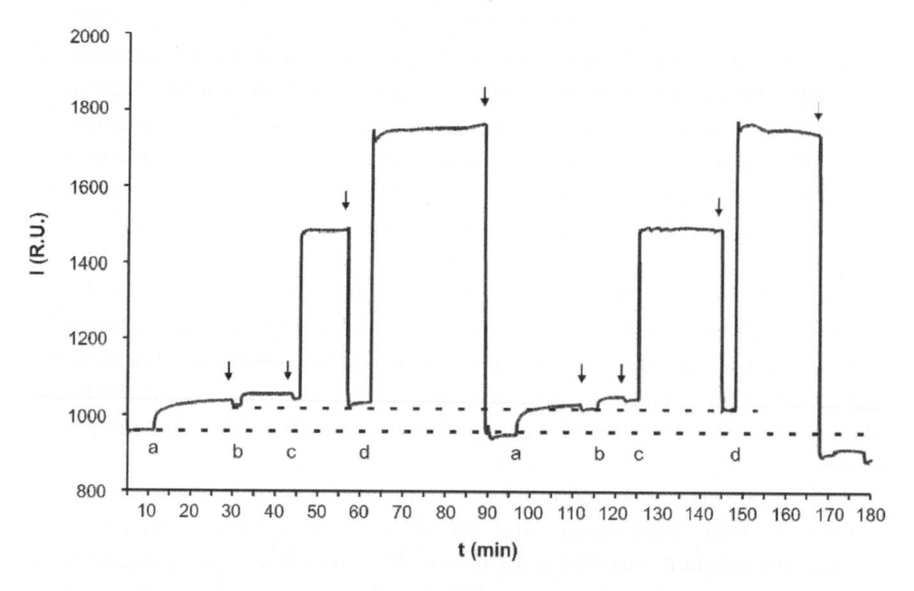

Figure 11-7. SPR sensogram showing the stepwise assembly and the subsequent stepwise disassembly of the molecular capsule **5@6** at the molecular printboard. The arrows (↓) indicate a background change to 10 mM aqueous KCl; *a* indicates adsorption of **5** (0.1 mM in 4.0 mM β-CD + 10 mM KCl); *b* indicates adsorption of **6** (0.1 mM in 4.0 mM β-CD + 10 mM KCl); *c* indicates desorption of **6** by 1 M KCl and *d* indicates desorption of **5** by 2-propanol.

In conclusion, these results show that multivalency can result in such strong binding that weaker, orthogonal interactions can be employed in subsequent steps to make more complex assemblies. Furthermore, it emphasizes the versatility of the molecular printboards as a building platform onto which assemblies can be constructed and removed again at will.

4. CREATING PATTERNS OF ASSEMBLIES AT THE MOLECULAR PRINTBOARD

As discussed above, thermodynamically and kinetically stable assemblies at molecular printboards can be created using the concept of multivalency. This bears the implication that also individual molecules or small groups can be firmly immobilized at predetermined positions. In this section, we discuss the creation of patterns based on supramolecular interactions at the molecular printboard using microcontact printing (μCP) and dip-pen nanolithography (DPN).[7]

μCP is parallel technique, developed by Whitesides *et al.* for the replication of relatively small patterns on surfaces, which is cost effective and fast.[29] Patterns are created by the transfer of molecules at the contact areas of a polymeric relief stamp with the substrate. Often poly(dimethyl siloxane) (PDMS) is used as a stamp material. In DPN, which is a serial technique, an AFM tip is inked, and the ink is transferred upon contact of the tip with the surface.[30,31] The patterns that are written can have a high resolution, and the transport of ink is controlled by a water meniscus between the tip and the sample.

Both techniques were applied for local transfer of **4** to the molecular printboard **2**. μCP of **4** yielded a pattern that was visible directly after printing (Fig. 8a), and remained visible after extensive rinsing with water (Fig. 8b). These results confirm the SPR studies described in section 2. Also extensive rinsing with 10 mM β-CD in solution did not completely remove the pattern, but led to loss of contrast due to partial removal of **4** induced by competition (Fig. 8c). The presence of **4** at the patterned areas of the substrate was also confirmed by secondary ion mass spectrometry (SIMS) (Fig. 8f). Analogously, **4** was printed onto a 11-mercapto-1-undecanol SAM. As can be seen in Figure 9.8d, **4** is transferred upon printing, but rinsing with water (Fig. 8e) was sufficient to remove **4** from the surface, attributable to the lack of host-guest interactions in this case.

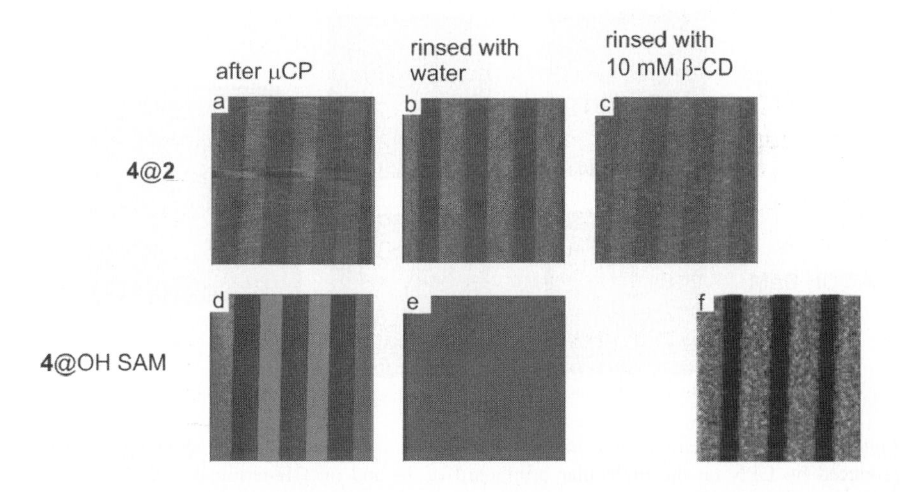

Figure 11-8. AFM friction force image (a-e; image size: 50 x 50 μm²) of patterns obtained by μCP of **4** (brighter areas) on the molecular printboard (a-c) and on OH-terminated SAMs (d,e) before rinsing (a,d), after rinsing with water (b,e), or after rinsing with 10 mM aqueous β-CD (e), respectively. TOF-SIMS image (f; image size: 56 x 56 μm²) of a β-CD SAM after μCP of **4**: the bright areas indicate the presence of the molecular ion peak of **4** at *m/z* = 1418.

In order to perform DPN nanolithography, silicon nitride AFM tips were dipped in aqueous solutions of **4**, and subsequently scanned across the molecular printboard, as well as on the OH-terminated reference SAMs. In both cases patterns were transferred (Figs. 9a and 9c). However, after rinsing with aqueous solutions, patterns were only visible at the molecular printboard, again indicating the need of multivalent host-guest interactions for pattern stability. To confirm that these patterns are not due to the mechanical force that was applied by the AFM tip to the surface, a reference experiment was performed in which a bare silicon nitride tip was used to write on the molecular printboard under the same conditions. These experiments did not show a visible pattern. The resolution that can be obtained by this method is below 100 nm, as shown by the line patterns in Fig. 9e. The line width is also in this case a function of contact time, tip radius, ink concentration, and ink-transfer mechanism.

Supramolecular multivalent interactions between the molecular printboard and a divalent guest can thus be employed in the patterning of surfaces. Stable, reversible, features are obtained upon printing and writing. The advantage of the use of supramolecular, multivalent interactions is the tunability and the reversibility of the system.

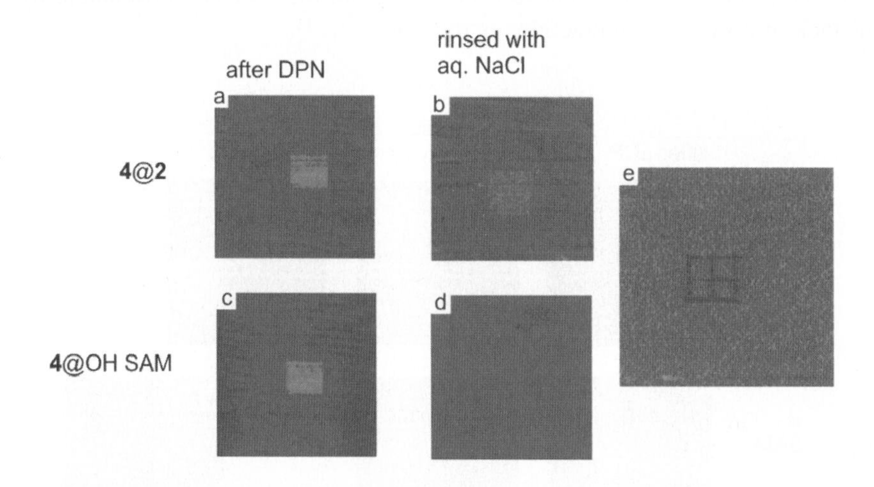

Figure 11-9. AFM friction force images (a-d; image size: 30 x 30 μm^2) showing patterns produced by DPN on the molecular printboard (a, b) and on OH-terminated SAMs (c, d) using **4** as the ink before rinsing (a, c), after rinsing in situ with with 50 mM aqueous NaCl (b, d), respectively. AFM friction force image (e; 4 x 4 μm^2; in air) showing an array of lines with mean widths of 60 ± 20 nm, produced by DPN of **4** at the molecular printboard.

5. CONCLUSIONS

Calix[4]arenes are versatile building blocks because of the possibility to modify the lower rim with multiple guest moieties in order to obtain multivalent systems. In section 2, a model is shown that describes multivalent binding processes in solution and at the surface. The binding of the bis(adamantyl)-calix[4]arene **4** to the molecular printboard has a binding affinity that is two to three orders of magnitude higher than the binding affinity for the divalent interaction between **4** and β-CD dimer **3** in solution. The model shows that this binding enhancement is due to a higher effective concentration at the surface.

Multivalent host-guest interactions can be employed to build larger noncovalent structures at surfaces. It was shown in section 3 that orthogonal interactions can be employed to create capsules at a surface. The bottom half of the capsule is a calix[4]arene modified at the lower rim with four guest moieties and at the upper rim with four guanidinium groups. This calix[4]arene is very strongly bound at the surface through four adamantyl-β-CD interactions. The top part of the capsule is a calix[4]arene modified at the upper rim with four sulfonate groups to enable capsule formation via ionic interactions with the guanidinium groups of the bottom half of the capsule. This capsule can be built up and broken down in subsequent steps at the molecular printboard.

Furthermore, it was shown in section 4 that supramolecular μCP and DPN can be used to create patterns based on multivalent host-guest interactions at a surface. These experiments showed that stable supramolecular features can be obtained upon printing, and that narrow line widths can be obtained.

Multivalent host-guest interactions as described in this chapter, can be used for the development of new nanofabrication schemes. All interactions can be tuned by changing the number and type of guest molecules, which makes the system very versatile, allowing the fabrication of 3D structures[32] as well as structures with interesting electrochemical properties.[4] Furthermore, the controlled attachment of biomolecules at the molecular printboard can be envisioned.

ACKNOWLEDGEMENTS

This work was financially supported by the Council for the Chemical Sciences of The Netherlands Organization for Scientific Research (CW-NWO; M.J.W.L.: Vidi Vernieuwingsimpuls Grant 700.52.423 to J.H.). Dr. Andrea Sartori, Prof. Allesandro Casnati, and Prof. Rocco Ungaro are gratefully acknowledged for the synthesis of calix[4]arenes **4** and **5**.

REFERENCES

1. Beulen, M. W. J., Bügler, J., Lammerink, B., Geurts, F. A. J., Biemond, E. M. E. F., Van Leerdam, K. G. C., Van Veggel, F. C. J. M., Engbersen, J. F. J., Reinhoudt, D. N., *Langmuir* **14**(22), 6424-6429 (1998).
2. Beulen, M. W. J., Bügler, J., De Jong, M. R., Lammerink, B., Huskens, J., Schönherr, H., Vancso, G. J., Boukamp, B. A., Wieder, H., Offenhäuser, A., Knoll, W., Van Veggel, F. C. J. M., Reinhoudt, D. N., *Chem. Eur. J.* **6**(7), 1176-1183 (2000).
3. Huskens, J., Deij, M. A., Reinhoudt, D. N., *Angew. Chem. Int. Ed.* **41**(23), 4467-4471 (2002).
4. Nijhuis, C. A., Huskens, J., Reinhoudt, D. N., *J. Am. Chem. Soc.* **126**(39), 12266-12267 (2004).
5. Rekharsky, M. V., Inoue, Y., *Chem. Rev.* **98**(5), 1875-1917 (1998).
6. Schönherr, H., Beulen, M. W. J., Bügler, J., Huskens, J., Van Veggel, F. C. J. M., Reinhoudt, D. N., Vancso, G. J., *J. Am. Chem. Soc.* **122**(20), 4963-4967 (2000).
7. Auletta, T., Dordi, B., Mulder, A., Sartori, A., Onclin, S., Bruinink, C. M., Péter, M., Nijhuis, C. A., Beijleveld, H., Schönherr, H., Vancso, G. J., Casnati, A., Ungaro, R., Ravoo, B. J., Huskens, J., Reinhoudt, D. N., *Angew. Chem. Int. Ed.* **43**(3), 369-373 (2004).
8. De Jong, M. R., Huskens, J., Reinhoudt, D. N., *Chem. Eur. J.* **7**(19), 4164-4170 (2001).
9. Mammen, M., Choi, S. K., Whitesides, G. M., *Angew. Chem. Int. Ed.* **37**(20), 2754-2794 (1998).
10. Rao, J. H., Lahiri, J., Isaacs, L., Weis, R. M., Whitesides, G. M., *Science* **280**(5364), 708-711 (1998).
11. Rao, J. H., Lahiri, J., Weis, R. M., Whitesides, G. M., *J. Am. Chem. Soc.* **122**(12), 2698-2710 (2000).
12. Gargano, J. M., Ngo, T., Kim, J. Y., Acheson, D. W. K., Lees, W. J., *J. Am. Chem. Soc.* **123**(51), 12909-12910 (2001).
13. Gestwicki, J. E., Cairo, C. W., Strong, L. E., Oetjen, K. A., Kiessling, L. L., J. *Am. Chem. Soc.* **124**(50), 14922-14933 (2002).
14. Gestwicki, J. E., Kiessling, L. L., *Nature* **415**(6867), 81-84 (2002).
15. Huskens, J., Mulder, A., Auletta, T., Nijhuis, C. A., Ludden, M. J. W., Reinhoudt, D. N., *J. Am. Chem. Soc.* **126**(21), 6784-6797 (2004).
16. Mulder, A., Auletta, T., Sartori, A., Del Ciotto, S., Casnati, A., Ungaro, R., Huskens, J., Reinhoudt, D. N., *J. Am. Chem. Soc.* **126**(21), 6627-6636 (2004)
17. Corbellini, F., Mulder, A., Sartori, A., Ludden, M. J. W., Casnati, A., Ungaro, R., Huskens, J., Crego-Calama, M., Reinhoudt, D. N., *J. Am. Chem. Soc.* **126**(51), 17050-17058 (2004).
18. Kitov, P. I., Shimizu, H., Homans, S. W., Bundle, D. R., *J. Am. Chem. Soc.* **125**(11), 3284-3294 (2003).
19. Kramer, R. H., Karpen, J. W., *Nature* **395**(6703), 710-713 (1998).
20. Mandolini, L., *Adv. Phys. Org. Chem.* **22**, 1-111 (1986).
21. Winnik, M. A., *Chem. Rev.* **81**(5), 491-524 (1981).
22. Ercolani, G., *J. Phys. Chem. B* **107**(21), 5052-5057 (2003).
23. Jacobson, H., Stockmayer, W. H., *J. Chem. Phys.* **18**(12), 1600-1606 (1950).
24. Kryschenko, Y. K., Seidel, S. R., Muddiman, D. C., Nepomuceno, A. I., Stang, P. J., *J. Am. Chem. Soc.* **125**(32), 9647-9652 (2003).
25. Warmuth, R., *J. Inclusion Phenom. Macrocyclic Chem.* **37**(1-4), 1-38 (2000).
26. Huisman, B. H., Rudkevich, D. M., Van Veggel, F. C. J. M., Reinhoudt, D. N., *J. Am. Chem. Soc.* **118**(14), 3523-3524 (1996).

27. Huisman, B. H., Rudkevich, D. M., Farrán, A., Verboom, W., Van Veggel, F. C. J. M., Reinhoudt, D. N., *Eur. J. Org. Chem.* **2**, 269-274, (2000).
28. Levi, S. A., Guatteri, P., Van Veggel, F. C. J. M., Vancso, G. J., Dalcanale, E., Reinhoudt, D. N., *Angew. Chem. Int. Ed.* **40**(10), 1892-1896 (2001).
29. Xia, Y., Whitesides, G. M., *Angew. Chem. Int. Ed.* **37**(5), 550-575 (1998).
30. Mirkin, C. A., Hong, S. G., Demers, L., *ChemPhysChem* **2**(1), 37-39 (2001).
31. Piner, R. D., Zhu, J., Xu, F., Hong, S., Mirkin, C. A., *Science* **283**(5402), 661-663 (1999).
32. Crespo-Biel, O., Dordi, B., Reinhoudt, D. N., Huskens, J., *J. Am. Chem. Soc.* **127**(20), 7594-7600 (2005).

Chapter 12

PEPTIDO- AND GLYCOCALIXARENES

Supramolecular and biomimetic properties

Laura Baldini, Alessandro Casnati, Francesco Sansone, and Rocco Ungaro
Dipartimento di Chimica Organica e Industriale, Università di Parma, Parco Area delle Scienze 17/a, 43100 Parma, Italy, and Consorzio INSTM, Via Giusti 9, 50121 Firenze, Italy

Abstract: Calixarenes adorned with peptide and carbohydrate units are called peptido- and glycocalixarenes, respectively. Peptidocalixarenes have been used as biomimetic receptors for the complexation of cations, anions, carbohydrates, oligopeptides and for protein surface recognition. Some of them are biologically active as antibiotics, enzyme inhibitors and even anticancer agents. Cleft-like, *C*-linked peptidocalix[4]arenes self-assemble in the solid state to give nanotubes, whereas *N,C*-linked derivatives give chiral dimeric capsules in apolar media, through the formation of β-sheets. Glycocalixarenes bind, in a multivalent fashion, to specific lectins and self-aggregate in water or in the presence of phosphate anions to give glycocluster nanoparticles (GNPs) able to function as cell-specific molecular delivery systems or as neutral cell transfectants. A divalent calix[4]arene ligand, containing a synthetic o-GM1 mimic as binding unit, shows a remarkable affinity for cholera toxin (CT), thus acting as efficient CT inhibitor.

Key words: Calixarenes, carbohydrates, peptides, capsules, multivalency, nanotubes, nanoparticles, cholera toxin.

1. INTRODUCTION

The name of calixarenes, given by Gutsche[1] to the macrocycles derived from the condensation of phenols and formaldehyde, clearly alludes to the main feature of these compounds, namely the presence of a molecular cavity, attractive for the inclusion of apolar guests. This concept was strengthened when we reported the X-ray crystal structure of the 1:1 inclusion complex of p-*tert*-butylcalix[4]arene and toluene[2] which is probably the first example of a neutral molecule encapsulated in the apolar, intramolecular cavity of a neutral synthetic receptor. Since then, the study of

J. Vicens and J. Harrowfield (eds.), Calixarenes in the Nanoworld, 233–257.

the molecular recognition of small organic molecules by calixarenes has flourished,[3] leading to interesting applications in gas sensors.[4]

About ten years ago, we decided to investigate the functionalisation of the *upper rim*[5] of calix[4]arenes with sugar[6] or amino acid[7,8] units and we synthesised the first glyco- (general formula **1** in Fig. 1) and peptidocalixarenes[9] (general formula **2** in Fig. 1). These hybrid receptors are characterised by the presence of polar, hydrogen bonding moieties in close proximity to the calixarene apolar cavity, which, in principle, should shift the molecular recognition properties of native calixarenes towards more polar neutral guests and ions. Moreover, they could also interact, in a multivalent fashion,[10] with biological targets or self-assemble through hydrogen bonding, giving novel supramolecular architectures and enlarging the scope of calixarenes in Supramolecular Chemistry.

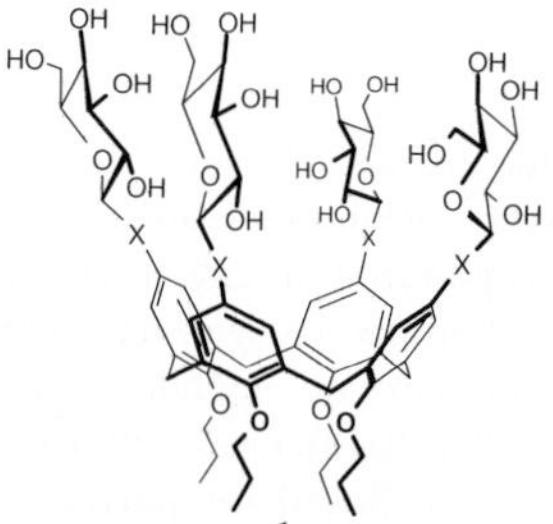

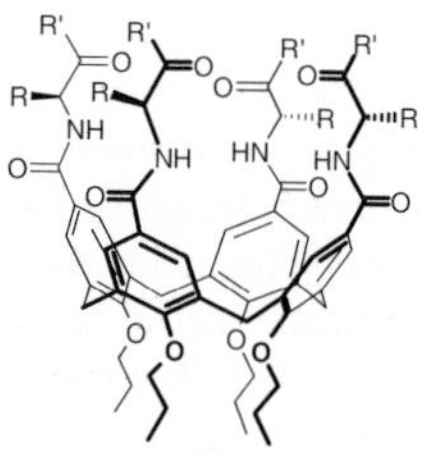

Figure 12-1. Upper rim tetrafunctionalised glyco- (**1**) and peptidocalix[4]arenes (**2**).

The present chapter will focus on the supramolecular properties of these novel synthetic receptors, highlighting their potential applications in biomimetic chemistry and nanoscience.

2. PEPTIDOCALIXARENES

α-Amino acids or peptides can be linked to the calixarene scaffold through the terminal amino or carboxylic acid group, leading to *N*-linked or *C*-linked peptidocalixarenes, respectively. Moreover, conjugation can occur at the *upper* or at the *lower rim* of the calixarenes. All these synthetic possibilities have been explored and some representative examples are reported in Fig. 2.

Figure 12-2. Examples of *N*-linked (**3**[7] and **5**[11]) and *C*-linked (**4**[12] and **6**[13]) peptidocalix[4]arenas.

2.1 *Upper rim* Peptidocalixarenes as Molecular Receptors

The *upper rim*, cleft-like peptidocalix[4]arene podands in the *cone* conformation, both di- (*e.g.* **3** and **4**) and tetrafunctionalised (*e.g.* **7** and **8**) are, in general, poor receptors, since they experience a residual conformational mobility between two flattened C_{2v} structures or undergo intermolecular self-association. The *C*-linked compounds have a tendency to form weak complexes with anionic species,[12] whereas the *N*-linked analogues are able to complex primary ammonium cations, showing shape selectivity.[7]

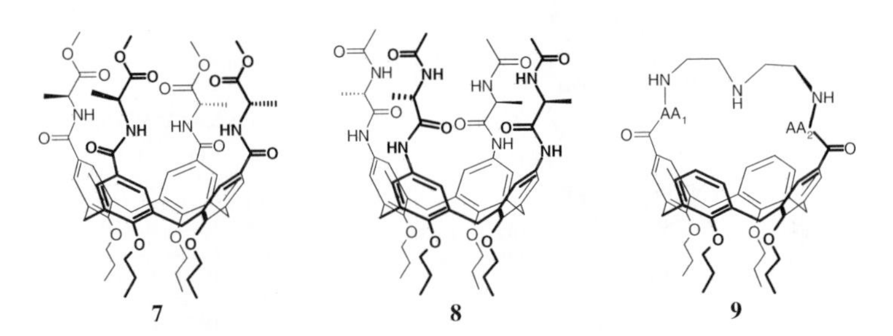

In order to increase the molecular recognition properties of peptidocalix[4]arenes, we rigidified their structure by synthesising a series of *upper rim* bridged peptidocalix[4]arenes, either *N*-linked (*e.g.* **9** and **10a,b**), or *C*-linked (**12a,b**).

The members of a small library of *N*-linked compounds having formula **9** were designed to act as vancomycin mimics.[8,9,14] Vancomycin is a glycolpeptide antibiotic, which performs its biological task by binding to the terminal -L-Lys-D-Ala-D-Ala sequence of the cell wall mucopeptide precursors of Gram-positive bacteria, thus inhibiting the growth of the cell wall and causing the cell lysis.[15] The model compounds **9** possess a pseudopeptide bridge in 1,3-position at the *upper rim* of a *cone* calix[4] arene, consisting of two α-amino acids AA_1 and AA_2 of different structure and configuration linked through a 1,3,5-diethylenetriamine spacer. The amino acid units should interact with the peptide substrate by hydrogen bonding, while the hydrophobic cavity of the calix[4]arene could take part in the recognition process and stabilise the complex through hydrophobic interactions. As a matter of fact, some members of the library have biological activity very similar to vancomycin: they show an anti-Gram-positive activity from moderate to good, while no activity was observed toward Gram-negative bacteria (*Escherichia coli*), yeast (*Saccharomyces cerevisiae*), or cell-wall-lacking bacteria (*Acholeplasma laidlawii*). NMR diffusion experiments performed in $CDCl_3$ + 3% DMSO-d_6 showed that the calixarene macrocycle **9** (AA_1= AA_2= L-Ala) binds the *N*-Ac-Ala-Ala dipeptide more strongly ($\mathrm{Log}K_{ass}$ = 3.4) than the simple amino acid derivative *N*-Ac-Ala ($\mathrm{Log}K_{ass}$ = 2.4).[14] Evidence was collected that the carboxylic group of the guest transfers its proton to the amino group of the host and the supramolecular complex is stabilised by the electrostatic interaction between the charged ammonium and carboxylate groups, and by hydrogen bonding between the pseudopeptide bridge and the guest (Fig. 3a).[8]

A similar general strategy was used in the design of macrobicyclic, *N*-linked peptidocalix[4]arene receptors **10a,b** for carbohydrate recognition.[16]

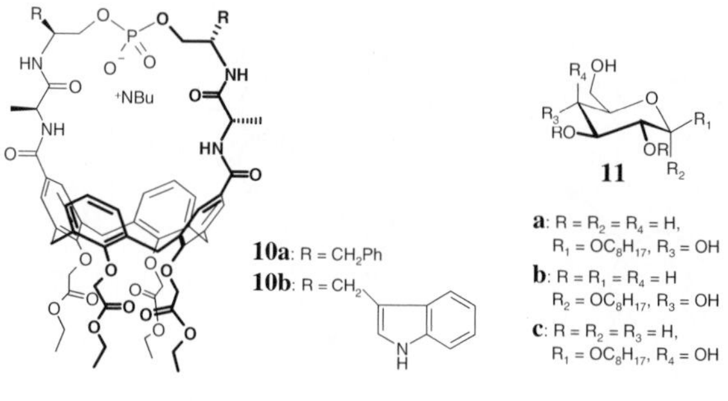

10a: R = CH_2Ph
10b: R = CH_2-(indol-3-yl)

11

a: R = R_2 = R_4 = H, R_1 = OC_8H_{17}, R_3 = OH
b: R = R_1 = R_4 = H R_2 = OC_8H_{17}, R_3 = OH
c: R = R_2 = R_3 = H, R_1 = OC_8H_{17}, R_4 = OH

Carbohydrates are known to interact with phosphate anions and several receptors incorporating phosphate or phosphonate groups in their structure have been proposed for carbohydrate recognition.[17,18] In general, it has also been pointed out that sugar binding can be strengthened through hydrogen bonding and other weak noncovalent interactions.[19,20] Several of these potential binding sites were incorporated in the "synthetic lectin" prototypes **10a,b**: i) hydrogen bonding donor and acceptor sites are provided by the amino acid units of the bridge; ii) a charged phosphate group is present in the middle of the pseudopeptide loop; iii) an apolar cavity is offered by the calix[4]arene in the *cone* conformation. In addition, receptors **10a,b** possess four ester groups at the *lower rim*, which can be used as connecting units to solid surfaces or can be hydrolysed to provide water soluble receptors. Carbohydrate recognition studies by ^{1}H NMR titration experiments were performed in organic media using the Phe derivative **10a** as host and the lipophilic sugar derivatives **11a-c** as guests.[16] The stability constant values in $CDCl_3$ point to a good selectivity of this receptor for the β-anomer **11a** of octylglucoside with respect both to its α-anomer **11b** and to the β-galactoside derivative **11c**. The association constant drops dramatically if the anionic phosphate group in receptor **10a** is transformed into its corresponding acid or methyl ester. The experimental data collected indicate that the phosphate anionic centre is the most important binding site and that additional hydrogen bonds and steric effects determine the observed preference for β-octylglucoside **11a**, as depicted in Fig. 3b.

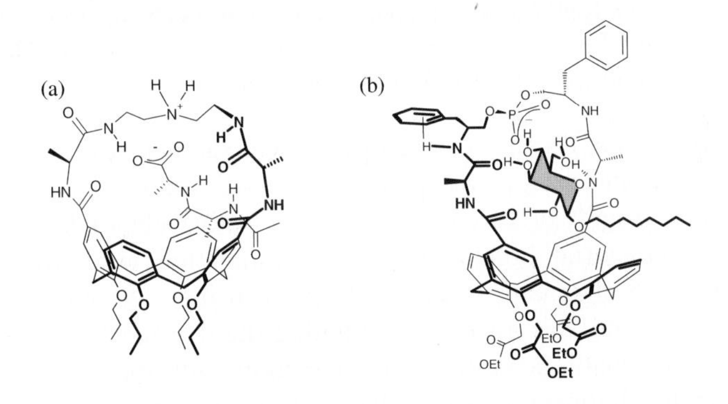

Figure 12-3. Proposed mode of binding of (a) *N*-acetyl-D-Ala-D-Ala by receptor **9** and (b) of β-octylglucoside **11a** by receptor **10a.**

Bridged *C*-linked peptidocalix[4]arenes (**12a,b**), containing an aromatic ring in the pseudopeptide loop were also synthesised.[21] The X-ray crystal structure of the pyrido derivative **12b** (Fig. 4b) shows that two molecules of acetone (used as solvent) are coordinated to the host through three hydrogen bonds, one with a calixarene NH and two with the Ala NH groups, thus suggesting that these compounds could be attractive hosts for the binding of polar organic molecules and anions. Solution studies performed in acetone–d_6 indicate that **12a,b** strongly bind carboxylate anions, showing association constants ($2.6 < \mathrm{Log}K_{ass} < 4.6$) two orders of magnitude larger than the cleft-like *C*-linked peptidocalix[4]arenes. The data obtained reveal that both ligands preferentially bind carboxylate anions having aromatic groups in their structure and that the amide NH groups of the host are involved in the complexation. We hypothesised that the anionic guest (*e.g.* benzoate) replaces the solvent molecule linked to the calixarene NH groups and that, beside the $COO^{-}\cdots HN$ hydrogen bonding interaction, additional stabilisation of the complexes results from π–π stacking between the host and guest aromatic nuclei.

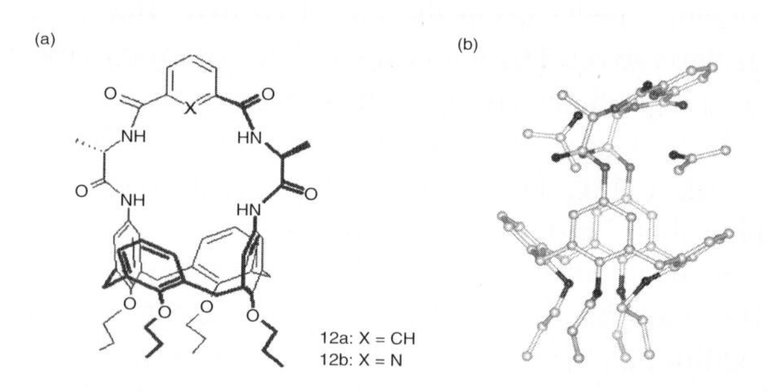

Figure 12-4. (a) Bridged *C*-linked peptidocalix[4]arenes **12a,b** and (b) ball-and-stick representation of the X-ray crystal structure of the complex of macrocycle **12b** with two acetone molecules (hydrogen atoms are omitted for clarity).

Other *upper rim*, *C*- and *N*-linked peptidocalixarenes have been reported in the literature and used as receptors for fluoride anions,[22] as catalysts for ester bond cleavage[23] or as carriers in ion selective electrodes.[24] A 3375-member combinatorial library (Fig. 5) of difunctionalised, fluorescence-labelled peptidocalixarenes was synthesised using 15 amino acids as building blocks and was assayed for binding the dye-labelled oligopeptide **13**, a Leu5 enkephalin derivative.[25] Interestingly, only four members of the library bind the oligopeptide, and all have a L-Tyr as the second amino acid.

Figure 12-5. Combinatorial library of fluorescent peptidocalixarenes.

Peptidoresorc[4]arenes were synthesised by linking amino acid moieties to a resorcarene scaffold.[26,27,28] Some of the resulting receptors were chemically bonded to a polymeric matrix to develop a new chiral stationary phase in capillary gas chromatography.[26,27]

2.2 *Upper rim* Peptidocalixarenes in Protein Surface Recognition

A family of calix[4]arenes functionalised at the *upper rim* with four cyclic peptides (**14a-d**, Fig. 6) has been designed by the group of Hamilton as synthetic agents with the challenging task of targeting protein surfaces to eventually disrupt clinically important protein-protein interactions.[29] The arrangement of the peptide loops around a hydrophobic cavity is based on the observation that a common feature of many natural protein-protein interfaces is, on each partner, a hydrophobic region surrounded by a ring of polar, charged and hydrogen bonding residues. The hydrophobic interactions between the binding partners are responsible for the affinity, while the electrostatic interactions provide the selectivity. The peptide loops are based on a cyclic hexapeptide in which two residues are replaced by the dipeptide analogue 3-aminomethylbenzoyl which contains a 5-amino substituent for linkage to the calixarene scaffold. Changing the amino acids $AA_1 - AA_4$ of the cyclic peptide results in a modification of the recognition characteristics of the receptors.[30]

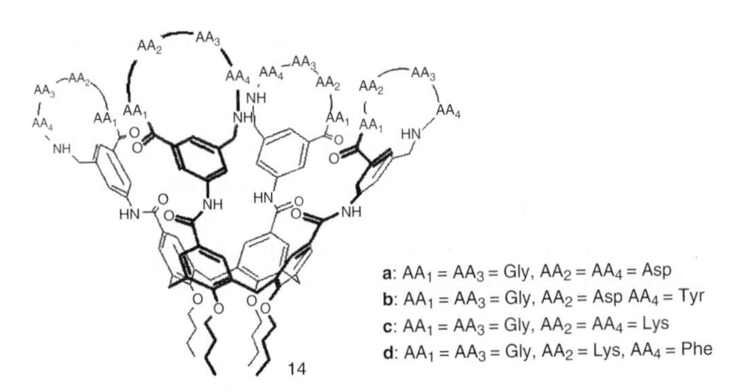

Figure 12-6. Hamilton's peptidocalix[4]arenes for protein surface recognition.

The effectiveness of this approach was confirmed by the successful binding of different protein targets. Compound **14a**, which has the anionic Gly-Asp-Gly-Asp peptide loops, strongly interacts with the surface of cytochrome c near the heme edge, the same region where the protein interacts with its natural partners (cytochrome c peroxidase and cytochrome oxidase).[31,32] This region contains a large hydrophobic area, surrounded by positively charged Lys residues, which is complementarily matched by the hydrophobic and anionic features of the calixarene ligand.

The Gly-Asp-Gly-Asp receptor **14a** was also identified from a pool of receptors having different anionic, hydrophobic and cationic amino acid residues to be a potent slow binding inhibitor (with submicromolar activity) of chymotrypsin, an important serine protease, presumably by binding to a patch of several cationic residues which is found near the active site cleft of the enzyme.[33]

Some members of this family of peptidocalixarenes have shown anti-cancer and antiangiogenic activity both *in vitro* and *in vivo* by targeting the surface of certain growth factors involved in tumour growth and angiogenesis. In fact, the binding of the synthetic ligand to the growth factor prevents the interaction of the protein with its membrane bound receptor. This inhibits the cell signalling pathways which are triggered by the growth factor-receptor interaction and eventually lead to cell growth and angiogenesis. In particular, **14b** (peptide sequence Gly-Asp-Gly-Tyr) is able to bind (the dissociation constant, K_d, is in the nanomolar range) the platelet-derived growth factor, whose receptor-binding region is composed of cationic and hydrophobic residues. *In vivo* studies showed that the treatment of nude mice bearing human tumours with **14b** resulted in a significant inhibition of the tumour growth and angiogenesis.[34,35] In a similar way, compound **14c**, which has the amino acid sequence Gly-Lys-Gly-Lys, selectively disrupts the binding of the vascular endothelial growth factor to its receptor and inhibits angiogenesis both *in vitro* and *in vivo*, and tumorigenesis and metastasis *in vivo*.[36]

Taking inspiration from Hamilton's work, Cunsolo *et al.* synthesised a *C*-linked octalysine calix[8]arene which binds the surface of tryptase near its active site[37] and Neri and co-workers prepared a small library of calix[4] arenes functionalised at the *upper rim* with tetrapeptides which were evaluated as transglutaminase inhibitors.[38]

2.3 *Lower rim* Peptidocalixarenes in Molecular Recognition

When the amino acids are attached to the *lower rim*, the calixarene acts as a scaffold to organise the chiral groups in space and no important role in guest binding is played by the apolar cavity. Nevertheless some useful systems have been obtained.

Calix[4]arenes functionalised with *N*-linked amino acid esters (**15**) were synthesised as receptors for alkali metal ions and Ag^+.[11,39,40] This family of compounds is characterised by the presence of a circular array of intramolecular hydrogen bonds between the amide groups. In presence of metal ions, the CO groups undergo a change in orientation and converge on the cation, which is accommodated in the polar niche formed by the four ether and the four carbonyl oxygens (Fig. 7). Due to this competition between the cation and the amide NHs for the carbonyl oxygen sites, the metal binding ability of these receptors is diminished with respect to the analogous calixarenes which do not form intramolecular hydrogen bonds, such as a simple tetraethyl ester[39] or the calix[4]arene *N,N*-diethylacetamide.[40,41]

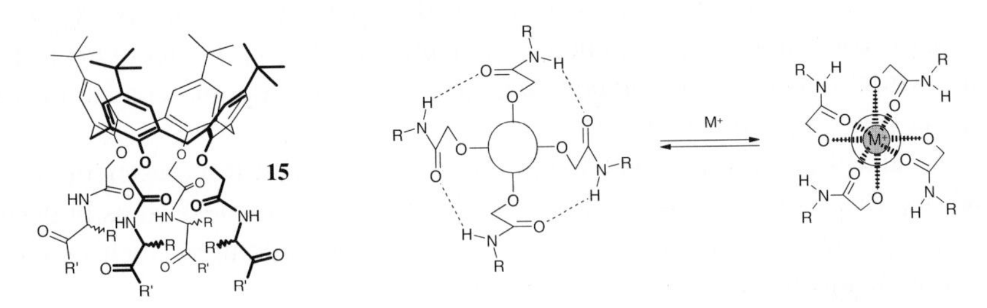

Figure 12-7. Interplay between hydrogen bonding and cation complexation in *lower rim N*-linked peptidocalix[4]arenas.

Analogous compounds functionalised with ValOBn (**15**, R = *i*-propyl, R' = OBn), Val-GlyOBn (**15**, R = *i*-propyl, R' = GlyOBn) and Val-Gly-TyrOBn (**15**, R = *i*-propyl, R' = Gly-TyrOBn) were synthesised by Lippard *et al.*, and studied as structural mimics of the selectivity filter of the K^+ channel from *Streptomyces lividans* (KcsA), which is formed by a 4-fold symmetrical arrangement of Thr-Val-Gly-Tyr-Gly polypeptides.[11] The authors were interested in getting evidences for the cation binding-release mechanism of the protein. By ^1H NMR and solid state IR studies on the model systems, they showed that the strength of the intramolecular hydrogen bonding between the amide groups of the first amino acid (Val) increases with the presence of additional amino acids (Gly and Gly-Tyr) and competes more strongly with the formation of the metal ion complex. As a consequence, the cation binding is weakest in the compounds bearing the longer chains and the cation release is easiest. The authors suggest that this ion-binding destabilisation mechanism may operate also in the natural K^+ channels, helping to lower the energy barrier for ion translocation.

As part of a project devoted to investigate the mechanism of biomineralisation, and in particular the interaction between inorganic material and organic molecules, two peptidocalixarenes functionalised with Asp (**15**, R = CH_2COOH, R' = OH) or Glu (**15**, R = CH_2CH_2COOH, R' = OH) were synthesised and studied as crystal growth modifiers.[42,43] Although these compounds have a different effect on the inorganic material, both were found to profoundly influence the crystal growth and morphology of calcite, barium sulphate and calcium oxalate. In particular, the aspartic acid derivative suppresses the growth of calcite and induces chiral morphologies and growth features on the crystal surface at concentrations as low as 2 ppm.

Peptidocalixarenes functionalised at the *lower rim* with amino acids or peptides have also been studied as receptors for anions such as Cl^-, Br^-, HSO_4^-, $H_2PO_4^-$ and N-Tosyl(L)alaninate,[13,44] 4-nitrophenyl phosphate dianion,[45] and lactate[46] and two large combinatorial libraries of peptidocalixarenes (50000[47] and 3375 members[48]), on polystyrene beads, having two different dipeptides or two identical tripeptides, respectively, were screened for binding dye-labelled oligopeptides. A certain selectivity in the interaction between the analyte and the different peptide chains of the macrocycles was observed.

A calix[4]arene functionalised at the *lower rim* with three ester moieties and one Lys unit linked through its ε-NH_2 group was synthesised as building block for the construction of multifunctional nanostructures, such as calix peptide dendrimers.[49]

2.4 Peptidocalixarenes in Self-assembly

If two calix[4]arenes in the *cone* conformation self-associate through the interaction of the functional groups on their *upper rims* in a "head-to-head" manner, a capsular structure is formed, whose interior can accommodate a guest. Through the self-assembly of two identical, self-complementary subunits a homodimer is formed, while the association of two differently functionalised calix[4]arenes gives rise to a heterodimer.

The presence of amino acids and peptides on the *upper rim* of calixarenes makes them good candidates for self-assembly through hydrogen bonding or electrostatic interactions. A number of papers have been published where the amino acids and peptides either linked through functional spacers or directly connected to the aromatic cavity participate or sometimes exclusively drive the noncovalent dimerisation process. In particular, we can distinguish between two main groups of synthetic receptors capable of forming self-assembled superstructures based on peptidocalixarenes. The first (i) comprises well-studied self-assembling systems to which amino acids or peptides have been added to convey additional properties such as chirality or supplementary hydrogen bonding groups. In the second group (ii), which is more relevant to this chapter, we find peptidocalixarenes able to self-assemble only through noncovalent interactions between the amino acid or peptide chains.

(i) Calix[4]arenes **16a,b** functionalised with urea groups were reported by Rebek[50] and Böhmer[51] to form in apolar solvents highly stable, noncovalent homodimers (see also Chapter 2 in this book) held together by a seam of 16 hydrogen bonds (Fig. 8).

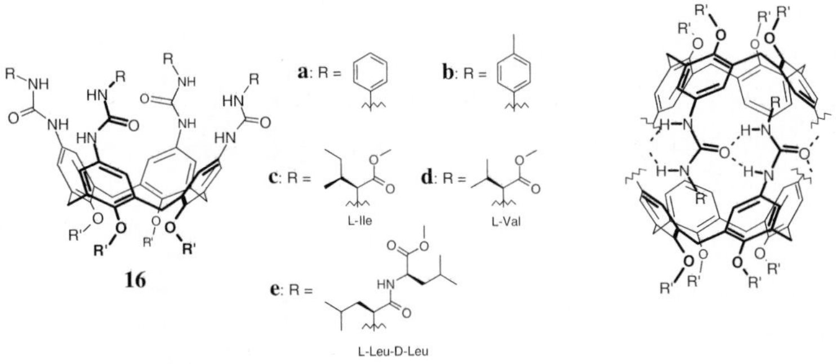

Figure 12-8. Dimeric capsules from ureidopeptidocalix[4]arenas.

The attachment of β-branched amino acids such as Ile (**16c**) and Val (**16d**) to the urea moieties markedly affects the behaviour of the calixarenes, which do not self-associate. Instead, the ureidopeptidocalixarenes **16c,d** form heterodimers with calix[4]arenes functionalised with arylurea groups (**16b**).[52] Moreover, by ^{1}H NMR spectroscopy it was demonstrated that the chirality of the amino acids is transferred to the capsule lining: in each heterodimer only one direction of the head-to-tail arrangements of the urea hydrogen bonding is observed (*e.g.* clockwise or counterclockwise), giving rise to an optically active capsule. Chiral guests are able to sense the chirality of the capsule and a modest enantioselection is observed.

Subsequently, the homodimerisation of a calix[4]arene ureidopeptide (**16e**), where four units of L-Leu-D-Leu-OMe are linked to the *upper rim* ureas was achieved by de Mendoza.[53] The alternation of L- and D- amino acids has the effect of projecting the bulky side chains outside the cavity, thus leaving more space for encapsulation. The capsule structure, elucidated by ROESY NMR, consists of a classical urea-urea dimer reinforced by an additional seam of hydrogen bonds provided by the peptide side chains, including the ester carbonyl groups.

(ii) The first example of a peptidocalixarene dimeric structure held together exclusively by hydrogen bonding between the amino acids (without participation of different functional groups) is provided by calix[6]arene **17**, which contains three leucyl amide residues at the *lower rim*.[54] In apolar solvents the dimerisation was confirmed by GPC. By molecular mechanics and dynamics calculations, together with NMR studies, it was established that the interaction is driven by the formation of up to eighteen hydrogen bonds between the amino acids in an extended conformation. By ^{1}H NMR dilution experiments, the dimerisation constant was measured as K_{dim} = 640

M^{-1} in $CDCl_3$. The inclusion of guests was not observed, probably because the methoxy groups and one of the amino acid flexible chains occupy the free space within the peptide cage.

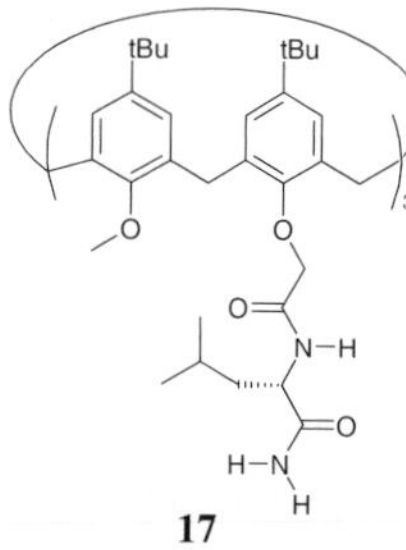

Reinhoudt's group reported an interesting water-soluble heterodimeric capsule, the first example of a dimer self-assembled in water without the driving force of metal-ligand interactions.[55] The assembly is based on multiple electrostatic interactions between the negatively charged carboxylate groups of calixarene **18**, functionalised at the *upper rim* with L-Ala moieties, and the positively charged amidinium groups of calixarene **19**. The system was studied by means of ^{1}H NMR spectroscopy, ESI-MS, and isothermal titration calorimetry (ITC), which allowed evaluation of the association constant $K_{ass} = 3.3 \times 10^4$ M^{-1} in buffered solution. More importantly, in water the dimer is able to encapsulate small molecules, such as *N*-methylquinuclidinium.

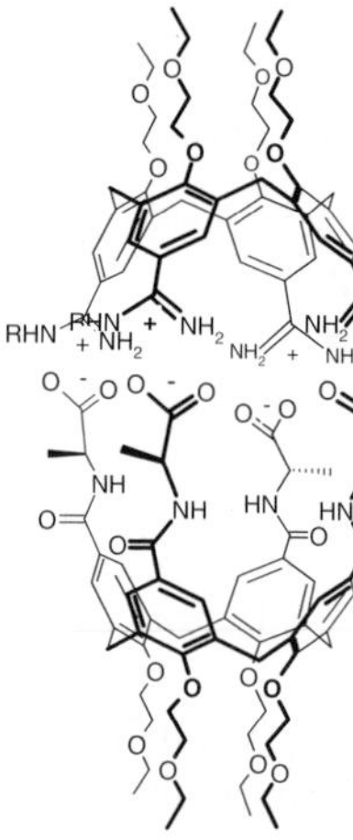

We have been interested in expanding the scope of peptidocalixarenes and we have synthesised the first examples of *N,C*-linked peptidocalix[4]arenes (**21**-**23**) by incorporating the calix[4]arene amino acid **20a** into a pseudopeptide sequence.[56] The two phenolic hydroxyl groups present at the *lower rim* block the calixarene in the *cone* conformation and the amino acid chains can be arranged in an extended orientation, which allows the formation of intermolecular β-sheets. In fact, all the *N,C*-linked peptidocalixarenes synthesised (including the simple diamide **20b**) showed peculiar self-assembling properties in apolar media, that were not observed for the corresponding *N*- or *C*-linked peptidocalixarenes. The concentration-dependence of ^{1}H NMR spectra in $CDCl_3$, together with the presence in the ESI-MS spectra of a molecular ion corresponding to a dimer, proved that these derivatives are able to form dimeric species.

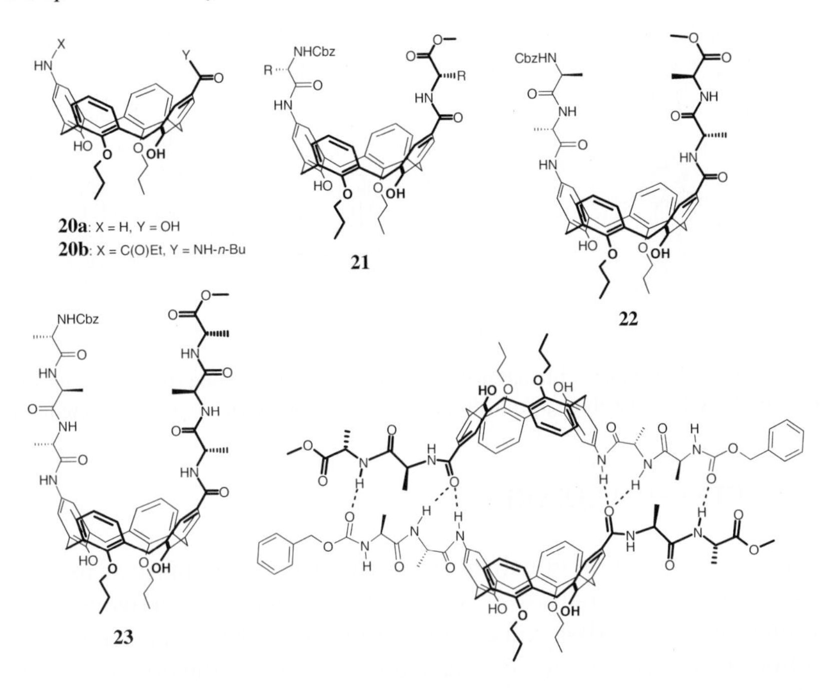

Figure 12-9. Self-assembled dimeric capsule from *N*,*C*-linked peptidocalix[4]arenas.

NOESY NMR spectroscopy and molecular modelling studies indicated that the dimer has a capsule-like structure held together by an array of hydrogen bonds in the form of an antiparallel β-sheet (Fig. 9). Dilution NMR experiments allowed the measure of the dimerisation constant, which increases with the number of CO and NH groups (K_{dim} = 74, 105, 776 and 1460 M^{-1} in $CDCl_3$ for compounds **20b**, **21** (R = CH_3), **22** and **23**, respectively). The introduction of bulkier side chains such as Phe (**21**, R = Bn) or Leu (**21**, R = *i*-butyl) causes a decrease in the dimerisation constant (K_{dim} = 63 and 82 M^{-1}, respectively).[57]

If two amino acids are *C*-linked through a methylene spacer to the opposite aromatic rings of a tetrapropoxycalix[4]arene, the resulting peptide-calixarenes (*e.g.* **24**) do not show any self-assembling ability in solution because the amino acid chains are involved in intramolecular hydrogen bonding. However, in the solid state, these derivatives form self-assembled nanotubes through a two-dimensional network of hydrogen bonds between the amide chains of adjacent molecules (Fig. 10b).[58] The conformation of the amino acid chains depends on the *N*-protecting group, but the interlocked array of hydrogen bonds which directs the nanotube formation is a common motif found in the crystal structure of all these derivatives.[59]

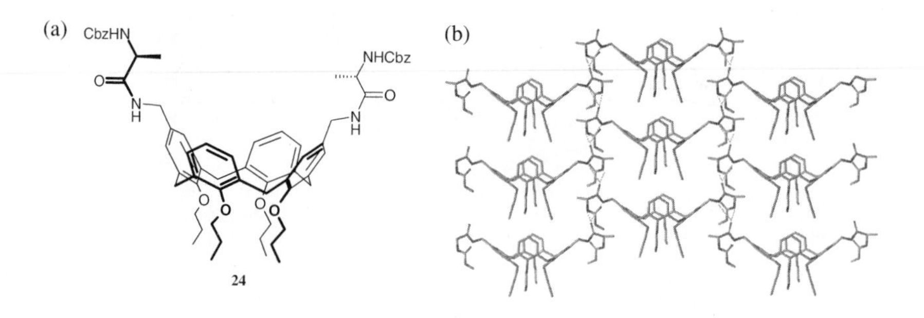

Figure 12-10. (a) A *C*-linked, difunctionalised peptidocalix[4]arene and (b) a side view of its packing in the solid state through a hydrogen bonding network (hydrogens are omitted for clarity).

3. GLYCOCALIXARENES

The functionalisation of calixarenes with a variable number of the same molecular unit to give polyvalent systems is particularly attractive when the repeating units are carbohydrates. Saccharides have been considered for a long time as a chemical energy source and structurally important elements, but they are now identified as fundamental substrates for specific receptors in a wide range of biological processes.[60] Intercellular communication, cell trafficking, immune response, infections by bacteria and viruses, growth and metastasis of tumour cells all occur thanks to the binding of sugar residues by saccharide receptors on the cell surface.[61] Since the affinity of a single carbohydrate unit for its receptor is usually low, the strong binding observed in these recognition events is determined by the simultaneous complexation of several identical glycoside residues, exposed at the substrate surface, by receptors (often proteins) bearing several equivalent binding sites. This phenomenon has been named *multivalency*,[10,62] or *glycoside cluster effect*,[63] and has inspired the design and synthesis of polyglycosylated, multivalent systems,[64] some of them based on calixarene scaffolds. Calixarenes offer special advantages in this context, since their molecular cavity can be exploited as an additional binding site for the complexation of guest molecules and the resulting glycocalixarenes have the potential of acting as site-directed molecular delivery systems. This is just one of the many attractive features of glycocalixarenes in Bionanotechnology (Fig. 11).

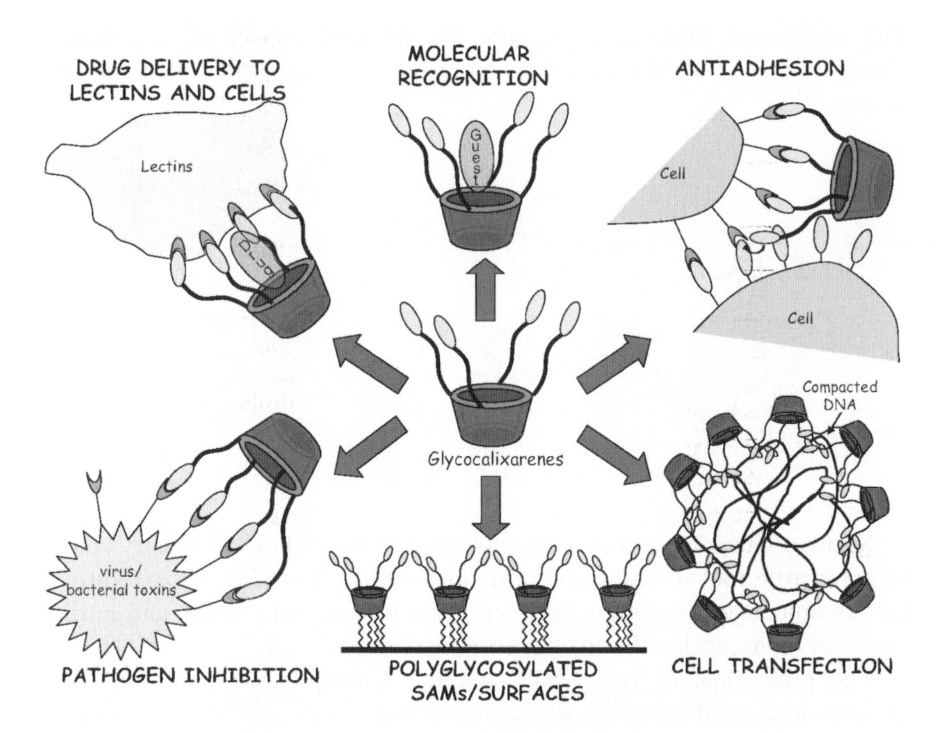

Figure 12-11. Glycocalixarenes in Bionanotechnology.

In the last decade, several groups have developed synthetic procedures for the linkage of glycoside units to the calixarene macrocycle. In a pioneering work, we reported[6,65,66] the first examples of calix[4]arene *O*-glycoconjugates bearing glycoside units at the *lower* or *upper rim* (*e.g.* **25**). Together with the *O*-neoglycoconjugates subsequently prepared by Dondoni *et al.*,[67] they remain until now the only examples of polyglycosylated calixarenes synthesised through "classical" glycosidation reactions, where bonds are formed at the anomeric carbon. However, we and others also explored alternative conjugation reactions, which do not affect the stereochemistry of the anomeric carbon, in some cases very hard to control. Coleman *et al.*,[68] for example, exploited a Suzuki type reaction to synthesise calix[4]arene-*O*-gluco-, *O*-galacto- (**26**) and *O*-maltoside, while phosphorus ylides of different monosaccharides were condensed by Dondoni *et al.*[67,69] with *p*-formyl calix[4]arenes to achieve *upper rim* *C*-glycoconjugates. Ester[70] and amide[71] bond formation between calixarenes functionalised with carboxylic moieties and saccharides presenting free hydroxy or amino groups has also been successfully employed as a conjugation method. We synthesised, for instance, the difunctionalised glycocalixarene **27** by condensation of the aspartylaminoglucoside with a dicarboxycalix[4]arene.[72]

The compound obtained represents the minimal model of a pseudo-glycopeptide and could lay the basis for developing cell surface glycoprotein mimics.

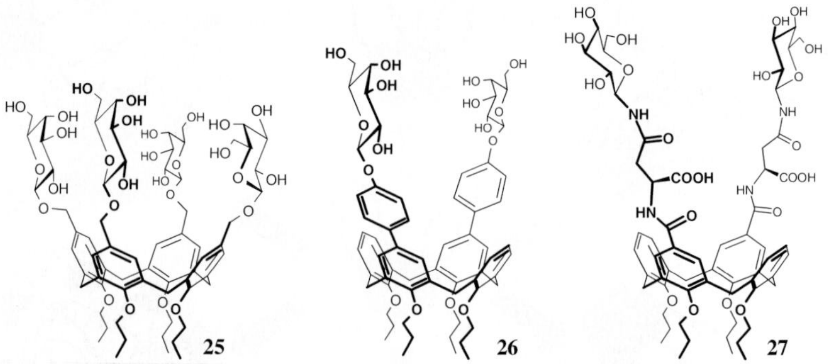

Santoyo-González and co-workers[73] showed that "click" chemistry could be a versatile tool to build-up multivalent derivatives (**28**), by exploiting the 1,3-dipolar cycloaddition between azide-bearing calixarenes and nitrile oxides of monosaccharides like mannose and glucose.

The polysialosides (*e.g.* **29**)[74] and the corresponding poly-*N*-acetyl-galactosides[75] synthesised by Roy and co-workers, through the conjugation of sugar units to a spacer present at the *lower rim* of calix[4]arenas, are the first glycocalixarenes studied as multivalent systems. They are actually able to efficiently crosslink and agglutinate carbohydrate binding proteins specific for the saccharide units present on the calixarene.

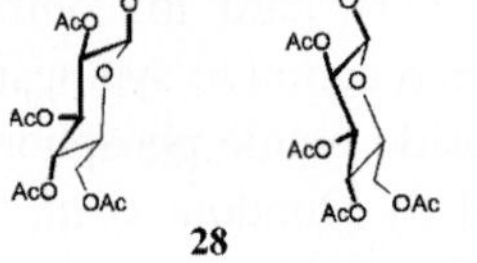

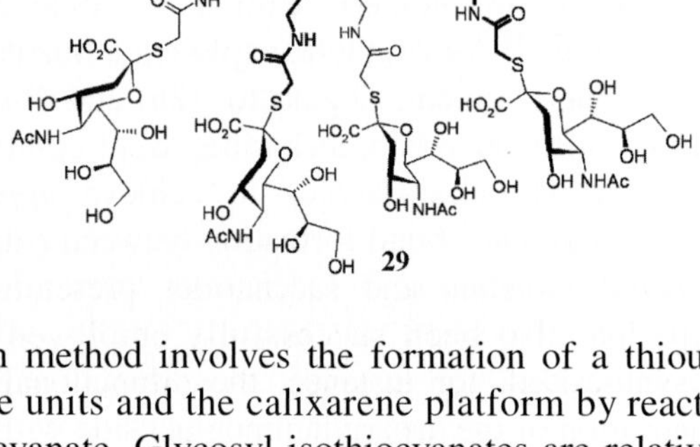

An attractive conjugation method involves the formation of a thiourea group between the glycoside units and the calixarene platform by reaction of an amine with an isothiocyanate. Glycosyl isothiocyanates are relatively simple to obtain and their conjugation with an amino containing counterpart occurs in high yields, reasonable time and mild conditions. We first exploited[76] the formation of a thiourea spacer at the *upper rim* of calix[4]arenes in the *cone* conformation (*e.g.* **30**) by condensation between aminocalixarenes and *β*-glycosylisothiocyanates, or between calixarene isothiocyanates and aminoglycosides. Then we extended this synthetic protocol to *1,3-alternate* (**31**) and *mobile* (**32**, n = 4) calix[4]arenes and to calix[6]- and -[8]arenes (**32**, n = 6, 8), linking both mono- (Glc and Gal) and disaccharides (Lac).[77]

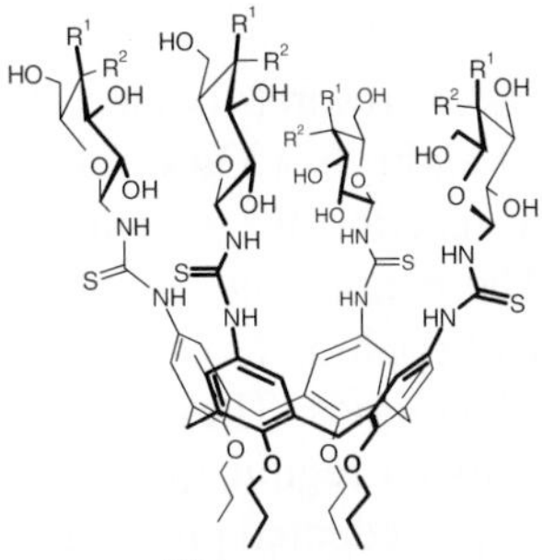

30: R^1 = OH, R^2 = H or R^1 = H, R^2 = OH or R^1 = O-1-β-Gal, R^2 = H *cone*

31: R^1 = OH, R^2 = H or R^1 = H, R^2 = OH or R^1 = O-1-β-Gal, R^2 = H *1,3-alt*

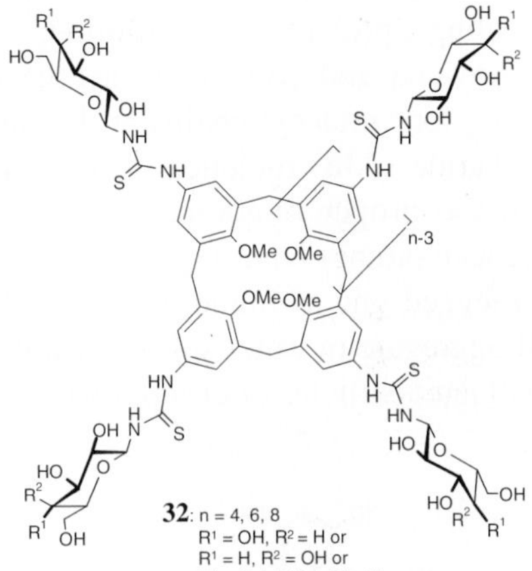

32: n = 4, 6, 8 R^1 = OH, R^2 = H or R^1 = H, R^2 = OH or R^1 = O-1-β-Gal, R^2 = H

The combination of saccharides with a calixarene scaffold confers a significant amphiphilic character to the resulting glycocalixarenes. Their solubility in water is up to 10^{-3} M, but they tend to self-assemble, as evidenced by the broadening of the ^{1}H NMR signals in D_2O and by the unambiguous detection by AFM on mica surfaces of discoid-like aggregates.[77] As expected, these multivalent compounds showed selectivity in binding to lectins. Simple and widely used turbidimetric experiments evidenced the specific agglutination ability of the glucose-bearing macrocycles towards Concanavalin A (ConA) and of those exposing galactose towards Peanut Agglutinin. Additionally, they were shown by NMR spectroscopy and MS to interact with anionic substrates through the thiourea units.[76] This is an important, preliminary result which establishes the basis for the design and synthesis of site-directed drug delivery systems, exploiting carbohydrate recognition for high organ or tissue specificity.

Calix[4]- and calix[8]arene glycoclusters were also prepared by Consoli *et al.*[78,79] Glycocalixarenes exposing *N*-acetylglucosamine (GlcNAc) units[79] were successfully tested as ligands of the wheat germ agglutinin, a GlcNAc binding protein, and inhibitors of its agglutination properties towards human erythrocytes, showing an amplified lectin affinity with respect to the simple monosaccharide.

Calixarenes conjugated with sugars at the *lower rim* through thiourea units were prepared by Santoyo-González[80] and Yan *et al.*[81] but no recognition properties were reported.

Resorc[4]arenes were adorned with glycoside units connected through a thiourea containing linker to the resorcine oxygens[81] or to the *upper rim* aromatic nucleus of the cavitands.[82,83] The recognition properties of this second series of receptors towards anions were studied in detail by ESI-MS, revealing a preference for chloride.

Aoyama and co-workers synthesised a series of resorc[4]arenes **33a-d** having four undecyl chains at the *lower rim* and eight di- (**33a-c**) or oligosaccharide (**33d**) moieties at the *upper rim*, by reacting the octamine **33e** with the proper oligosaccharide lactone.[84-87] These nonionic, amphiphilic glycocalixarenes have a quite unusual four legged lipophilic tail rather than a two legged one as found in phospholipids and glycolipids. They are able to self-aggregate in water solution even in the micromolar region, giving rise to small nanoparticles of exceptional stability and unprecedented properties.[88]

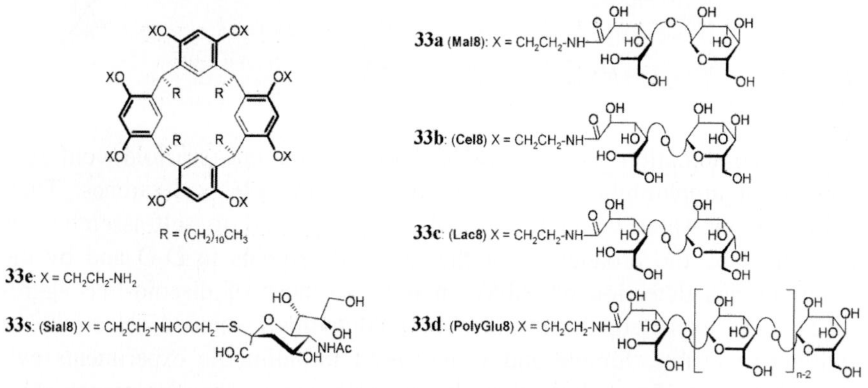

$R = (CH_2)_{10}CH_3$

33e: $X = CH_2CH_2\text{-}NH_2$

33s: (Sial8) $X = CH_2CH_2\text{-}NHCOCH_2\text{-}S\text{-}$

33a (Mal8): $X = CH_2CH_2\text{-}NH\text{-}$

33b: (Cel8) $X = CH_2CH_2\text{-}NH\text{-}$

33c: (Lac8) $X = CH_2CH_2\text{-}NH\text{-}$

33d: (PolyGlu8) $X = CH_2CH_2\text{-}NH\text{-}$

The micellar assemblies (Fig. 12), named glycocluster nanoparticles (GNPs), were characterised by NMR, gel permeation chromatography (GPC), dynamic light scattering (DLS), TEM and AFM microscopies and showed no notable oligosaccharide chain-length dependent size (2.4-3.5 nm). The molecular weights of the GNPs formed by **33a** and **33d** (n = 7) indicate, for example, that they are composed of 4 or 6 molecules, respectively.[87] The unusual stability of GNPs comes from lateral side-by-side inter-saccharide interactions, which immobilize the hydrophobically associated and otherwise labile micelles. Moreover, glycocalixarenes of type **33** form complexes in water with a large variety of anionic and cationic dye molecules having association constants in the range of $10^2\text{-}10^6\ M^{-1}$.[84,85,89-91]

Compound **33a** proved to be effective as molecular delivery system to surfaces such as quartz[89,92] or gels coated with ConA.[90] The octagalactose **33c** is able to deliver fluorescent guests, such as phloxine B or calcein, to hepatocytes, which have receptors for the terminal galactose residues of asialoglycoproteins. The selectivity for different biological receptor sites can be finely tuned by changing the saccharide units present on the cluster, as further confirmed by the glycocluster **33s**, having eight sialic acid residues successfully tested as ligand for influenza viruses.[93] This highly specific saccharide-mediated cell recognition event suggests a possible application of glycoclusters in anticancer therapy. If functionalised with the appropriate sugar, they could create a "cancer net" around a solid tumour able to retard its growth and/or inhibit metastasis. A first attempt to obtain such "cancer net" drug was performed by Menger *et al.*[94] who prepared a double resorcarene having 14 galactose units. However, no evidence could be collected for aggregation of cancer cell suspensions and the authors proposed to correct the design by rigidifying the linker between the two-calixarene units.

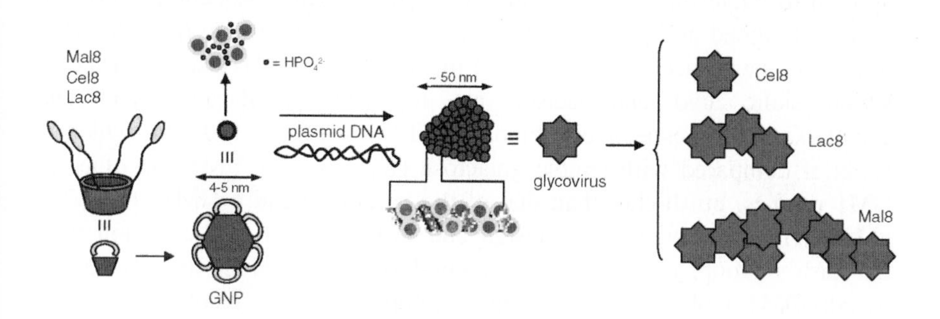

Figure 12-12. Assembly of GNPs and inclusion of plasmid DNA.

Another quite remarkable peculiarity of these resorcarene glycoclusters is their ability to interact with anionic substrates and in particular with phosphates,[87] nucleotides[85] and nucleic acids.[86,95] While guanosine-5'-monophosphate (GMP) and adenosine mono-, di- or triphosphate (AMP, ADP or ATP) form 1:1 complexes with **33a**,[85] HPO_4^{2-} induces formation of large aggregates with a guest/host ratio higher than 10. The collected evidence clearly indicates that micellar GNPs originating from **33a-d** in pure water are agglutinated or cross-linked in the presence of HPO_4^{2-} to give aggregates whose dimensions (50-140 nm) depend on the oligosaccharide chain length and where the anions (H-acceptors) act as glue for the oligosaccharide (H-donors) units (Fig. 12). The ability of GNPs to complex phosphate anions suggested their application as non-viral gene vectors. Indeed, GNPs are able to coat and compact a plasmid DNA (7040 base pairs) into viral (50 nm) sized particles, named artificial glycoviruses.

The aggregation tendency increases in the order Cel8 (**33b**) << Lac8 (**33c**) < Mal8 (**33a**) showing that the alteration in stereochemistry of a single glycoside linkage (Mal *vs.* Cel) or a single OH group (Mal *vs.* Lac) may result in a significant change in size of the glycovirus (Fig. 12). These glycoviruses are able to transfect Hela cells with a remarkable size-regulated activity, the smaller (50 nm) cellobiose glycoviruses being more potent than the larger lactose (200 nm) or maltose (300 nm) ones.[95] Interestingly, for HepG2 cells, which possess receptors for β-galactose, lactose glycoviruses showed a transfection activity 100-fold higher than expected on the basis of size.

Cholera toxin (CT) and its homologous heat-labile toxin of *E. Coli* are toroid-like AB_5 toxins. The B proteins simultaneously bind to five units of their natural ligand, the GM1 ganglioside, contained on the cell membrane. The recognition process primarily involves the galactose and the sialic acid molecules of the GM1 oligosaccharide (o-GM1, **34**), kept in the correct orientation by a lactose scaffold. Several multivalent glycoclusters have been applied to bacterial targets, including cholera toxin.[96] Recently, Bernardi *et al.*[97] designed and prepared an interesting synthetic mimic (pseudo-GM1 **35**) of the pentasaccharide contained in the natural GM1. Pseudo-GM1, in which sialic acid and lactose are replaced by R-lactic acid and cyclohexanediol, respectively, binds to CT with a K_d = 190 μM, which is lower if compared with simple galactose ($K_d = 4 \times 10^4$ μM) and other o-GM1 mimics, but higher than that of the natural ligand o-GM1 (K_d = 219 nM).[97] This prompted us to design and synthesise, in collaboration with Bernardi's group, a series of calixarene-based multivalent ligands having pseudo-GM1 units as toxin binding epitopes. So far, results have been obtained with the divalent ligand **36**, which possesses two pseudo-GM1 units linked in diametral position at the *upper rim* of a calix[4]arene blocked in the *cone* conformation. A long spacer based on a polyethylenglycole bis-amine and squaric acid is present between the saccharide and the cavity to allow the simultaneous interaction of the two pseudo-GM1 units with two non adjacent binding sites of CT.[98] This divalent system proved, from spectrofluorimetric titration experiments and ELISA inhibition tests, to be a highly efficient ligand for CT, showing a 50% saturation concentration of 48 nM, slightly better than that measured for the natural o-GM1. This value also corresponds to a remarkable affinity enhancement factor close to 4000 fold (2000 fold per sugar mimic) with respect to the single pseudo-GM1, which is one of the highest ever observed for CT inhibitors.

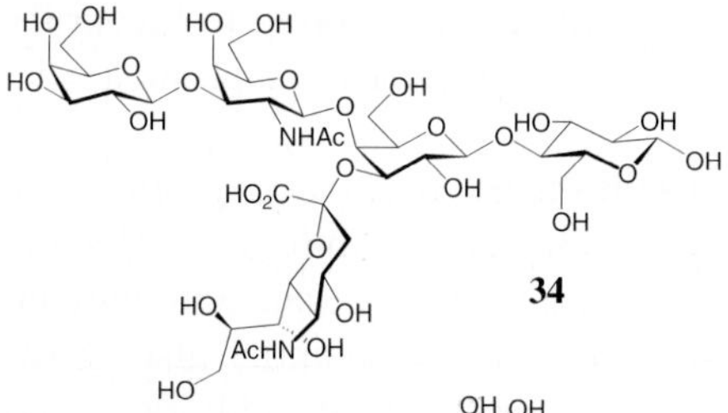

34 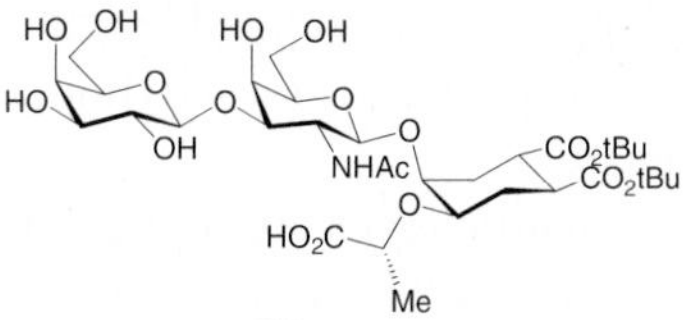 **35**

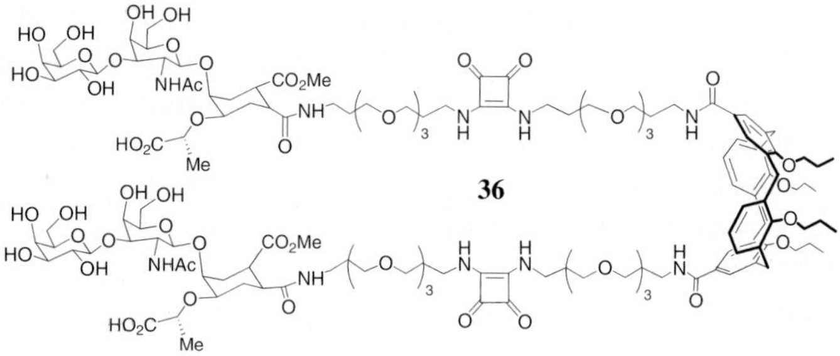

36

4. OUTLOOK

Almost 30 years of Supramolecular Chemistry have contributed to break, or at least to loosen, the barriers between classical scientific disciplines (Chemistry, Physics, Biology and Molecular Medicine) in such a way that the results achieved in one context often deeply affect research in the others. This is particularly true when studying molecular recognition processes involving carbohydrates and peptides, including the use of peptido- and glycocalixarenes, since the results obtained can be of interest to biomedicine and material science as well.

Several prospects, which the design and synthesis of peptido- and glycolcalixarenes were based on, have already come true, but many others are at the horizon, waiting for discovery. The finding that some cleft-like peptidocalix[4]arenes self-assemble in the solid state, to give nanotubes, opens the way to design novel porous materials useful for molecular storage and separation. The synthesis of the first example of calix[4]arene amino acid offers the possibility of incorporating this unit in cyclopeptide receptors for molecular recognition or to design novel conformationally constrained peptidomimetics. The anchoring on surfaces of peptidocalix[4]arenes able to form dimeric capsules makes it attractive, *inter alia*, to explore the possibility of pattern replication *via* hydrogen bonding, using the techniques and principles of nanofabrication.

For the formation of dimeric capsules through antiparallel β-sheet-like self-assembly we have used short L-Ala oligopeptides. An attractive feature is to change the nature and length of the peptide chains in order to favour the formation of α-helices, amyloid fibres or novel nanoscale supramolecular architectures.[99]

Glycocalixarenes are interesting molecules in the fields of Glycomics and Glycobiology, where they can be used as selective inhibitors or effectors of biological processes based on carbohydrate-carbohydrate and protein-carbohydrate recognition. Moreover, they have great potential also in the novel and rapidly growing field of Bionanotechnology. In addition to the results already achieved outlined in Fig. 11, an interesting perspective is to design carbohydrate microarrays based on multivalent glycocalixarenes, which could confer robustness, efficiency and selectivity to this type of glycan-based device, very promising for the study of carbohydrate-cell interactions, to detect pathogens and for drug discovery.[100,101] Finally, the cluster glycoside effect of glycocalixarenes can be also useful for the design of novel organ specific contrast agents or other supramolecular devices.

5. ACKNOWLEDGEMENTS

The authors thank the Ministero dell'Istruzione, Università e Ricerca (MIUR) (FIRB Project RBNEO19H9K: Manipolazione molecolare per macchine nanometriche).

6. REFERENCES

1. C. D. Gutsche and R. Muthukrishnan, *J. Org. Chem.* **43**, 4905-4906 (1978).
2. G. D. Andreetti, R. Ungaro and A. Pochini, *J. Chem. Soc., Chem. Commun.*1005-1007 (1979).
3. A. Pochini and A. Arduini, in *Calixarenes in Action,* Imperial College Press (Eds.: L. Mandolini, R. Ungaro), London, **2000**, pp. 37-61.
4. A. Arduini, A. Casnati, E. Dalcanale, A. Pochini, F. Ugozzoli and R. Ungaro, in *Supramolecular Science: Where it is and Where it is going,* KAP, (Eds.: R. Ungaro, E. Dalcanale), Dordrecht, **1999**, NATO ASI *Ser. C.* **527**, pp. 67-94.
5. C. D. Gutsche, *Calixarenes*, The Royal Society of Chemistry (Ed.: J. F. Stoddart), Cambridge, **1989**.
6. A. Marra, M. C. Scherrmann, A. Dondoni, A. Casnati, P. Minari and R. Ungaro, *Angew. Chem. Int. Engl. Ed.* **33**, 2479-2481 (1994).
7. F. Sansone, S. Barboso, A. Casnati, M. Fabbi, A. Pochini, F. Ugozzoli and R. Ungaro, *Eur. J. Org. Chem.* 897-905 (1998).
8. A. Casnati, M. Fabbi, N. Pelizzi, A. Pochini, F. Sansone, R. Ungaro, E. Di Modugno and G. Tarzia, *Bioorg. Med. Chem. Lett.* **6**, 2699-2704 (1996).
9. A. Casnati, F. Sansone and R. Ungaro, *Acc. Chem. Res.* **36**, 246-254 (2003).
10. M. Mammen, S.-K. Choi and G. M. Whitesides, *Angew. Chem. Int. Engl. Ed.* **37**, 2755-2794 (1998).

11. J. C. M. Rivas, H. Schwalbe and S. J. Lippard, *Proc. Natl. Acad. Sci. U. S. A.* **98**, 9478-9483 (2001).
12. M. Lazzarotto, F. Sansone, L. Baldini, A. Casnati, P. Cozzini and R. Ungaro, *Eur. J. Org. Chem.* 595-602 (2001).
13. S. Ben Sdira, C. P. Felix, M. B. A. Giudicelli, P. F. Seigle-Ferrand, M. Perrin and R. J. Lamartine, *J. Org. Chem.* **68**, 6632-6638 (2003).
14. L. Frish, F. Sansone, A. Casnati, R. Ungaro and Y. Cohen, *J. Org. Chem.* **65**, 5026-5030 (2000).
15. D. H. Williams and B. Bardsley, *Angew. Chem.Int. Ed.* **38**, 1173-1193 (1999).
16. M. Segura, B. Bricoli, A. Casnati, E. M. Munoz, F. Sansone, R. Ungaro and C. Vicent, *J. Org. Chem.* **68**, 6296-6303 (2003).
17. G. Das and A. D. Hamilton, *Tetrahedron Lett.* **38**, 3675-3678 (1997).
18. A. S. Droz, U. Neidlein, S. Anderson, P. Seiler and F. Diederich, *Helv. Chim. Acta* **84**, 2243-2289 (2001).
19. A. P. Davis and R. S. Wareham, *Angew. Chem. Int. Engl. Ed.* **38**, 2979-2996 (1999).
20. E. Klein, M. P. Crump and A. P. Davis, *Angew. Chem. Int. Ed.* **44**, 298-302 (2005).
21. F. Sansone, L. Baldini, A. Casnati, M. Lazzarotto, F. Ugozzoli and R. Ungaro, *Proc. Natl. Acad. Sci. U. S. A.* **99**, 4842-4847 (2002).
22. R. Miao, Q. Y. Zheng, C. F. Chen and Z. T. Huang, *Tetrahedron Lett.* **45**, 4959-4962 (2004).
23. R. E. Brewster, B. G. A. Dalton and S. B. Shuker, *Bioorg. Chem.* **33**, 16-21 (2005).
24. E. A. Shokova, A. E. Motornaya, A. K. Shestakova and V. V. Kovalev, *Tetrahedron Lett.* **45**, 6465-6469 (2004).
25. H. Hioki, Y. Ohnishi, M. Kubo, E. Nashimoto, Y. Kinoshita, M. Samejima and M. Kodama, *Tetrahedron Lett.* **45**, 561-564 (2004).
26. J. Pfeiffer and V. Schurig, *J. Chromatog., A* **840**, 145-150 (1999).
27. C. Berghaus and M. Feigel, *Eur. J. Org. Chem.* 3200-3208 (2003).
28. A. Ruderisch, J. Pfeiffer and V. Schurig, *Tetrahedron Asymmetry* **12**, 2025-2030 (2001).
29. R. Jain, J. T. Ernst, O. Kutzki, H. S. Park and A. D. Hamilton, *Molecular Diversity* **8**, 89-100 (2004).
30. Q. Lin, H. S. Park, Y. Hamuro and A. D. Hamilton, in *Supramolecular Science: Where it is and Where it is going,* KAP, (Eds.: R. Ungaro and E. Dalcanale), Dordrecht, **1999**, NATO ASI Series C **527**, pp. 197-204.
31. Y. Hamuro, M. Crego-Calama, H. S. Park and A. D. Hamilton, *Angew. Chem. Int. Engl. Ed.* **36**, 2680-2683 (1997).
32. Q. Lin, H. S. Park, Y. Hamuro, C. S. Lee and A. D. Hamilton, *Biopolymers* **47**, 285-297 (1998).
33. H. S. Park, Q. Lin and A. D. Hamilton, *J. Am. Chem. Soc.* **121**, 8-13 (1999).
34. M. A. Blaskovich, Q. Lin, F. L. Delarue, J. Sun, H. S. Park, D. Coppola, A. D. Hamilton and S. M. Sebti, *Nat. Biotech.* **18**, 1065-1070 (2000).
35. S. M. Sebti and A. D. Hamilton, *Oncogene* **19**, 6566-6573 (2000).
36. J. Z. Sun, M. A. Blaskovich, R. K. Jain, F. Delarue, D. Paris, S. Brem, O. M. Wotoczek, Q. Lin, D. Coppola, K. H. Choi, M. Mullan, A. D. Hamilton and S. M. Sebti, *Cancer Res.* **64**, 3586-3592 (2004).
37. T. Mecca, G. M. L. Consoli, C. Geraci and F. Cunsolo, *Bioorg. Med. Chem.* **12**, 5057-5062 (2004).
38. S. Francese, A. Cozzolino, L. Caputo, C. Esposito, M. Martino, C. Gaeta, F. Troisi and P. Neri, *Tetrahedron Lett.* **46**, 1611-1615 (2005).
39. E. Nomura, M. Takagaki, C. Nakaoka, M. Uchida and H. Taniguchi, *J. Org. Chem.* **64**, 3151-3156 (1999).
40. L. Frkanec, A. Visnjevac, P. B. Kojic and M. Zinic, *Chem. Eur. J.* **6**, 442-453 (2000).

41. F. Arnaud-Neu, S. Barboso, F. Berny, A. Casnati, N. Muzet, A. Pinalli, R. Ungaro, M. J. Schwing-Weill and G. Wipff, *J. Chem. Soc., Perkin Trans. 2* 1727-1738 (1999).
42. M. J. Bartlett, M. Mocerino, M. I. Ogden, A. Oliveira, G. M. Parkinson, J. K. Pettersen and M. M. Reyhani, *J. Mater. Sci. Tech.* **21**, 1-5 (2005).
43. F. Jones, M. Mocerino, M. I. Ogden, A. Oliveira and G. M. Parkinson, *Crys. Growth Des.* **5**, 2336-2343 (2005).
44. S. Ben Sdira, R. Baudry, C. P. Felix, M. B. Giudicelli and R. J. Lamartine, *Tetrahedron Lett.* **45**, 7801-7804 (2004).
45. X. B. Hu, A. S. C. Chan, X. X. Han, J. Q. He and J. P. Cheng, *Tetrahedron Lett.* **40**, 7115-7118 (1999).
46. W. Guo, J. Wang, C. Wang, J. Q. He, X. W. He and J. P. Cheng, *Tetrahedron Lett.* **43**, 5665-5667 (2002).
47. H. Hioki, T. Yamada, C. Fujioka and M. Kodama, *Tetrahedron Lett.* **40**, 6821-6825 (1999).
48. H. Hioki, M. Kubo, H. Yoshida, M. Bando, Y. Ohnishi and M. Kodama, *Tetrahedron Lett.* **43**, 7949-7952 (2002).
49. H. Xu, G. R. Kinsel, J. Zhang, M. L. Li and D. A. Rudkevich, *Tetrahedron* **59**, 5837-5848 (2003).
50. K. D. Shimizu and J. Rebek, *Proc. Natl. Acad. Sci. U. S. A.* **92**, 12403-12407 (1995).
51. O. Mogck, M. Pons, V. Böhmer and W. Vogt, *J. Am. Chem. Soc.* **119**, 5706-5712 (1997).
52. R. K. Castellano, C. Nuckolls and J. Rebek, *J. Am. Chem. Soc.* **121**, 11156-11163 (1999).
53. A. M. Rincon, P. Prados and J. de Mendoza, *J. Am. Chem. Soc.* **123**, 3493-3498 (2001).
54. A. M. Rincon, P. Prados and J. de Mendoza, *Eur. J. Org. Chem.* 640-644 (2002).
55. F. Corbellini, L. Di Costanzo, M. Crego-Calama, S. Geremia and D. N. Reinhoudt, *J. Am. Chem. Soc.* **125**, 9946-9947 (2003).
56. F. Sansone, L. Baldini, A. Casnati, E. Chierici, G. Faimani, F. Ugozzoli and R. Ungaro, *J. Am. Chem. Soc.* **126**, 6204-6205 (2004).
57. L. Baldini, F. Sansone, A. Casnati and R. Ungaro, *unpublished results.*
58. L. Baldini, F. Sansone, A. Casnati, F. Ugozzoli and R. Ungaro, *J. Supramol. Chem.* **2**, 219-226 (2002).
59. L. Baldini, F. Sansone, A. Casnati, C. Massera, F. Ugozzoli and R. Ungaro, *unpublished results.*
60. A. Varki, *Glycobiology* **3**, 97-130 (1993).
61. H. Lis and N. Sharon, *Chem. Rev.* **98**, 637-674 (1998).
62. P. I. Kitov and D. R. Bundle, *J. Am. Chem. Soc.* **125**, 16271-16284 (2003).
63. Y. C. Lee and R. T. Lee, *Acc. Chem. Res.* **28**, 321-327 (1995).
64. J. J. Lundquist and E. J. Toone, *Chem. Rev.* **102**, 555-578 (2002).
65. A. Marra, A. Dondoni and F. Sansone, *J. Org. Chem.* **61**, 5155-5158 (1996).
66. A. Dondoni, A. Marra, M. C. Scherrmann, A. Casnati, F. Sansone and R. Ungaro, *Chem. Eur. J.* **3**, 1774-1782 (1997).
67. A. Dondoni, M. Kleban, X. Hu, A. Marra and H. D. Banks, *J. Org. Chem.* **67**, 4722-4733 (2002).
68. C. Felix, H. Parrot-Lopez, V. Kalchenko and A. W. Coleman, *Tetrahedron Lett.* **39**, 9171-9174 (1998).
69. A. Dondoni, M. Kleban and A. Marra, *Tetrahedron Lett.* **38**, 7801-7804 (1997).
70. A. Dondoni, X. B. Hu, A. Marra and H. D. Banks, *Tetrahedron Lett.* **42**, 3295-3298 (2001).
71. J. Budka, M. Tkadlecova, P. Lhotak and I. Stibor, *Tetrahedron* **56**, 1883-1887 (2000).
72. U. Schädel, F. Sansone, A. Casnati and R. Ungaro, *Tetrahedron* **61**, 1149-1154 (2005).

73. F. G. Calvo-Flores, J. Isac-García, F. Hernandez-Mateo, F. Perez-Balderas, J. A. Calvo-Asin, E. Sanchez-Vaquero and F. Santoyo-González, *Org. Lett.* **2**, 2499-2502 (2000).
74. S. J. Meunier and R. Roy, *Tetrahedron Lett.* **37**, 5469-5472 (1996).
75. R. Roy and J. M. Kim, *Angew. Chem. Int. Engl. Ed.* **38**, 369-372 (1999).
76. F. Sansone, E. Chierici, A. Casnati and R. Ungaro, *Org. Biomol. Chem.* **1**, 1802-1809 (2003).
77. F. Sansone, A. Casnati and R. Ungaro, *unpublished results.*
78. G. M. L. Consoli, F. Cunsolo, C. Geraci, T. Mecca and P. Neri, *Tetrahedron Lett.* **44**, 7467-7470 (2003).
79. G. M. L. Consoli, F. Cunsolo, C. Geraci and V. Sgarlata, *Org. Lett.* **6**, 4163-4166 (2004).
80. C. Saitz-Barria, A. Torres-Pinedo and F. Santoyo-González, *Synlett* 1891-1894 (1999).
81. Y. Ge, Y. H. Cai and C. G. Yan, *Synth. Commun.* **35**, 2355-2361 (2005).
82. G. V. Oshovsky, W. Verboom, R. H. Fokkens and D. N. Reinhoudt, *Chem. Eur. J.* **10**, 2739-2748 (2004).
83. G. V. Oshovsky, W. Verboom and D. N. Reinhoudt, *Collect. Czech. Chem. Commun.* **69**, 1137-1148 (2004).
84. T. Fujimoto, C. Shimizu, O. Hayashida and Y. Aoyama, *Gazz. Chim. Ital.* **127**, 749-752 (1997).
85. O. Hayashida, M. Kato, K. Akagi and Y. Aoyama, *J. Am. Chem. Soc.* **121**, 11597-11598 (1999).
86. Y. Aoyama, T. Kanamori, T. Nakai, T. Sasaki, S. Horiuchi, S. Sando and T. Niidome, *J. Am. Chem. Soc.* **125**, 3455-3457 (2003).
87. O. Hayashida, K. Mizuki, K. Akagi, A. Matsuo, T. Kanamori, T. Nakai, S. Sando and Y. Aoyama, *J. Am. Chem. Soc.* **125**, 594-601 (2003).
88. Y. Aoyama, *Chem. Eur. J.* **10**, 588-593 (2004).
89. Y. Aoyama, Y. Matsuda, J. Chuleeraruk, K. Nishiyama, K. Fujimoto, T. Fujimoto, T. Shimizu and O. Hayashida, *Pure Appl. Chem.* **70**, 2379-2384 (1998).
90. T. Fujimoto, C. Shimizu, O. Hayashida and Y. Aoyama, *J. Am. Chem. Soc.* **120**, 601-602 (1998).
91. K. Fujimoto, T. Miyata and Y. Aoyama, *J. Am. Chem. Soc.* **122**, 3558-3559 (2000).
92. T. Fujimoto, C. Shimizu, O. Hayashida and Y. Aoyama, *J. Am. Chem. Soc.* **119**, 6676-6677 (1997).
93. K. Fujimoto, O. Hayashida, Y. Aoyama, C. T. Guo, K. I. P. J. Hidari and Y. Suzuki, *Chem. Lett.* 1259-1260 (1999).
94. F. M. Menger, J. W. Bian, E. Sizova, D. E. Martinson and V. A. Seredyuk, *Org. Lett.* **6**, 261-264 (2004).
95. T. Nakai, T. Kanamori, S. Sando and Y. Aoyama, *J. Am. Chem. Soc.* **125**, 8465-8475 (2003).
96. S.-K. Choi, *Synthetic Multivalent Molecules*, John Wiley & Sons, Inc., Hoboken, New Jersey, **2004**, pp. 23-78.
97. A. Bernardi, D. Arosio and S. Sonnino, *Neurochem. Res.* **27**, 539-545 (2002).
98. D. Arosio, M. Fontanella, L. Baldini, L. Mauri, A. Bernardi, A. Casnati, F. Sansone and R. Ungaro, *J. Am. Chem. Soc.* **127**, 3660-3661 (2005).
99. S. Gilead and E. Gazit, *Supramol. Chem.* **17**, 87-92 (2005).
100. E. W. Adams, D. M. Ratner, H. R. Bokesch, J. B. McMahon, B. R. O'Keefe and P. H. Seeberger, *Chemistry & Biology* **11**, 875-881 (2004).
101. I. Shin, S. Park and M. R. Lee, *Chem. Eur. J.* **11**, 2894-2901 (2005).

Chapter 13

MODELS OF METALLO-ENZYME ACTIVE SITES

Calixarenes and enzyme mimicry

Olivia Reinaud,[a] Yves Le Mest,[b] and Ivan Jabin[c]

[a]Laboratoire de Chimie et Biochimie Pharmacologiques et Toxicologiques, UMR CNRS 8601, Université René Descartes, 45 rue des Saints-Pères, 75270 Paris cedex 06, France ; [b]Laboratoire de Chimie, Electrochimie Moléculaires et Chimie Analytique, UMR CNRS 6521, Université de Bretagne Occidentale, 6 avenue Le Gorgeu, 29238 Brest cedex 3, France; [c]URCOM, Université du Havre, 25 rue Philippe Lebon, BP 540, 76058 Le Havre cedex, France

Abstract: The unique capacity of the calix[6]arene cavity to encapsulate small molecules has been used to explore the influence of the environment of a protein-like pocket on the biomimetic reactivity of metal ions bound within that cavity. Increasingly sophisticated functionalisation of calix[6]arene has provided three generations of enzyme models, the most advanced providing accurate mimicry for selective substrate binding and guest-controlled redox properties. These supramolecular models provide useful understanding of the detailed mechanisms of the function of both Zn and Cu dependent enzymes in hydrolytic and redox processes, opening thereby the route to a "synthetic enzyme" (synzyme) system.

Key words: Metalloenzyme models, funnel complexes, tris(aza-donor) sites, zinc hydrolases, copper oxidases.

In the biological world, particles of nanometre dimensions – proteins, nucleic acids and polysaccharides, for example – might be considered the norm, rather than the exception. The diverse functions of these materials are characterized by their selectivity, a property which can be seen as deriving from their complex macromolecular structure and its organization so as to control the numerous factors which may influence chemical reactivity. Recognition processes are based on non-covalent interactions such as hydrogen bonding, electrostatic and the weaker but multiple CH/π and cation/π interactions. Many proteins also contain metal ions, which may

J. Vicens and J. Harrowfield (eds.), Calixarenes in the Nanoworld, 259–285.

have a structural role. They can also be directly involved in recognition processes or even act as a catalyst leading to the selective transformation of an organic substrate. In enzymes, the control of the metallo-site is due to the structuring of the protein backbone in order to not only preorganize the coordination site for the metal ion, but also to provide a cavity and a corridor connecting it to the bulk that controls the binding of exogenous molecules and the reactivity of the metal ion. A few examples are given in Fig. 1.

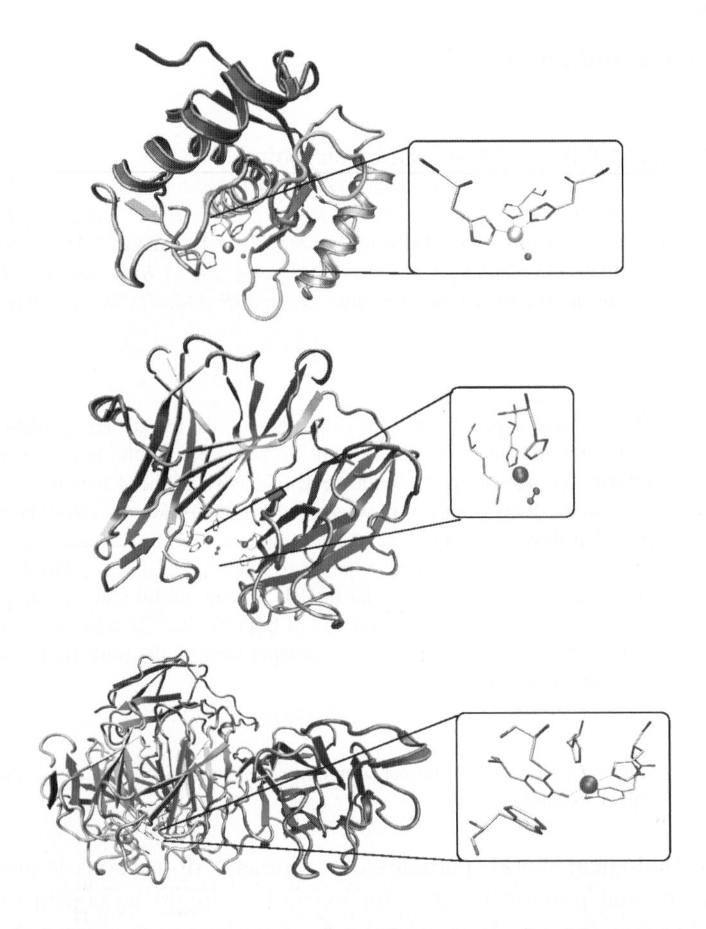

Figure 13-1. Schematic representations of the 3D structures of some mononuclear metallo-enzymes with their active sites in insets. From top to bottom: Adamalysine (a Zn-matrix metallo-peptidase from snake venom), Cu-Peptidylglycyl α-hydroxylating mono-oxygenase and Cu-galactose oxidase, PDB codes : 1IAG, 1SDW and 1GOF, respectively.

Among the great variety of metallo-enzymes, interesting sub-families present a mononuclear active site where a single metal ion is coordinated to a polyimidazole core.[1] For example, carbonic anhydrase, human collagenases, snake venom matrix metalloproteinases and β-lactamase II, all possess in their active site a mononuclear zinc center coordinated by three histidine residues and a water molecule. The latter is either displaced by the substrate or deprotonated.[2] A mono-copper site coordinated to a polyimidazole core is found in important copper enzymes.[3,4] Some display a mono-oxygenase activity and hydroxylate a CH_2 moiety of the glycinamide terminal group of peptides (in peptidyl amidating mono-oxygenase)[5] or of dopamine (in dopamine β-mono-oxygenase). Other enzymes are oxidases and dehydrogenate alcohols (galactose oxidase), amines (amine oxidase) or aldehydes (glyoxal oxidase).

A good chemical model of such metallo-enzymes is a key to the understanding of the fundamental mechanisms of the chemistry involved in the bio-catalytic cycles and to the design of efficient and selective new tools for the synthetic chemist. For that purpose, three important elements are required :

- a polydentate ligand to mimic the geometry and the chemical nature of the amino-acid residues involved in the coordination sphere of the metal;
- an appropriate environment to isolate the metal center and control the nuclearity of the system;
- a protected vacant site to allow coordination and exchange of an external ligand.

The *tris*(histidine) motif, frequently encountered in the case of zinc and copper enzymes, has guided the conception of tri-aza ligands able to reproduce the first coordination sphere of the metal. Among the best known artificial tripodal systems are the *tris*(pyrazolyl)borates, also called scorpionate ligands,[6] and triazacyclononanes.[7,8] The classical strategy to protect the reactive metal center against deactivating interactions (extra-coordination, dimerization,...), consists in tuning the steric environment provided by a tripodal ligand. This, however, presents a dilemma: high steric hindrance will prevent the metal ion from reacting with an external substrate molecule.[9-11]

Few biomimetic complexes have combined a metal ion and a hydrophobic cavity but a supramolecular approach is very important for the detailed understanding of the functioning of natural systems. Recognition mechanisms, chemio-selectivity and allosteric control are indeed moderated by the interactions of the substrate with the protein. We were interested in developing a supramolecular model that would allow the control of not only the first but also the second coordination sphere as well as the substrate access channel that selects and drives the substrate to the metal center, and expels the product. Our strategy consists in the covalent linkage of coordination arms, which reproduces the natural core provided by the amino acid residues found in the enzymes, to a hydrophobic cavity playing the role of the enzyme active site pocket (see Fig. 2).

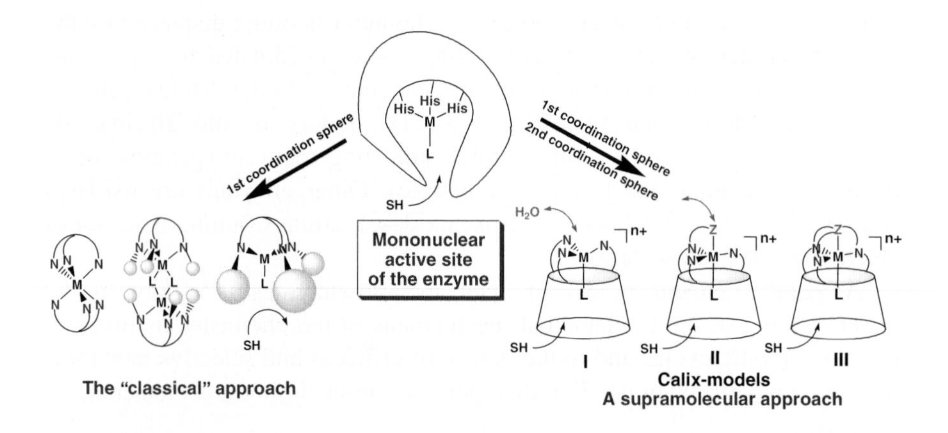

Figure 13-2. Modeling the active site of mononuclear enzymes with tripodal systems.

A possible way to construct a supramolecular system that combines a biomimetic coordination core with a cavity is to use readily available building blocks such as cyclodextrins,[12] cyclotriveratrylenes,[13] resorcinarenes,[14] or calixarenes.[15,16] In comparison to the others, relatively little has been reported on the use of the calixarene cavity, although these cyclic phenolic oligomers can easily be functionalized. This stems from the fact that, on the one hand, the conical cavity provided by calix[4]arenes is too small to play the role of a good host for organic molecules. On the other hand, the increased flexibility of higher oligomers such as calix[6]arenes, due to the facile ring inversion of their phenolic units, constitutes an obstacle for obtaining a receptor with a cavity. Our first hope was that, upon coordination to a metal center by adequately functionalized arms grafted to the narrow rim of the calix[6]arene, the macrocyclic structure would become constrained in the cone conformation required to play the role of a host, and orient a free coordination site toward the inside of the calixarene hydrophobic cavity. On this basis, three generations of calix[6]arene-based ligands have been developed.

1. FIRST GENERATION OF CALIX[6]-MODEL COMPLEXES

Tridentate *N*-ligands presenting either aromatic amines or alkylamines are synthesized from p-*t*-Bu-calix[6]arenes *O*-protected in alternate positions by three alkyl groups R. They form a large family of biomimetic N_3-ligands with different electronic and steric properties (see Fig. 3).[17,18]

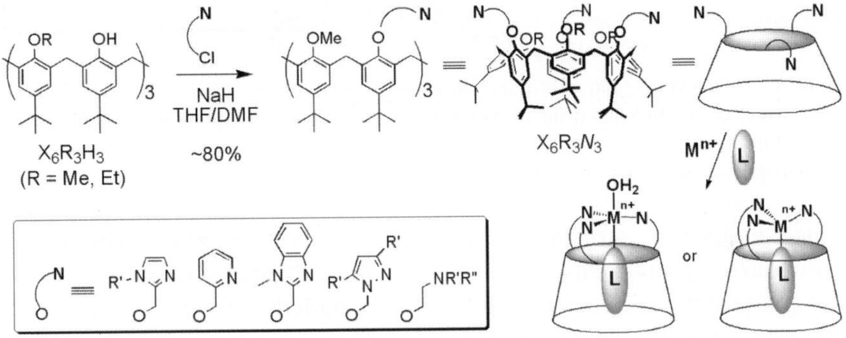

Figure 13-3. First generation of calix[6]arene-based N_3-ligands.

1.1 Zn(II) *Funnel* Complexes with the Calix[6]N_3 Ligands

The capacity of these N_3 ligands to coordinate Zn(II) in acetonitrile was evaluated.[17-21] Tertiary amines were too basic and sterically hindered, leading to precipitation of $Zn(OH)_2$. Ligands with pyrazole, benzymidazole and imidazole donor units, all form stable Zn(II) complexes under stoichiometric conditions. The calixarene functionalized by three pyridine groups, on the other hand, did not appear to be a good ligand for Zn(II), which stands in contrast to its remarkable ability at stabilizing copper(I) (*vide infra*).

1.1.1 Supramolecular Stabilization of a *tris*(imidazole) Zn-aqua Complex: A Structural Model for Mono-zinc Active Sites of Enzymes

Several chemical systems have been developed by various groups to reproduce the $[Zn(His)_3(OH_2)]^{2+}$ coordination core encountered in many hydrolytic Zn enzymes. Surprisingly, zinc aqua model complexes have proven extremely difficult to stabilize and most classical models only succeeded in stabilizing Zn-hydroxo species because of the high Lewis-acidity of the

Zn(II) center. In strong contrast, the reaction of the calix[6] *tris*(imidazole) ligand with $Zn(H_2O)_6(ClO_4)_2$ in THF, readily yielded a very stable dicationic zinc-aqua complex (Fig. 4).[20]

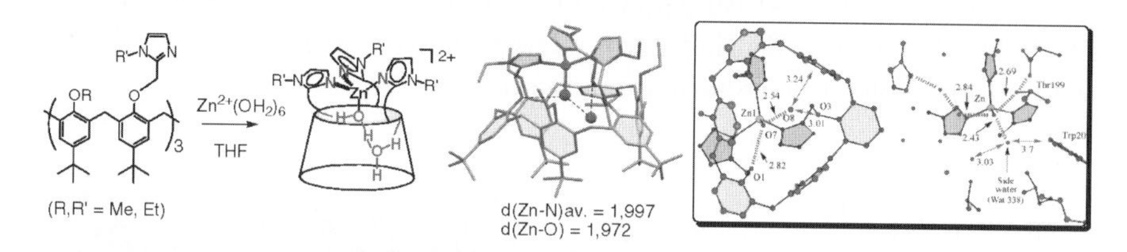

Figure 13-4. Synthesis and molecular structure (XRD) of the first dicationic Zn-aqua complex. Insert: comparison of the hydrogen bond networks in the calixarene-based complex (left) and in the carbonic anhydrase (right). Alkyl substituents in the calixarene structure have been omitted for clarity. Distances are given in Å.

Its molecular structure shows a tetrahedral zinc within the *tris*(imidazolyl) environment provided by the calixarene-based ligand in a cone conformation. Two water molecules are buried in the calixarene cavity, with one of them coordinated to Zn. The second water molecule is suspended in the heart of the cavity by a very strong hydrogen bond to the aqua ligand and an OH/π stabilizing interaction. Each water molecule is also hydrogen-bonded to an oxygen atom linked to the calixarene nitrogen arms. A comparison with the active site of carbonic anhydrase[22] shows surprising similarities. In the acidic form of the enzyme, the aqua ligand is strongly hydrogen-bonded to a threonine residue and to two water molecules. One of them which is displaced by the substrate CO_2, is situated at a very short distance from the aqua ligand. It also stands next to another H_2O molecule and to a tryptophan residue. Hence, the analogies between this side water and the (O8) water molecule of the calixarene-based complex are quite remarkable. The exceptional stability of this calixarene-based Zn-aqua complex is best illustrated by its reluctance to deprotonation in the presence of one molar equivalent of an amine. Instead, the aqua ligand is displaced by a primary amine yielding the 4-coordinate species $[Zn(N_3)(NH_2R')]^{2+}$ (*vide infra*). Such a behavior stands in contrast to the strong acidity expected from a water molecule coordinated to a dicationic tetrahedral zinc center.

1.1.2 XRD Characterization of Ternary Complexes $[Zn(N_3)(L)]^{2+}$ with Various Organic Guests L

When crystallized out of a solution containing small organic coordinating molecules (L), ternary complexes were isolated.[18,19,21] In each case, X-ray diffraction analysis showed a Zn(II) center in the regular tetrahedral environment provided by the *tris*(imidazole) core and the guest ligand L. With protic guests such as amines, alcohols or primary amides, hydrogen bonds connect their acidic protons to one or two calixarene phenoxyl units (as in the case of the aqua-complex). The guest conformations, often undergoing gauche interactions, seem to correspond to an optimized filling of the calixarene cavity with stabilizing CH/π interactions between the guest alkyl chain and the aromatic walls of the host (Fig. 5).

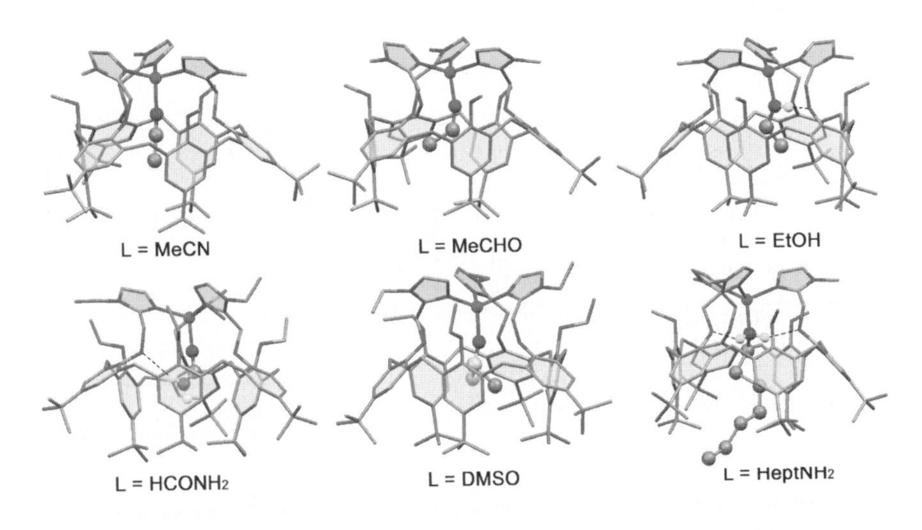

Figure 13-5. XRD structures of ternary dicationic Zn(II) complexes based on calix[6]*tris*-(imidazole) ligands with various organic guests L.

The ethanol[21] and acetaldehyde[18] ternary complexes provide interesting models for substrate binding in Liver Alcohol Dehydrogenase (LADH), a zinc enzyme that catalyses the reversible dehydrogenation of alcohols to aldehydes *via* hydride transfer with NAD^+.[20]

1.1.3 Solution Characterization (^{1}H NMR) of the Ternary Complexes $[Zn(N_3)(L)]^{2+}$

^{1}H NMR spectroscopy has proven to be a powerful tool to monitor the presence of a coordinating molecule (L) inside the cavity.[18,19,21] Studies in $CDCl_3$ showed the easy exchange of the guest ligand with only little change in the C_{3v} flattened cone conformation of the calixarene, in agreement with the XRD data. In most cases, the peaks of the included coordinated molecule are very sharp and well defined (see Fig. 6). This indicates that chemical exchange at the metal centre was slow on the NMR timescale. The proton resonances of the NH_2 and OH groups were also identified, displaying a coupling with the adjacent methylene protons. Hence, the exchange of the protons at the level of the coordinating heteroatom is substantially slower in comparison with the free molecules (L = RNH_2, ROH). This reflects the high hydrophobicity of the cavity around the guest molecule. The up-field shifts ($\Delta\delta$) measured for the guest protons were dependent on their spatial position in the aromatic cavity of the calixarene. An excellent correlation between their relative position to the coordinating heteroatom (Y = *N* or *O*) and the corresponding $\Delta\delta$ is observed for all molecules.

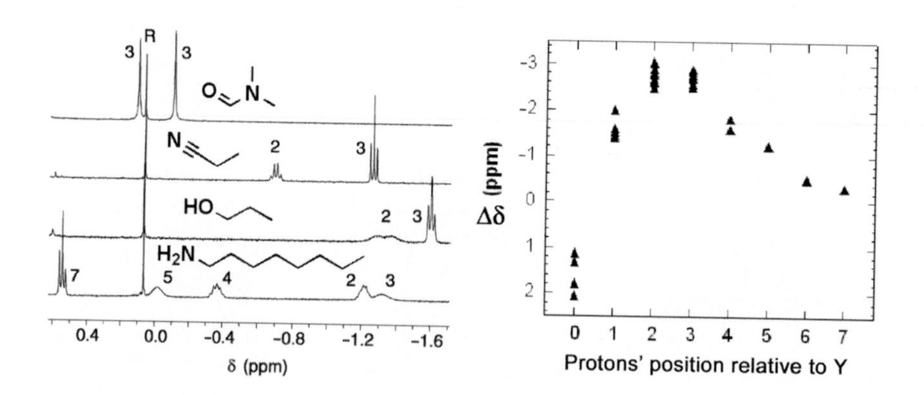

Figure 13-6. Representative ^{1}H pattern of included guests L ($CDCl_3$, 298 K) and mapping of the corresponding $\Delta\delta$ shifts (Y stands for guest coordinating atom).

1.1.4 The Zn *Funnel* Complexes Behave as Highly Selective Receptors for Small Neutral Guests

The equilibrium constants $K_{L/2H_2O}$ were shown to be first and second order relative to the L and water guests concentrations, respectively. With L = DMF, the positive equilibrium entropy of $\Delta S = 91(3)$ $J.K^{-1}.mol^{-1}$ associated with the enthalpy value of $\Delta H = 33.5(7)$ $kJ.mol^{-1}$ accords well with an exchange process in which two water molecules are replaced by one molecule of DMF.

The relative capacity of organic molecules to bind to the Zn(II) center is evaluated by the equilibrium constants $K_{L'/DMF}$ (see Fig. 7). These show that the selectivity of the binding in the cavity is based on both, the affinity of the donor atom Y of the guest ligand for the zinc ion and the relative host/guest geometries.

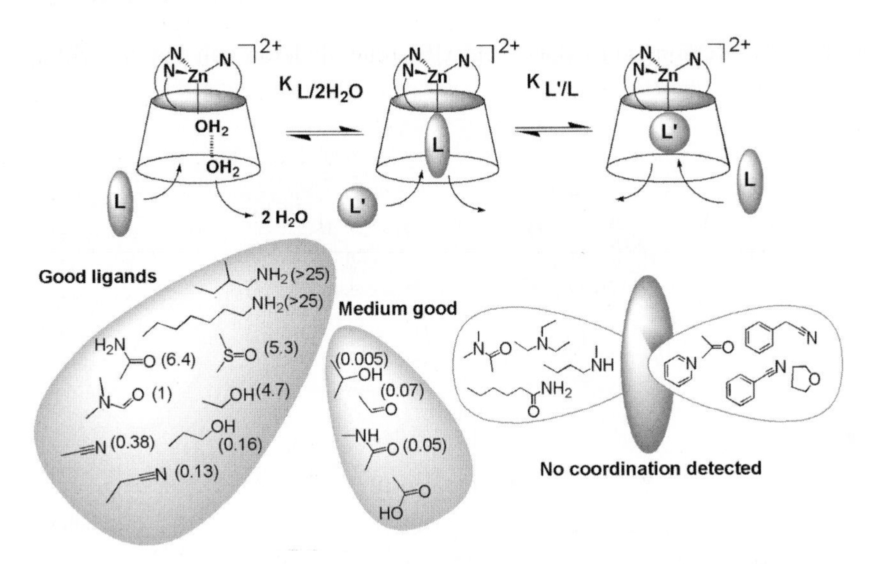

Figure 13-7. Ligand exchange at the Zn(II) center of the *funnel* complexes. The equilibrium constants $K_{L'/DMF}$ (exchange of L = DMF for L' at 298 K in $CDCl_3$) are given in parenthesis.

- With primary amines, coordination to the metal center is stoichiometric and quantitative. Amides and alcohols are also excellent guests, better than nitriles. Coordination of aldehydes and carboxylic acids was much weaker, although detected. Neither ether nor ketone yielded detectable coordinated species.
- Steric hindrance at the level of the coordinating atom (Y) and at its α position is a major factor of selectivity: whereas primary amines are the best ligands, secondary amines do not coordinate the metal center at all. Coordination of 1-propanol is 30 times stronger than 2-propanol.
- Either a methyl substituent in 2-position (as in DMF) or a long alkyl chain does not preclude coordination at the metal center. However, benzo- and benzyl-nitrile are too sterically encumbered to yield a stable adduct.

1.1.5 Chirality

These *funnel complexes* are chiral due to their helical shape. In solution, both enantiomers are in equilibrium.[18] Sterically hindered *N*-donors have the highest enantiomerization barrier (>70 kJ/mol). The helicity, which originates from the metal binding of the three amino arms, is efficiently transmitted to the calixarene cavity that is twisted, hence providing a chiral environment that ultimately is experienced by the guest (see Fig.8). Conversely, a chiral guest can control the equilibrium between the two helical forms of the complexes, thereby transmitting its own chirality to the system. This demonstrates that a calix[6]arene skeleton can convey chiral information.

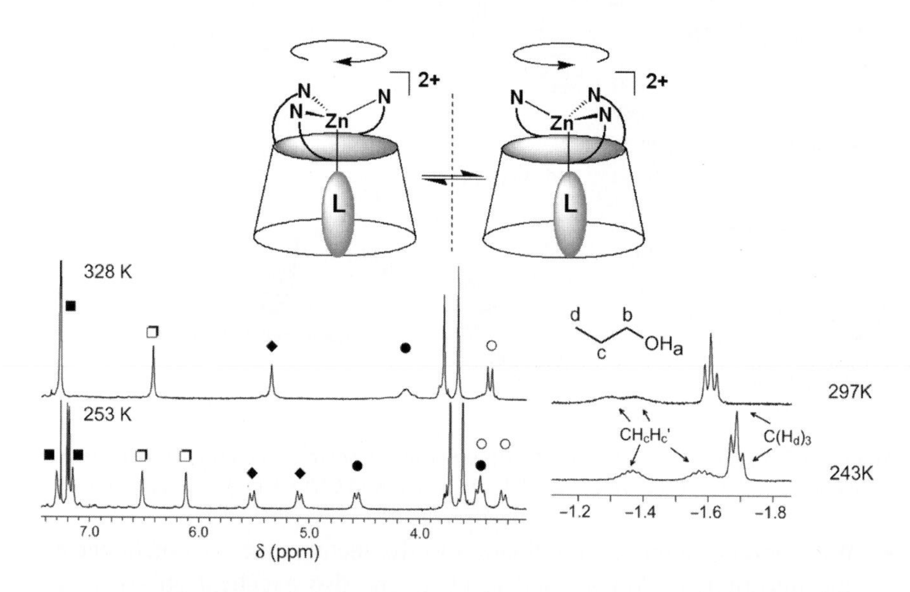

Figure 13-8. Diastereotopic differentiation of the host {left, calix[6]*tris*(imidazole)} and guest (right, L = EtOH) ^{1}H NMR resonances (400 MHz) upon freezing the equilibrium between the helical enantiomers in $CDCl_3$.

1.2 Calix[6]N_3 Cu(II) and Cu(I) *Funnel* Complexes

The ability of the calix[6]N_3 ligands to stabilize Cu complexes was explored. As the function of copper enzymes is based on Cu(II)/Cu(I) redox processes, the calix[6]arene-based complexes of both Cu(I)[23-27] and Cu(II)[28-30] have been synthesized and characterized in solution (NMR, EPR, UV-vis and electrochemistry) and in the solid state (XRD). Classically, the Cu(I) complexes preferentially adopt a tetrahedral environment, whereas the Cu(II) complexes are 5-coordinate in a distorted square-based pyramidal (SBP) environment. For Cu(II), an additional water molecule binds outside the cavity and is stabilized by a hydrogen bonding network (see Fig. 9). For both redox states however, the hydrophobic protein-like pocket of the calixarene controls an exchangeable guest. These Cu complexes displayed quite different binding and redox properties, depending on the nature of the coordinating arms. On one hand, unlike Zn(II), tetrahedral Cu(I) complexes appear to be best stabilized by the *tris*(pyridine) ligand with a bound RCN guest buried inside the cavity.[23] On the other hand, the *tris*(imidazole) ligand yields highly stable Cu(II) complexes with either nitrile, alcohol or amide guests.[28,29] In all cases, the Cu(I) complexes are resistant to air oxidation in spite of their labile site. This stands in contrast to the classical model systems that undergo fast dimerization into Cu(II) dinuclear complexes upon exposure to dioxygen.

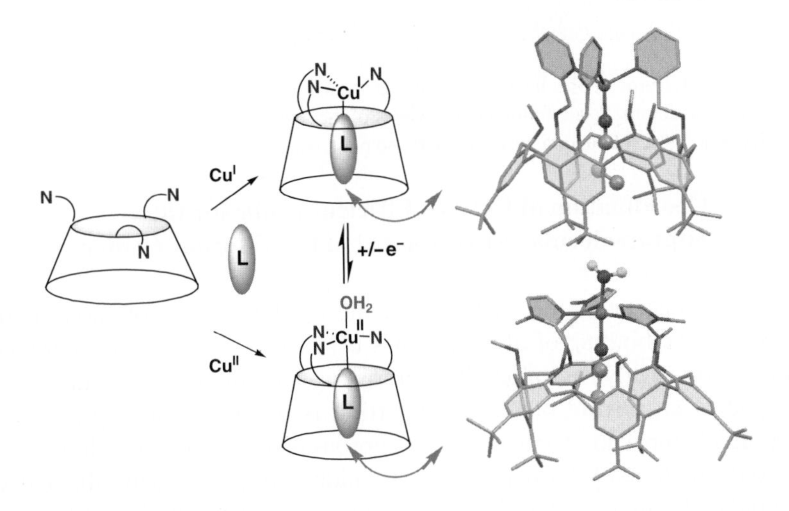

Figure 13-9. General scheme for the complexation of Cu(I) and Cu(II) by the calix[6]N_3 ligands and XRD structures of complexes $[Cu(I)N_3(EtCN)]^+$ and $[Cu(II)N_3(MeCN)(H_2O)]^{2+}$ stabilized by a *tris*(pyridine) and a *tris*(imidazole) core, respectively.

1.2.1 Supramolecular Control of the Thermodynamics and Kinetics of RCN Guest Exchange

The mechanism of the guest exchange at the cuprous center has been thoroughly explored with *tris*(pyridine)-based calixarene ligands.[26] For this redox state, thermodynamic and kinetic constants could be extracted from NMR experiments, which indicated a dissociative process. With this system, the $(t\text{Bu})_6$-calixarene-based model (**a**) was compared to the one having three *t*Bu substituents removed at the wide rim (**b**) (see Fig. 10). As the latter presents an enlarged cavity with a wider opening, the recognition pattern for MeCN *vs.* PhCN is inverted, the relative affinity constants differing by three orders of magnitude.

The removal of the *t*Bu door (b) also leads to a 100 fold increase of the MeCN exchange rate due to both a lower activation enthalpy and a higher activation entropy. This shows that bumping into the *t*Bu door is part of the selective process. Hence, these supramolecular systems provide a rare and interesting model for the hydrophobic substrate channel giving access to a metallo-enzyme active site.

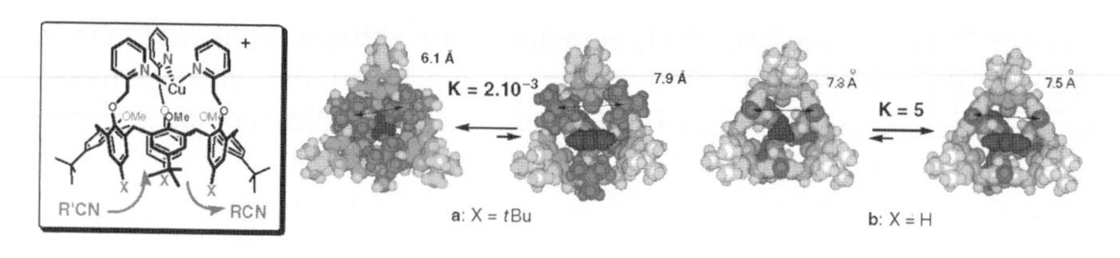

Figure 13-10. Control of the ligand exchange at the cuprous center by the "X_3 door" of the calix[6]arene wide rim. Space-filling bottom views of the *tris*(pyridine) Cu(I) complexes modeled with MeCN and PhCN bound in the cavity (left and right, respectively, for each equilibrium).

1.2.2 Coordination of CO: An Efficient Probe for the Supramolecular Environment of the Cuprous Center

In most instances, the first step of the catalytic cycle of mono-copper enzymes is the binding of dioxygen by Cu(I). However, mononuclear Cu-O_2 adducts are highly unstable. Therefore, carbon monoxide, which is an analogue of O_2 devoid of redox properties, is often used as a probe in the substrate chamber to characterize the cuprous state of enzymes. In our case, the conformationally ill-defined Cu(I) adduct obtained with the calix[6] *tris*(imidazole) ligand when RCN is not present as a fourth ligand for metal, leads to a very stable tetrahedral complex upon CO coordination. This adduct was characterized by IR and ^{1}H-NMR spectroscopies. Interestingly, a VT ^{1}H NMR study revealed the partial inclusion of one *t*Bu group into the calixarene pocket (see Fig. 11).[25]

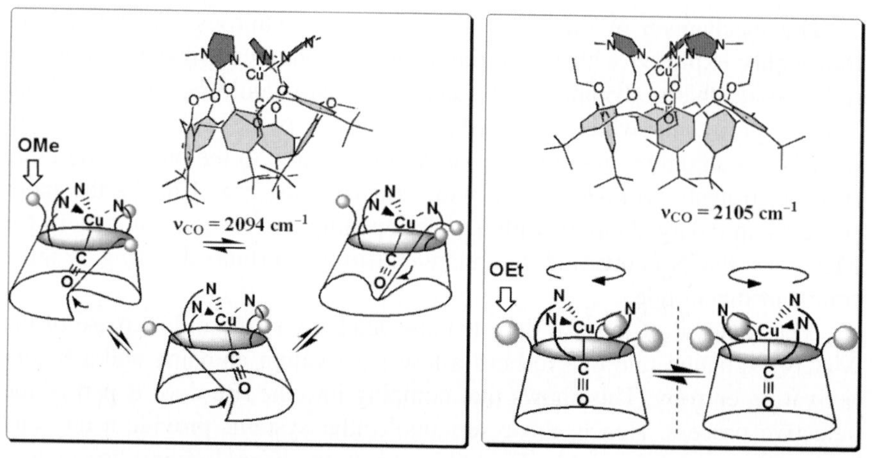

Figure 13-11. Coupled electronic and conformational properties of the calixarene-based *tris*(imidazole) Cu(I)(CO) complexes controlled by the narrow rim OR substituents.

As the CO ligand is not large enough to fill the pocket, the Cu(I)(CO) moiety actually wobbles between three dissymmetric but equivalent conformations with a self-included *t*Bu group. At RT, their fast interconversion results in an averaged *pseudo*-C_{3v} symmetrical structure, the Cu complex undergoing a fast "three-step dance". This process is actually controllable by the *O*-R substituents at the narrow rim of the calixarene: if it is bulkier than methyl, such as ethyl for example, the C_{3v} symmetrical flattened cone conformation is maintained even at low T and the complex is restricted to a "two-step dance" between two helical enantiomers, as in the case of Zn(II) complexes.

Although the accountable small ligand change occurs relatively far away from both the cavity and the metal ion, the resulting differences in the steric demand also induces significant modifications of the electronic properties of the metal center (ν_{CO} = 2094 and 2105 cm^{-1} for R = Me and Et, respectively). Hence, as in copper proteins, the vibrational spectrum of the bound CO is a sensitive gauge of the coordination geometry of the Cu(I) ion. For example, a FT IR study of a mitochondrial cytochrome c oxidase showed a Cu-CO absorption split into two bands (ν_{CO} = 2055 and 2065 cm^{-1}). As in the present model, this observation was attributed to the flexibility of the Cu-CO adduct embedded in a very non-polar pocket.[31]

1.2.3 Redox Behavior of the Calix[6]N_3 Cu *Funnel* Complexes

In simple species, the Cu(II)/Cu(I) redox process involves a structural reorganisation between five- and four-coordinate species. The electrochemical behaviour of the Cu-calix[6]areneN_3 complexes, especially in the case of the pyridine-donors, is chemically and electrochemically irreversible, indicating major structural restrictions, except in the case of solutions in CH_3CN.[32] Binding of CH_3CN as a ligand appears to be necessary to achieve control of the cone conformation of the calixarene and to result in sufficient flexibility for the macrocycle to adapt not only to the coordination requirements of Cu(II) and Cu(I) but also of the intermediates states involved in electron transfer. This role of the solvent molecule is the basis for considering CH_3CN as a *shoetree* molecule (Fig. 12), inducing a conformational adaptation of the calixarene analogous to *induced fit* behaviour in biological systems.

The *tris*(pyridine) and *tris*(imidazole)-based Cu(II)/Cu(I) systems have very different redox potentials with Ep_{ox} = 0.9 V and 0.2 V, respectively. This difference can be accounted for by a square-scheme mechanism involving the route of electron transfer through whichever is the more stable intermediate. This This defines the value of the redox process E as equal to either E^0_A or E^0_B ($E^0_A < E^0_B$). For pyridine donors, their relatively strong π-acceptor character, combined with the retracting effect of the *shoetree* towards Cu, strongly favours tetrahedral geometry even for Cu(II) and this strained, "entatic" state of the metal is strongly oxidising ($E = E^0_B$). Thus, the complex will electrocatalytically oxidise phenols, for example.[29] Conversely, with the better σ-donors provided by imidazole units, the retracting effect of the CH_3CN ligand has less influence and five-coordination is possible, leading to a much weaker oxidant ($E = E^0_A$).

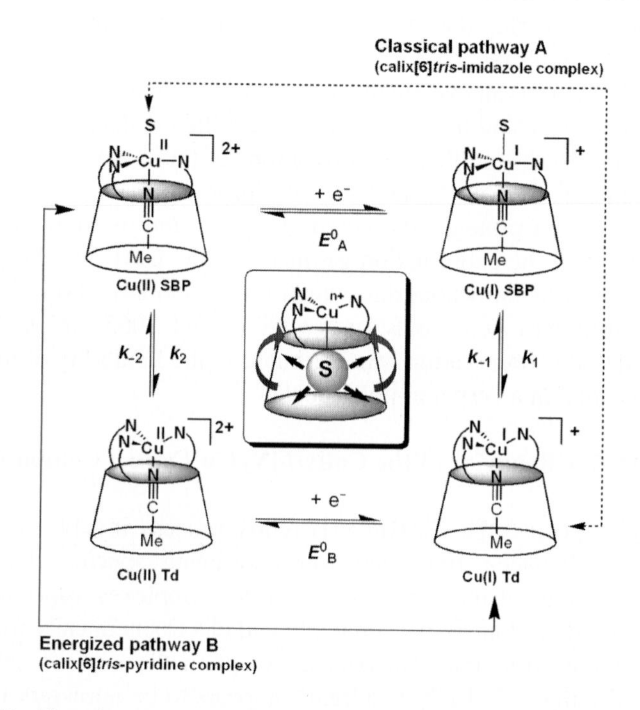

Figure 13-12. The *shoetree* supramolecular concept illustrated by the square-scheme mechanism for rationalizing the Cu(II)/Cu(I) topological reorganization through the two possible pathways.

Hence, the supramolecular control of the coordination of copper allows the tuning of the redox properties of the *funnel complexes*. This aspect is highly reminiscent of the blue copper proteins involved in electron transfer as well as mono-copper enzymes. Indeed, for the former, the first and second coordination spheres control the redox potential and electron pathway through an entatic state. For the latter, stepwise reaction with O_2, sequential electron inputs and substrate oxidation are redox key steps that are highly controlled by the supramolecular environment at the active site.

1.3 Towards a Supramolecular Control of a Hydrophobic Coordination Sphere in an Aqueous « Biological » Medium

The binding of a metal ion to a calix[6]arene bearing three coordinating arms allows the shaping of a molecular receptor. The resulting calix[6]N_3 mononuclear complexes are chiral supramolecular edifices. They act as selective receptors for neutral molecules thanks to the establishment of weak stabilizing interactions between host and guest. As a result, the Zn(II) system turns out to be highly reminiscent of the Michaelis complexes in enzymes. The remarkable stability of the *funnel* complexes is attributable to the control of the second coordination sphere. Hence, while the Cu(I) and Cu(II) complexes present different coordination numbers and environment, for both redox states, a supramolecular connection links one copper labile site and the calixarene cavity. This allows the control of the geometry and the interaction with the substrate as well as the electron exchange pathway, as in copper proteins. Also, the nature of the *N*-donors, imidazole or pyridine, allows the tuning of the electronic properties, the latter giving rise to an enhanced reactivity for the oxidation of exogenous substrates. Lastly, these studies show that the calixarene structure efficiently conveys the conformational modifications from one end of the molecule to the other. Such supramolecular behavior, promoted by the strong conformational coupling between the metal center and the host structure, is reminiscent of allosteric effects common in biological systems, such as protein folding or substrate binding.

Subsequently, we began exploring a possible route to transform the organo-soluble calix[6]N_3 derivatives into water-soluble compounds. The synthesis of the first water-soluble biomimetic cuprous complex was successfully carried out with a calix[6] *tris*(imidazole) ligand having hydrophilic substituents at the wide rim in place of the *t*Bu groups.[27] Unless coordinatively saturated, Cu(I) complexes are usually not stable in water and readily disproportionate. This is a limitation to the study (and applications) of copper model complexes in biological media. Like its organo-soluble analog, the water-soluble Cu(I) complex proved to be resistant to air autoxidation and reacted with carbon monoxide in water. Hence, careful functionalization of the calixarene-based ligand allows the retention of the most important properties of the complex (Fig. 13), *i.e.* the stabilization of the cuprous state despite the presence of a free valence, the protection of the metal site in the concave area of a hydrophobic cavity, and the host behavior of the complex toward small coordinating guests.

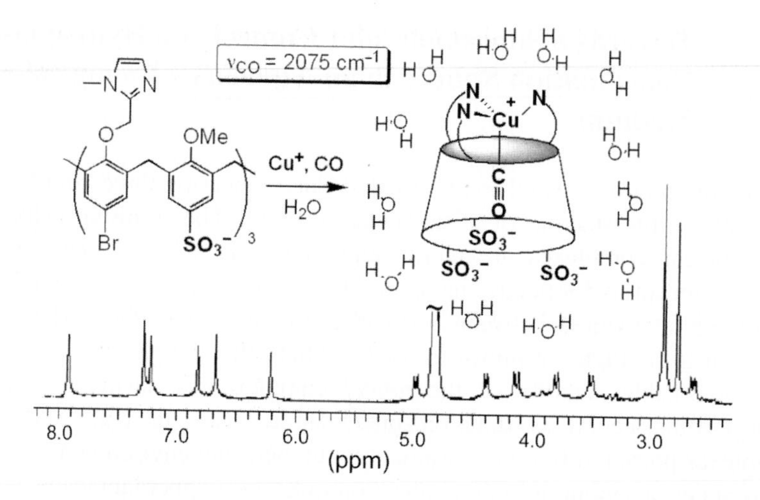

Figure 13-13. Synthesis and characterization (^{1}H NMR at 500 MHz in D_2O and IR) of the first water-soluble biomimetic Cu(I)CO complex.

This demonstrates the successful transposition of an organo-soluble biomimetic system into water. Such a supramolecular system reproduces not only the first coordination sphere encountered in many enzymes and the frequently hydrophobic microenvironment of the active site, but also the aqueous macro-environment of a physiological medium.

2. SECOND GENERATION OF CALIX[6]-MODELS: THE CALIX[6]N_3Z COMPLEXES

Wanting to introduce a different donor group into the coordination sphere of the metal ion, we took advantage of the *out* coordination site that was shown to be accessible for water in the case of divalent metal ions within the first generation of calix[6]N_3 ligands. Hence, a second generation of ligands, calix[6]N_3Z, was developed.[33,34] It presents a fourth donor group *Z* that is covalently linked to one of the nitrogenous arms and replaces the former aqua ligand, leaving a unique coordination site for an exogenous binding controlled by the calixarene pocket. This fourth donor can play the role of a semi-labile arm and control the inner binding.[35] It can also be a redox-active function, such as a phenoxide, and participate to the oxidation of a substrate, hence providing a good functional model of the copper enzyme, galactose oxidase.[36]

The key step for the ligand synthesis is the mono-alkylation at the narrow rim of the C_{3v} symmetrical trimethyl ether derivative of *t*Bu-calix[6]arene with *N*-Boc-2-chloroethylamine to yield a novel calix[6]arene synthon (see

Fig. 14). The subsequent grafting of the two imidazole arms and of a salicylaldehyde derivative yields the calix[6]arene-based ligands with mixed *N/O* donors.

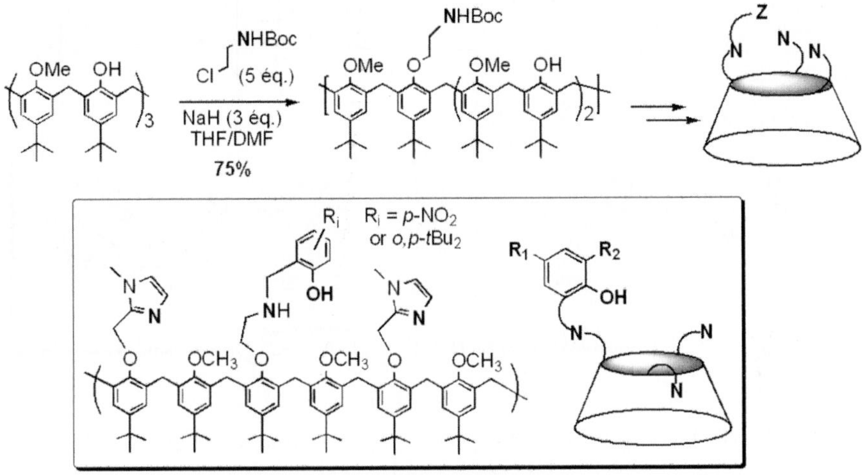

Figure 13-14. Synthetic strategy and 2nd generation ligand design.

2.1 Calix[6]N_3ArO Cu(II) Complexes: A Supramolecular Model for Galactose Oxidase

The intended role of a coordinated phenoxide (ArO^-) was to mimic the redox-active tyrosinate residue which is oxidizable into the tyrosinyl radical and found in some metal-radical enzymes, such as the copper enzyme galactose oxidase (GAO).[37]

Hence, the calix[6]N_3ArOH ligand was used to coordinate Cu(II) in the presence of a base.[36] The crystal structure of the Cu(II) complex displays a Cu(II) ion coordinated in a distorted square-based pyramidal N_4O environment. The main structural characteristics of this complex are quite similar to those described above with the *tris*(imidazole)-based calixarene system as the phenoxide group occupies the external site, thereby capping the cupric complex and leaving a single accessible site for the coordination of an exogenous ligand. The Cu complex undergoes fully reversible oxidation system, $E°' = 0.32$ V (*vs* Fc^+/Fc) ($\Delta Ep = 60$ mV, $i_{pa}/i_{pc} \approx 1$). Upon electro- or chemical [CAN (Ceric Ammonium Nitrate), 1 equiv/Cu] oxidation in the presence of MeCN at low T, the solution turned from purple to green with the formation the corresponding 1-electron oxidized species, namely a phenoxyl radical ($ArO^•$) bound to Cu(II) complex. The latter displayed spectroscopic features that are very similar to those reported for the oxidized active form of GAO: intense phenoxyl $\pi \rightarrow \pi^*$ transition at 405 nm and disappearance of the Cu(II) EPR signature due to a Cu(II)-ArO$^\bullet$ antiferromagnetic coupling (Fig. 15).

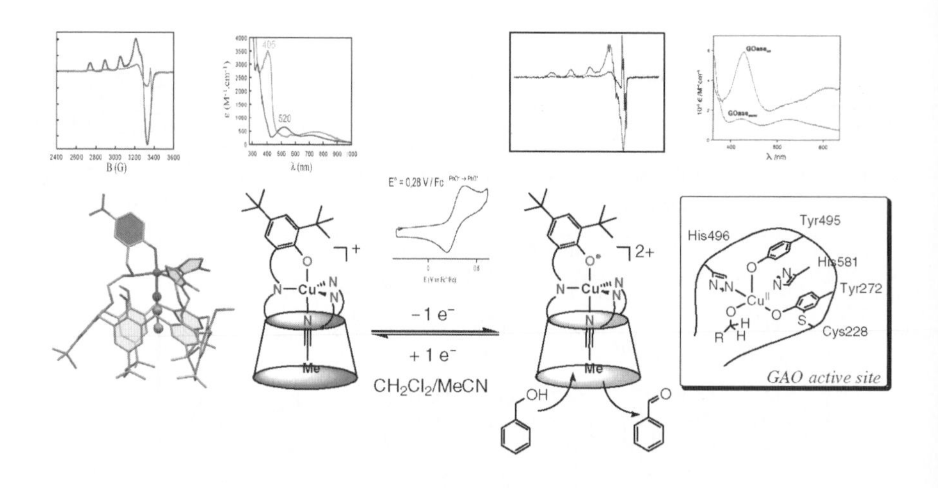

Figure 13-15. Modelling the active site of GAO. Left bottom XRD structure of the Cu(II) phenolate complex; top: EPR and UV-vis spectra of both phenate and phenoxyl radical Cu(II) centers for the model complex (left) and GAO (right).

A comparative study of the Cu(II) phenoxyl radical complex with its Zn(II) analog showed that, whereas the latter is stable for hours at RT, the former stoichiometrically oxidizes benzylic alcohol into benzaldehyde, which is consistent with an inner-cavity process through which only $[ArO^\bullet Cu(II)]^{2+}$, not $[ArO^\bullet Zn(II)]^{2+}$, can mediate the two-electron oxidation. Hence, this second generation of calix[6]N_3ArO systems is unique in that it associates a metal ion, an organic radical and a hydrophobic cavity. These specific features reproduce, as compared to the first generation, not only the protecting effect of the protein cavity relative to the substrate receptor site but also the redox activity of metal-radical enzymes.

2.2 Calix[6]N_3Z Zn(II) Complexes: An Acid-base Switch for Guest Binding

The coordination chemistry of Zn in the calix[6]N_3ArOH environment has also been studied (Fig. 16).[35] Three different protonation states for the corresponding Zn(II) complexes have been characterized: $[Zn(II)N_3ArOH]^{2+}$ $[Zn(II)N_3ArO]^{+}$ and $[Zn(II)(OH)N_3ArO]$. Whereas the dicationic 5-coordinate species is very sensitive to guest binding, the mono-cationic complex binds a guest ligand with a lower affinity.

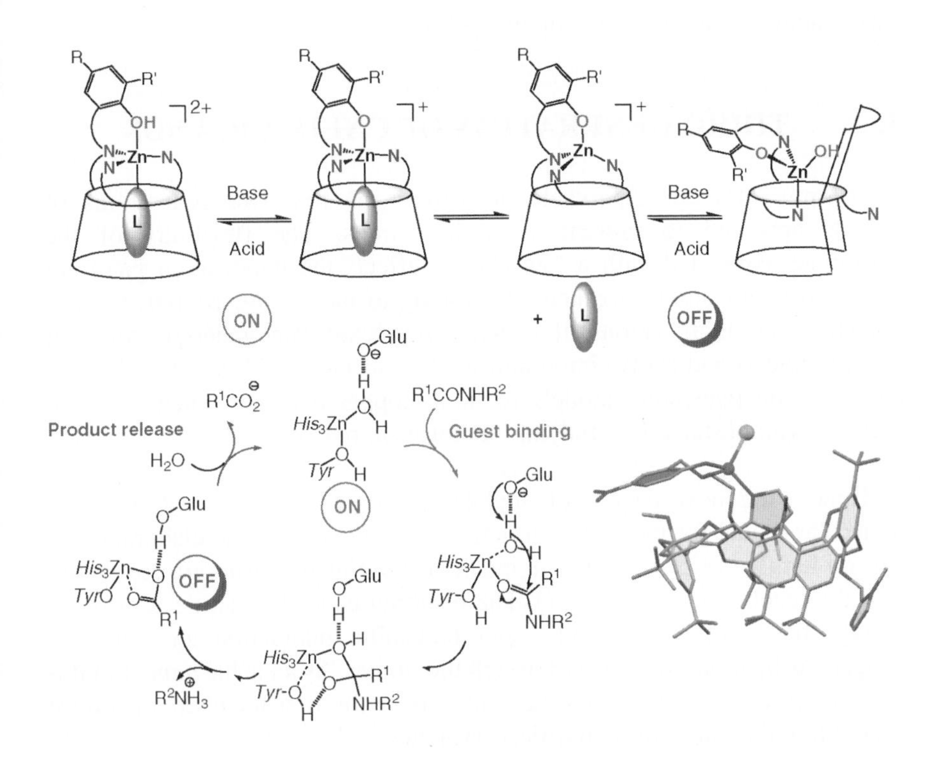

Figure 13-16. Acid-base switch for guest binding by the calix[6]N_3ArOH based Zn(II) complexes. Bottom right: XRD structure of the neutral chlorophenoxide complex; Bottom left: proposed mechanism for the peptidase activity of astacyn and serralysin Zn-enzyme families.

The neutral species can be obtained upon reaction with a base to yield a hydroxo complex or with an anion such as a chloride that coordinates the metal center from the outside of the calixarene cavity. The simultaneous binding of two anionic donors decreases the Zn Lewis acidity, allowing an impressive conformational reorganization of the system. One imidazole arm is released by the metal center. The other one undergoes self-inclusion into the π-basic calixarene cavity since the low affinity of the metal center for neutral ligands does not allow the endo-coordination of an exogenous guest. As a result, the calix[6]N_3Ar*O*H-based Zn complexes act as an acid-base switch for guest binding.

Several aspects of this system appear reminiscent of Zn-peptidases of the astacin and serralysin families.[2] For these enzymes, in addition to the three histidine residues, a side chain tyrosine coordinates the metal ion and its role has been questioned. This model system actually suggests that one of its roles could be to accurately control the activity of the enzymes as the pH varies, acting as an off-switch upon a pH rise.

3. THIRD GENERATION OF CALIX-LIGANDS

In spite of their remarkable properties, the first two generations of ligands appeared to present some limitations. The flexibility of the calixarene associated with a low chelate effect (the donor-atom pairs are separated by 17 atoms) weakens its binding to metal ions. Moreover, with potentially bridging anions, the metal complexes can undergo structural rearrangements and evolve into multi-nuclear species.[38-40] Even though these appear to be interesting models of multi-copper enzymes, such behavior represents an obstacle for studying the intrinsic reactivity of a mononuclear site.

Hence, the next degree of sophistication in the design of a supra-molecular model for mononuclear active site consisted in the elaboration of a novel family of ligands in which the three coordinating arms are covalently linked to each other (see Fig.17). The resulting capped structure provides a strong chelate effect, precludes any bimetallic interaction and enforces exogenous ligation exclusively through the funnel. As an achievement in this project, we recently described the synthesis of the first members of a third generation of ligands, the calix[6]azacryptands.[41-43]

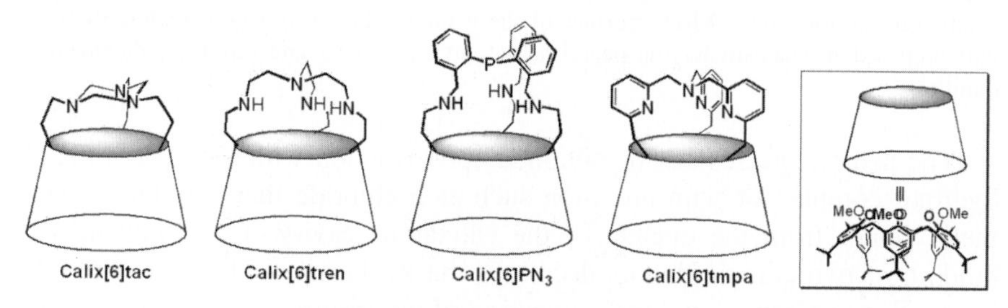

Figure 13-17. Calix[6]azacryptands.

An interesting feature of these compounds resides in the fact that coordination of a metal ion is not anymore a prerequisite to the rigidification of the calixarene core in a cone conformation. As a consequence, the calix-cryptands themselves display interesting tunable receptor properties.[44-46]

3.1 Calix[6]tren: A Versatile Receptor

We developed straightforward syntheses of these capped compounds mainly through [1+1] macrocylization reactions between tripodal partners. It is noteworthy that examples of calix[6]arenes bearing a tripodal cap are very rare in the literature. One of them, calix[6]tren, presents a calix[6]arene structure that is capped by a tren unit (Fig.18).[41] Calix[6]tren behaves as a quite remarkable receptor thanks to the combination of a calix[6]arene macrocycle constrained in a cone conformation and an aza cap that closes the narrow rim of the receptacle, leaving a single entrance controlled by the flexible *t*Bu door. The cap has a grid-like nitrogenous moiety that is highly basic. Hence, it can be used to polarize the edifice by protonation, offering in addition number of hydrogen bonding sites. The tren unit also offers a binding site for a metal ion that is firmly coordinated at the bottom of the concave cavity due to a strong chelate effect. As a result, three different host-guest systems have been described.

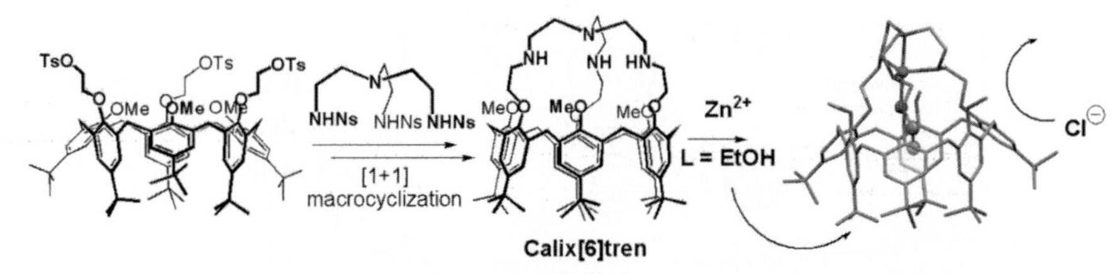

Figure 13-18. Synthesis of calix[6]tren and complexation of Zn(II) (XRD).

- Calix[6]tren itself complexes ammonium ions.[45] Here, the cap plays the role of a Brønsted base and recognition is based on multiple hydrogen bonding with the ammonium functionality of the guest and a good fit between its organic part and the calixarene cavity with stabilizing CH-π interactions.[42] In the absence of hydrophobic effect (requiring water as a solvent), non-polar interactions are not strong enough to allow the efficient binding of a neutral guest.
- In the *per*-protonated host, the cap presents four positive charges. This highly polarizes the receptor yielding strong charge-dipole interactions with the guest and offers multiple hydrogen bond donor sites.[45] The resulting remarkable binding properties toward neutral guests emphasize the efficiency of combining a poly-ammonium site and a hydrophobic cavity to build up a receptor for polar neutral molecules.
- The cavity can also be tuned by the coordination of a metal ion. A variety can be used. Zn(II) has been thoroughly explored.[44,45] X-ray and solution (NMR) structures show a dissymmetrical environment provided by the metallo-host and a host-guest selectivity based on the nature of the coordination link in the cap.

Hence, the affinity of the host varies according to the way it is polarized. For example, the protonated host does not bind benzylamine, whereas, due to the strength of the coordination bond, the Zn complex does. The versatility of the system is further illustrated by the reversible transformation of one form of the polarized receptor to another in solution (see Fig. 19). Such interconversion in solution allows the binding properties to be tuned by the environment (more or less basic, presence of metal ions etc...). The complexes are very stable and remarkably resistant to anion binding.

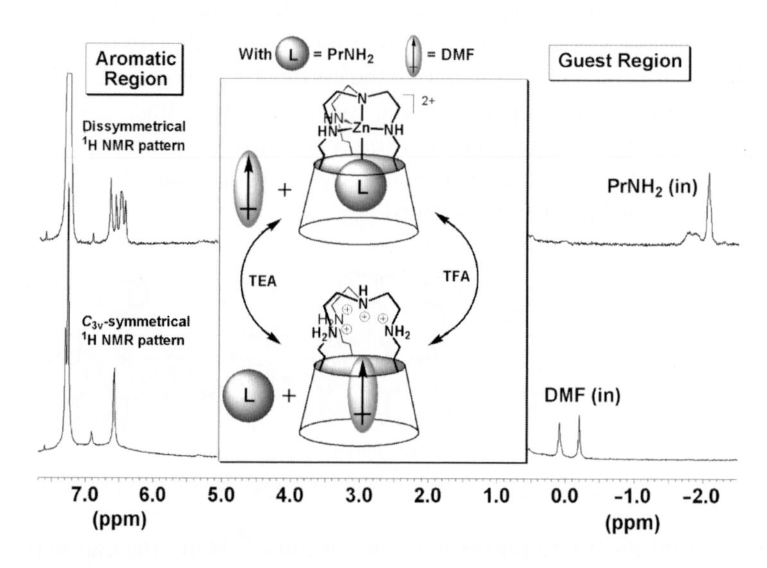

Figure 13-19. Guest switch through the reversible protonation of calix[6]tren-Zn(II) complex monitored by ^1H NMR in $CDCl_3$ (298 K); TFA and TEA stand for trifluoroacetic acid and triethylamine, respectively.

3.2 Redox-driven "Anti-thermodynamic" Ligand Exchange at the Calix[6]tren Cu Complexes

Upon coordination to a Cu(II) ion, the calix[6]tren ligand gives rise to a cupric complex that has a very strong affinity for a variety of small neutral guests (L = H_2O, EtOH, DMF, and MeCN).[46]

In contrast to the first- and second-generation models, the calix[6]-tren ligand forms a Cu(II) complex which undergoes well-defined redox chemistry that is at least chemically reversible. The metal is protected against influences from the surrounding medium by the ligand superstructure and the tren unit tends to enforce a trigonal array of three of its N-donors about the metal in both oxidation states. Coordination sphere reorganisation during oxidation state changes is limited to dissociation or binding of the axial guest. This on/off guest binding process in the calixarene pocket can be electrochemically controlled through the different affinities between Cu(II) and Cu(I) for the different guests L.

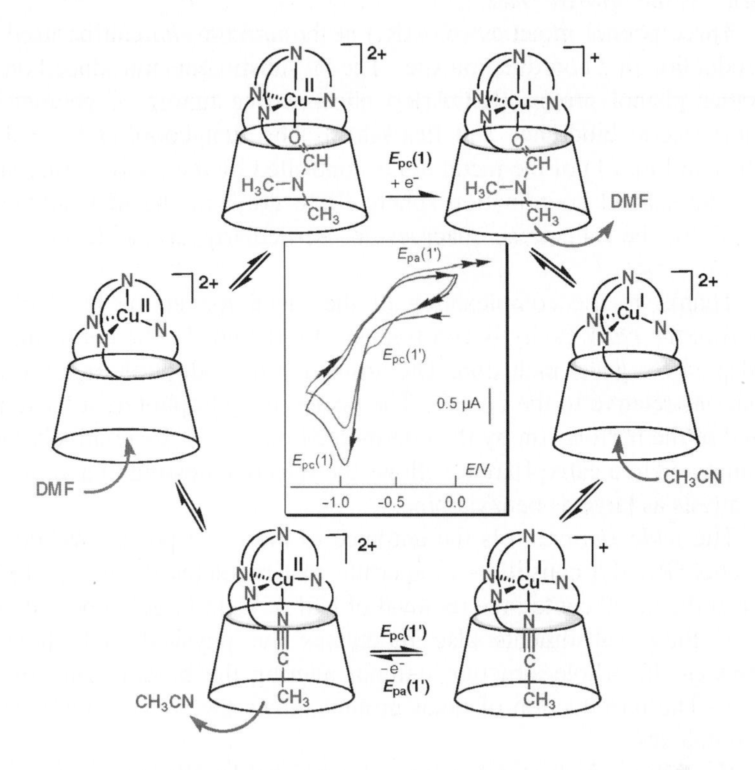

Figure 13-20. Redox driven ligand exchange within the calixtren Cu system.

Indeed, three types of redox driven guest-on/guest-off processes can be managed: guest binding/ejection, guest equilibrium and more remarkably a guest interconversion process. In this last process, a strong donor ligand (DMF) can be switched, after a reduction [Cu(II)(DMF) → Cu(I)(MeCN)]/ reoxidation process [Cu(I)(MeCN) → Cu(II)(MeCN)], for a weaker donor (MeCN) at the Cu(II) site. Such a conversion of a more stable Cu(II)(DMF) complex into a less stable Cu(II)(CH_3CN) complex may be viewed as a "remarkable antithermodynamic" redox-driven ligand exchange (Fig. 20).

In conclusion, it clearly appears that the calix[6]tren model closely approaches the concept of synzymes, both: i) in structural terms when considering the control of geometry and coordination sphere of the metal, its encapsulation in a protein like pocket with a free access channel, ii) in functional terms in regards to its redox behavior.

4. CONCLUSION

Calix[6]arene can be considered as a unique scaffold to model the active site pocket of metallo-enzymes.

- Three phenol moieties (-OArR_1) at the *narrow rim* can be used for the introduction of a coordination site. The R' substituents introduced on the three other phenol groups (R'OArR_2) allow a fine tuning of geometrical constrains (steric hindrance and flexibility). The first coordination sphere (dark grey in Fig. 21) of the metal ion is controlled by the coordinating arms whereas the second coordination sphere (light grey) of the M-L adduct is controlled by the calixarene macrocycle, particularly at the level of the narrow rim.
- Thanks to the complexation of the metal ion at the level of the narrow rim, the *calix* cavity is constrained in a flattened cone conformation well adapted for guest inclusion. The aromatic units adopt alternate *in* and *out* positions relative to the C_3 axis. The pocket is hydrophobic, π-basic and polarized at the narrow rim by the coordinated metal ion. Its relatively large size (compared to a calix[4]arene) allows the selective hosting of a variety of neutral guests as large as benzylamine.
- The *wide rim* controls the entrance to the guest pocket. While *t*Bu substituents (R_1, R_2) constitute an aperture that regulates the access to the free site at the metal center, the removal of half of them largely opens it. The nature of these substituents also modulates the physical and chemical properties of the whole structure without altering the binding site for the metal ion. The introduction of polar groups allows the water-solubilization of the complexes.

The flexibility of the structure can be restricted by the introduction of a cap (X) at the narrow rim. The resulting capped receptors strictly control the interactions with the guest molecules (L) through the large rim. They also better entrap these guest molecules from both a thermodynamic and a kinetic point of view.

The ability of this system to mimic specific aspects of metallo-enzyme active sites has been emphasized. In hydrolytic Zn enzymes, the metal ion plays the role of a Lewis acid for the activation of the water molecule and the substrate. In redox Cu enzymes, the metal ion either activates O_2 or mediates electron transfer reactions, hence swinging between two redox states. Our modeling ambitions have been directed toward mononuclear sites integrating a free access for exogenous binding, which are particularly difficult to model due to the propensity of the reactive species to undergo dimerization. Hence, another key role of the calix structure is to control the nuclearity of the complex.

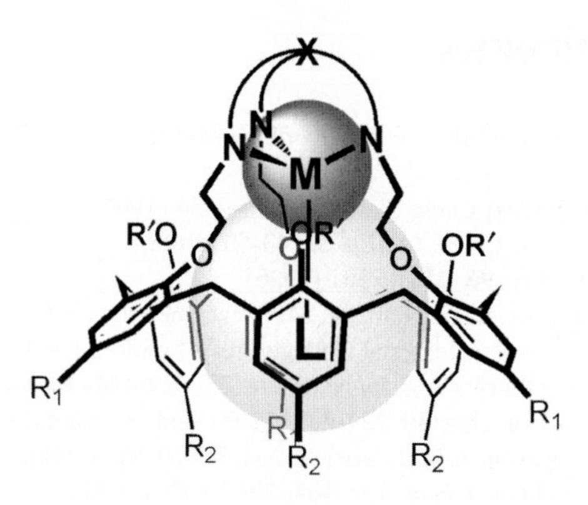

Figure 13-21. Schematic representation of the first and second coordination spheres of the funnel complexes.

- Our work with *Zn(II)* led to the characterization of the first dicationic tetrahedral Zn(II) complexes. Their exceptional stability stems from the second coordination core, which provides hydrogen bond acceptors (ArO) that release the Brønstedt acidity of the coordinated guest. It is also due to stabilizing interactions between the π-basic cavity and the alkyl group of the guest (CH-π interactions). These complexes led to the first model for the dicationic aqua complex in Zn-enzymes and mimicked nicely the Michaelis-Menten adducts, behaving as selective funnels for small neutral coordinating molecules. When covalently capped at the narrow rim, the rigidified system even better entraps the guest with a stronger bond between the Zn(II) complex and the substrate mimic.
- With *copper*, both oxidation states Cu(I) and Cu(II) were stabilized in spite of their different geometries and coordination numbers at the level of the first coordination sphere. The guest ligand plays the interesting role of a "*shoetree*" molecule that allows a continuum in the electron exchange process, hence controlling it. This can lead to an entatic state modeling bio-redox processes. Capping the system allows the full control of the guest binding through electrochemical monitoring; thus an "anti-thermodynamic" guest exchange was highlighted as a possible pathway. Lastly, introduction of redox active donor on the calix-ligand yielded a complex capable of a two-electron oxidation of an alcohol that is fully controlled by the cavity, hence providing a supramolecular nano-reactor, model of the natural system, the radical-enzyme GAO.

These supramolecular enzyme mimics open now a number of perspectives. Indeed, beside fundamental research for a better understanding of the mechanisms in biological systems, novel nano-devices for selective catalysis, sensing, electron transfer, redox mediation…, may be developed.

5. REFERENCES

1. I. Bertini, A. Sigel, H. Sigel (Eds.), *Handbook on Metalloproteins*, Marcel Dekker, New York, 2001.
2. W. N. Lipscomb, S. Sträter, *Chem. Rev.* **96**, 2375-2434 (1996).
3. W. Kaim, J. Rall, *Angew. Chem. Int. Ed.* **35**, 43-60 (1996).
4. J. P. Klinman, *Chem. Rev.* **96**, 2541-2561 (1996).
5. S. T. Prigge, B. A. Eipper, R. E. Mains, L. M. Amzel, *Science* **304**, 864-867 (2004).
6. S. Trofimenko, *Scorpionates,* Imperial College Press, London, 1999.
7. K. Wieghardt, P. Chaudhury, *Prog. Inorg. Chem.* **35**, 329-426 (1988).
8. W. B. Tolman, *Acc. Chem. Res.* **30**, 227-237 (1997), and ref. cited therein.
9. L. Q. Hatcher, K. D. Karlin, *J. Biol. Inorg. Chem.* **9**, 669-683 (2004).
10. E. A. Lewis, W. B. Tolman, *Chem. Rev.* **104**, 1047-1076 (2004).
11. L. M. Mirica, X. Ottenwalder, T. D. P. Stack, **104**, 1013-1045 (2004).
12. Reviews on cyclodextrins: *Chem. Rev.* **98**, 1741-2076 (1998).
13. A. Collet, *Tetrahedron* **43**, 5725-5759 (1987).
14. D. J. Cram, J. M. Cram, in J. F. Stoddart (Eds.), *Container Molecules and Their Guests, Monographs in: Supramolecular Chemistry*, The Royal Society of Chemistry, Cambridge, 1994.
15. C. D. Gutsche, in J. F. Stoddart (Eds.), *Calixarenes Revisited, Monographs in: Supramolecular Chemistry*, The Royal Society of Chemistry, Cambridge, 1998.
16. Z. Asfari, V. Böhmer, J. Harrowfield, J. Vicens (Eds.), *Calixarenes 2001*, Kluwer Academic Publishers, Dordrecht, 2001.
17. O. Sénèque, Y. Rondelez, L. Le Clainche, C. Inisan, M.-N. Rager, M. Giorgi, O. Reinaud, *Eur. J. Inorg. Chem.* 2597-2604 (2001).
18. O. Sénèque, M. Giorgi, O. Reinaud, *Supramol. Chem.* **15**, 573-580 (2003).
19. O. Sénèque, M.-N. Rager, M. Giorgi, O. Reinaud, *J. Am. Chem. Soc.* **122**, 6183-6189 (2000).
20. O. Sénèque, M.-N. Rager, M. Giorgi, O. Reinaud, *J. Am. Chem. Soc.* **123**, 8442-8443 (2001).
21. O. Sénèque, M. Giorgi, O. Reinaud, *J. Chem. Soc. Chem. Commun.* 984-985 (2001).
22. K. Håkansson, M. Carlsson, L. A. Svensson, A. Liljas, *J. Mol. Biol.* **227**, 1192-1204 (1992).
23. S. Blanchard, L. Le Clainche, M.-N. Rager, B. Chansou, J.-P. Tuchagues, A. F. Duprat, Y. Le Mest, O. Reinaud, *Angew. Chem. Int. Ed.* **37**, 2732-2735 (1998).
24. S. Blanchard, M.-N. Rager A. F. Duprat, O. Reinaud, *New J. Chem.* 1143-1146 (1998).
25. Y. Rondelez, O. Sénèque, M.-N. Rager, A. Duprat, O. Reinaud, *Chem. Eur. J.* **6**, 4218-4226 (2000).
26. Y. Rondelez, M.-N. Rager, A. Duprat, O. Reinaud, *J. Am. Chem. Soc.* **124**, 1334-1340 (2002).
27. Y. Rondelez, G. Bertho, O. Reinaud, *Angew. Chem. Int. Ed.* **41**, 1044-1046 (2002).
28. L. Le Clainche, M. Giorgi, O. Reinaud, *Inorg. Chem.* **39**, 3436-3437 (2000).
29. L. Le Clainche, Y. Rondelez, O. Sénèque, S. Blanchard, M. Campion, M. Giorgi, A. F. Duprat, Y. Le Mest, O. Reinaud, *C. R. Acad. Sci., Série IIc* **3** , 811-819 (2000).
30. O. Sénèque, M. Campion, M. Giorgi, Y. Le Mest, O. Reinaud, *Eur. J. Inorg. Chem.* 1817-1826 (2004).
31. J. O. Alben, P. P. Moh, F. G. Fiamingo R. A. Altschuld, *Proc. Natl. Acad. Sci. USA* **78**, 234-237 (1981).
32. N. Le Poul, M. Campion, G. Izzet, B. Douziech, O. Reinaud, Y. Le Mest, *J. Am. Chem. Soc.* **127**, 5280-5281 (2005).

33. O. Sénèque, O. Reinaud, *Tetrahedron* **59**, 5563-5568 (2003).
34. Y. Rondelez, Y. Li, O. Reinaud, *Tetrahedron Lett.* **45**, 4669-4672 (2004).
35. O. Sénèque, M.-N. Rager, M. Giorgi, T. Prangé, A. Tomas, O. Reinaud, *J. Am. Chem. Soc.* **127**, 14833-14840 (2005).
36. O. Sénèque, M. Campion, B. Douziech, M. Giorgi, Y. Le Mest, O. Reinaud, *Dalton Trans.* 4216–4218 (2003).
37. J. W. Whittaker, *Chem. Rev.* **103**, 2347-2364 (2003).
38. O. Sénèque, M. Campion, B. Douziech, M. Giorgi, E. Rivière, Y. Journaux, Y. Le Mest, O. Reinaud, *Eur. J. Inorg. Chem.* 2007-2014 (2002).
39. G. Izzet, Y. M. Frapart, T. Prange, K. Provost, A. Michalowicz, O. Reinaud, *Inorg. Chem.* **44**, 9743-9751 (2005).
40. G. Izzet, H. Akdas, N. Hucher, M. Giorgi, T. Prangé, O. Reinaud, *Inorg. Chem.* **45**, 1069-1077 (2006).
41. I. Jabin, O. Reinaud, *J. Org. Chem.* **68**, 3416-3419 (2003).
42. U. Darbost, M. Giorgi, O. Reinaud, I. Jabin, *J. Org. Chem.* **69**, 4879-4883 (2004).
43. X. Zeng, N. Hucher, O. Reinaud, I. Jabin, *J. Org. Chem.* **69**, 6886-6889 (2004).
44. U. Darbost, X. Zeng, M.-N. Rager, M. Giorgi, I. Jabin, O. Reinaud, *Eur. J. Inorg. Chem.* 4371-4374 (2004).
45. U. Darbost, M.-N. Rager, S. Petit, I. Jabin, O. Reinaud, *J. Am. Chem. Soc.* **127**, 8517-8525 (2005).
46. G. Izzet, B. Douziech, T. Prangé, A. Tomas, I. Jabin, Y. Le Mest, O. Reinaud, *Proc. Natl. Acad. Sci. USA* **102**, 6831-6836 (2005).

Chapter 14

AMINO-ACID, PEPTIDE AND PROTEIN SENSING

Calixarenes in monolayers

Reza Zadmard and Thomas Schrader
Department of Chemistry, Marburg University, Marburg, Germany

Abstract: Protein surface recognition by synthetic molecules remains a challenge because the directed systhesis of molecules with a predetermined pattern of solvent-exposed functional groups is still in its infancy. Self-assembly of small receptor units within the fluid environment of lipid mono- or bi-layers offers a very useful alternative to such difficult syntheses. This chapter describes the use of calixarenes as versatile species for the recognition of basic amino-acids and their incorporation into self-assembled monolayers for protein sensing at the nanomolar level.

Key words: Amino-acids, peptides, proteins, peptidocalixarenes, phosphonatocalixarenes, monolayers, recognition, sensing.

1. INTRODUCTION

Protein surface recognition is one form of nanoparticle binding and it is an area where Nature provides innumerable examples of highly stable and selective systems. The synthesis of artificial receptors for proteins, however, remains a difficult task because of the need to create large molecules with controllable stereochemistry but it is here that calixarenes provide particularly valuable starting points or "scaffolds" for the construction of giant receptors. Since proteins are polymers built from amino acids in passing through oligopeptides, it is useful to consider the development of protein receptors in terms of the elaboration of hosts for these smaller components.

J. Vicens and J. Harrowfield (eds.), Calixarenes in the Nanoworld, 287–309.

2. AMINO ACID RECOGNITION

2.1 Recognition of Amino-Acids by Calixarenes

Calixarenes[1] are well known for their capacity to encapsulate both neutral molecules and ionic species, the latter including quaternary ammonium ions and simpler ammonium and guanidinium species related to the protonated side chain species derived from basic amino-acids such as lysine and arginine. Early studies of quaternary ammonium ion binding were based upon anionic cavitands such as those derived from resorc[4]arenes[2] and sulfonated calixarenes[3] and were concerned largely with gaining an understanding of the binding modes involved in abiological systems.[4] Although the biological activity of sulfonated calixarenes, as ion-channel blockers, for example, has been characterised,[5] it is only relatively recently that their applications in biomedical science have begun to be explored.

A recently initiated research programme is specifically concerned with the use of NMR spectroscopy and microcalorimetry to characterise the interactions between sulfonated calixarenes and the basic amino acids lysine and arginine.[6] Arginine differs from lysine only in the replacement of the basic side-chain amino group by an even more basic guanidine unit and in acidic media, where both substituents are fully protonated, both amino acids form strong complexes with sulfonated calix[4]arene. They are much more strongly bound than any of the other common species, with association constants (for sulfonated calix[4]arene binding) being 10 - 100 times greater than those known for valine, leucine, phenylalanine, histidine and tryptophan.[7] A structural study[8] of the solid complex of lysine with sulfonated calix[4]arene has provided evidence that cation-π interactions involving the protonated side-chain amino group of lysine and the phenyl rings of the calixarene are important. For higher sulfonated calix[n]arenes, which adopt more flattened conformations, cation-π interactions with lysine and arginine appear to be weaker, though this is a misleading indicator of their interactions with proteins[9] (see ahead).

Initial extensions of this work to lysine- and arginine-containing di- and tri-peptides furnished intriguing results.[6] Complexation by sulfonated calix[n]arenes in water at pH 8 appears to be largely due to favourable enthalpy terms, attributed to van der Waals interactions between the guest and the hydrophobic cavity of the host, though entropy changes attributed to desolvation of the charged groups due to ion pairing are also important. NMR spectroscopy also provided evidence that even when bound to the smallest sulfonated calix[4]arene, lysyl-lysine adopts a compact, folded structure while arginyl-arginine has a more extended conformation.

2.2 Recognition of Amino Acids by Peptidocalix[4]arenes

Peptidocalix[4]arenes, discussed in detail in Chapter 12, are attractive as hosts because of their chirality and the wide range of functional groups which can be derived from the amino acid side-chains. Readily fixed in the cone or pinched-cone conformation (Fig. 1) with a well-defined cavity, such calixarenes have been shown to bind a variety of small molecules including amino acids.

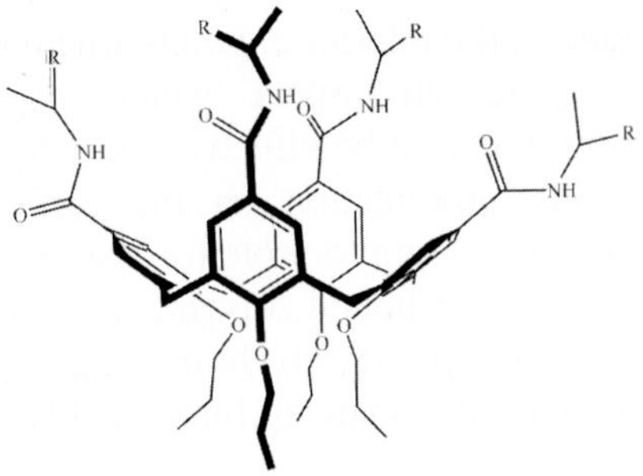

1 ($R = CO_2^-$)

Figure 14-1. The tetra-alaninato calix[4]arene **1**.

The alanine derivative **1**, for example, binds primary ammonium ions and protonated amino acids, though not amino acids in their anionic form.[10,11] Such simple derivatives, however, retain sufficient conformational flexibility to be relatively weak binders.

2.3 Oligopeptide Recognition

2.3.1 Sequence-selective Recognition of Short Peptides

Oligopeptide recognition represents an intermediate stage on the way to recognition of protein surfaces. Short peptide sequences (3 – 20 amino acid residues) are important targets in their own right, differing significantly from proteins because of their retention of a considerable degree of conformational flexibility as well as in their requirement for only a relatively small number of interaction sites in a host. Binding to a protein surface, in contrast, requires large arrays of interaction sites with limited conformational freedom.[12] For oligopeptides, various ditopic receptors based on rather different host components, such as cyclodextrins,[13] porphyrins[14] and peralkylammonium salts,[15] have been found to give strong binding in aqueous solution. The capacity of β-cyclodextrin cavities to accommodate phenyl rings leads here to selective binding of peptides with non-polar aromatic residues such as phenylalanine.[13,16]

2.4 Protein Surface Recognition by Synthetic Receptors

A protein in its native, folded conformation has a solvent-exposed (exterior) surface and a solvent-excluded (interior) surface. Enzyme active sites are most often found in the interior at points where the functional group presentation is convergent, an aspect which explains why highly functionalised small molecules can be effective inhibitors of the enzyme. In contrast, functional group presentation on the surface of a protein is predominantly divergent. In binding to protein surfaces, synthetic molecules must be able to displace surface-bound solvent, as well as complement the often highly irregular surface topology. Tight binding requires recognition of large areas and therefore multiple points of functionality.[12]

2.4.1 Recognition of Cytochrome-c

The general principles in design of a protein surface binding agent are usefully illustrated by the case of cytochrome-c recognition. Cytochrome-c is a well-characterised small protein that plays a key role in mitochondrial electron transfer and apoptosis.[17] The region of the protein that comes into closest contact with its natural redox partner proteins is a hydrophobic patch near the edge of the haem unit which is surrounded by several invariant positively charged lysine residues. The complementary domains on the partner proteins (such as cytochrome-c peroxidase or cytochrome-c oxidase) have a hydrophobic patch surrounded by anionic aspartate and glutamate residues. This simple matching of hydrophobic and charged domains (Fig. 3) inspired the design of artificial receptors **2** - **8** for cytochrome-c in which a hydrophobic centre, a tetra-arylporphyrin, was surrounded by negatively charged carboxylate group substituents (Fig. 3).[18]

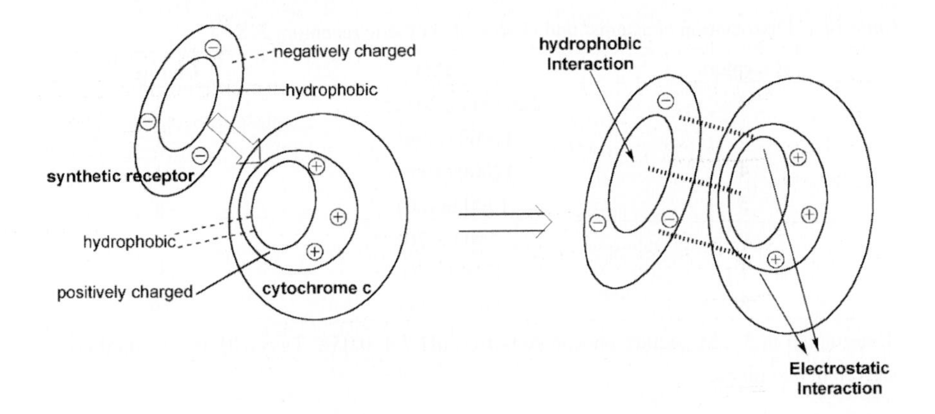

Figure 14-2. Schematic representation of a general concept for cytochrome-c recognition by synthetic receptors: complementary surface areas with hydrophobic cores and oppositely charged surroundings.[18]

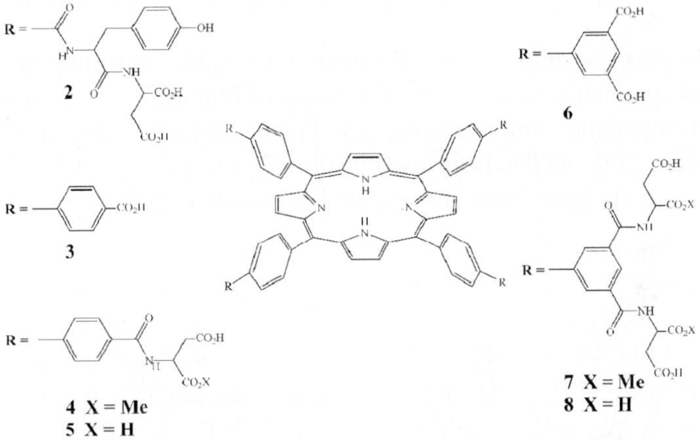

Figure 14-3. Structures of porphyrin-based cytochrome-c receptors **2-8**.[18]

The first results with receptor **2** indicated that better recognition might result by increasing both the area of the hydrophobic unit and the negative charge density around it, features incorporated into the receptors **3** - **8**. Binding affinities of all these receptors for horse heart cytochrome-c were investigated by fluorescence quenching measurements, which resulted in the equilibrium constants for 1:1 complexes given in Table 1.

Table 14-1. Dissociation constants[a] and charge of synthetic receptors **2–8** [18].

Receptors	K_d (nM)	Charge
2	220 (±17), 20 (±5)[b]	–8
3	17000 (±840)	–4
4	12000 (±950)	–4
5	1300 (±130)	–8
6	1500 (±170)	–8
7	1700 (±97)	–8
8	0.67(±0.34)	–16

[a)]Determined in 5 mM sodium phosphate buffer, pH 7.4, 0.05% Tween20, rt. Standard error is given in parentheses.

There is clearly a strong dependence of the binding affinity on the number of anionic groups on the receptor periphery, though receptor **2** is superior to any of **3** - **7**, indicating that flexibility in the porphyrin pendent groups may be important. The exceptionally high charge on the receptor **8** quite dramatically improves the binding, to an extent considerably greater than might be expected from the effect of charge doubling in passing from **3** and **4** to **5** - **7**.

A calix[4]arene in the cone conformation resembles a porphyrin in being a fourfold-symmetric scaffold for attachment of binding sites. Attachment of cyclic hexapeptide units known to favour β-loop structures[19] to a calix[4]arene (Fig. 4) has been used to provide receptors of very large area (>500Å^3) which have been investigated as another type of cytochrome-c binding agent.[20]

Figure 14-4. Synthetic antibody mimics based on calix[4]arene scaffolds and cyclic peptide loops. [20]

As mentioned above, there is a positive patch on the surface of cytochrome-c which appears to play a key role in the binding of redox partners, so that it was anticipated that binding of a synthetic receptor such as **9a** would inhibit electron transfer reactions of the protein (Figs 5 and 6). Indeed, addition of slightly more than one equivalent of **9a** to cytochrome-c in phosphate buffer causes a tenfold reduction in the rate of reduction of ascorbate and essentially complete inhibition results when two equivalents are added. These effects are very similar to those seen when cytochrome-c peroxidase binds (in a manner defined by a crystal structure determination[21]), leading to the conclusion that the interaction modes must also be similar. Molecular modelling of the cytochrome-c:**9a** complex confirmed that the four peptide loops can cover a large section of the protein surface and, in particular, contact four of the five lysine residues close to the haem edge. Experiment showed, from comparison of **9a** - **9c**, that the nature of the narrow rim substituent also influences the binding strength, presumably as a consequence of relatively minor conformational changes. More significantly, a receptor such as **9a** is capable of effectively competing with the natural protein substrates of cytochrome-c, including extremely strongly bound species such as the apoptosis protease activating factor-1.[22]

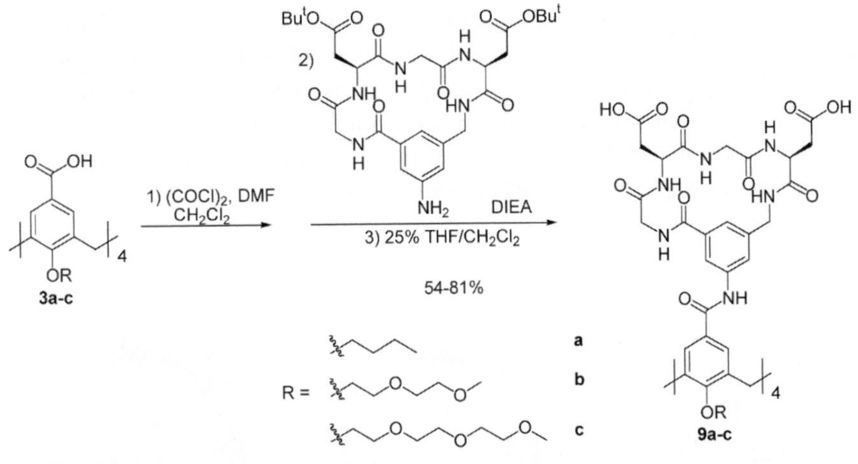

Figure 14-5. Synthesis of protein surface receptors **9a-c** from calixarene scaffold **3a-c**.

Figure 14-6. Calculated complex structures between cytochrome-c and receptors **8** and **9a.**[20]

Structurally simpler calixarene derivatives have also been shown to interact with cytochrome-c, sufficiently strongly, for example, to enable extraction of the protein into apolar solvents,[23] such extraction being of biotechnological interest.[24] Thus, the hexakis(oxycarbonylmethoxy) calix[6] arene **10** can be used to extract cytochrome-c into chloroform, the extraction efficiency being much greater than for the calix[4]arene analogue or the simple p-t-butylphenoxyacetic acid "monomer".[23] It is assumed that the association of the protein and the calixarenes involves lysine ammonium and carboxylate ion pairing (Fig. 7), though clearly there is a dependence upon the size and conformation of the macrocycle as well as upon its degree of deprotonation under the conditions of the extraction experiment. This is an interesting example of nanoparticle extraction.

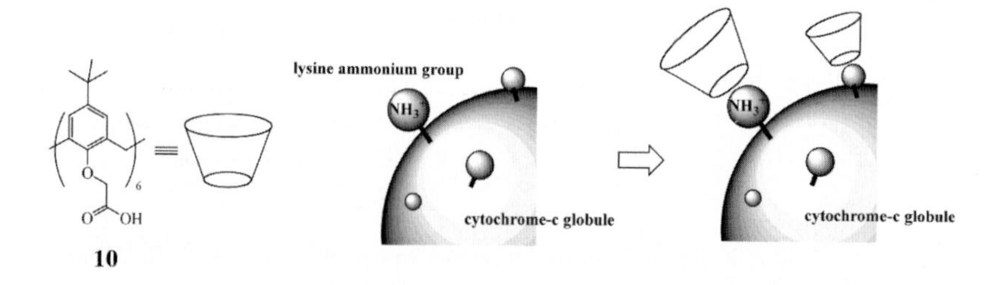

Figure 14-7. Left: Molecular structure of cytochrome-c extractant **10** . Right: schematic illustration of the cytochrome-c extraction with **10**.[23]

2.4.2 Complexation of Bovine Serum Albumin by Sulfonatocalix[n]arenes

The serum albumins are a class of globular proteins which have a major role in adhesion to surfaces as anchors for subsequent formation of protein films.[25] They are capable of binding both cations and anions, with multiple arginine and lysine residues being the known sites for anion interaction once again. As part of a general study of the formation of surface films by bovine serum albumin (BSA), it was found that BSA forms strong complexes with sulfonatocalix[n]arenes.[26] In extension of this work,[6,8,27] electrospray mass spectrometry has been used to show that calix[4]arene tetrasulfonate forms 1:1, 1:2 and 1:3 (BSA:ca1) complexes (Fig. 9) whereas calix[6]arene hexasulfonate forms 1:1 and 1:2 species and calix[8]arene octasulfonate only a 1:1 complex.

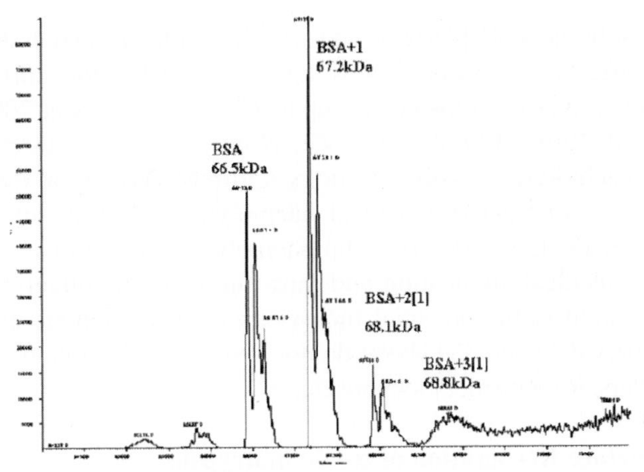

Figure 14-8. ESI-MS spectrum of the BSA complexes with sulfonatocalix[4[arene at a molar ratio of 1:4.[27]

For simple amino acids in solution, the stability constants for comparable species increase with the size of the macrocycle but both the strength and selectivity of complexation are influenced by the nature of the substituents on the narrow rim of the calixarene (*para* to the sulfonate groups) Both carboxylate and amino substituents enhance the binding of aspartic acid, for example. Ring size and substituent effects have thus both been studied as influences on the haemolytic activity of sulfonatocalixarenes,[28] the range of structures investigated being shown in Fig. 9.

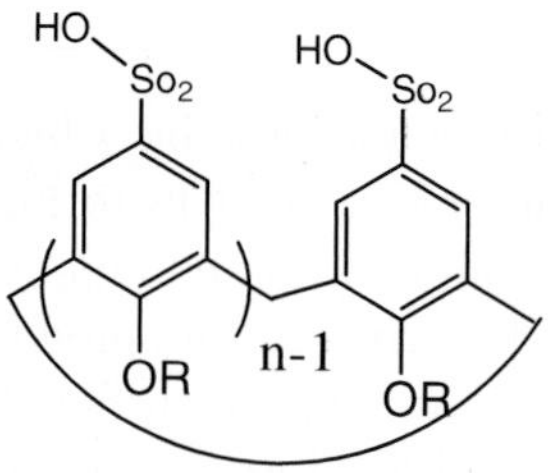

11a, **12a**, **13a**, R = H

11b,**12b**, **13b**, R = CH_2COOH

11c, **12c**, **13c**, R = CH_2CONH_2

11d, **12d**, **13d**, R = $CH_2CH_2NH_2$

11 n = 4, **12** n = 6, **13** n = 8

Figure 14-9. Schematic representation of various water-soluble *para*-sulphonatocalix[*n*]-arenes discussed above for their biological activity.[28]

While *p*-sulfonatocalix[4]arene and its ethers **11a-d** have essentially no haemolytic effects in concentrations up to 200 mM, the corresponding calix[6]arene derivatives **12a-d** cause up to 12% haemolysis at 200 mM and those of calix[8]arene **13a-d** up to 29%. None of these calixarenes causes appreciable haemolysis in concentrations <50 mM. While an aminoethoxy substituent causes an enhancement of haemolytic ability, a carboxymethyl substituent causes its diminution. The haemolytic activity of cyclodextrins delayed their medical application and thus this is an important property to evaluate in regard to the potential biotoxicity of water-soluble calixarenes. The preliminary data detailed above do not *a priori* exclude their application as drug-carriers, ion-channel blockers etc.

2.4.3 Surface Recognition of α-Chymotrypsin

To recognise different antigens, antibodies use changes in the sequence and conformation of six hypervariable loops of their protein structure. In analogy, the tetraloop peptide derivatives of calix[4]arene, briefly described above in consideration of cytochrome-c receptors, were designed to allow the possible variability in the peptide units to be used for the selective recognition of different proteins. Thus, using structural information for α-chymotrypsin, a selective receptor for this protein was designed and synthesised from calix[4]arene.[29] Again, it is a protein with surface arginine and lysine residues close to the entry to the active site and the calixarene shown in Fig. 10 was designed to interact with these residues and thus block access to the active site, acting as an inhibitor but not one directly bound to the active site.

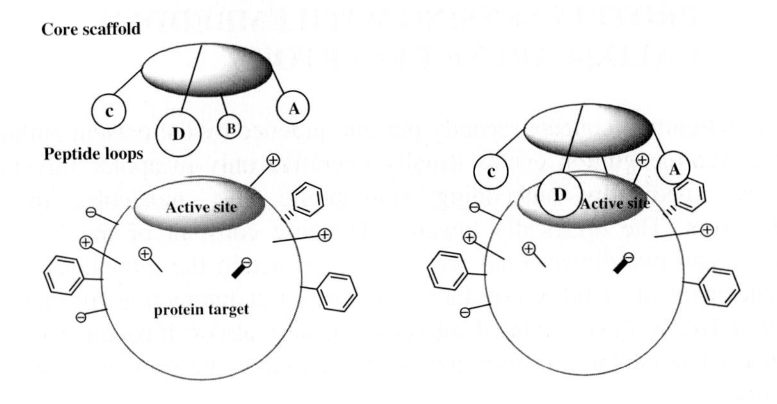

Figure 14-10. Schematic representation of enzyme inhibition by specific protein surface recognition with receptors carrying peptide loops on a central scaffold.

Indeed, receptor **14** is able to displace a trypsin inhibitor from its 1:1 complex with α-chymotrypsin but the unmodified receptor can be subsequently recovered, indicating that, as desired, it does not bind directly to the active site.

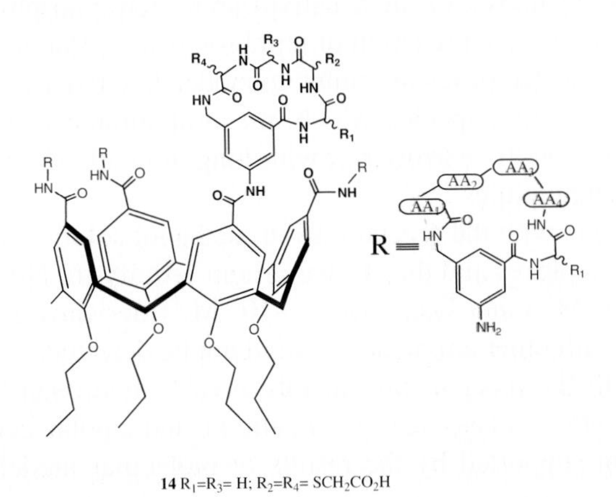

Figure 14-11. Artificial receptors based on 3-amino-5-aminomethylbenzoic acid for chymotrypsin surface recognition.[29]

3. PROTEIN SENSING WITH EMBEDDED CALIX[4]ARENE RECEPTORS

A recognition concept recently put into practice by the present authors is to transfer recognition events usually operative only in apolar solvents to aqueous media by embedding amphiphilic host molecules in lipid monolayers. The drastically lowered dielectric constant of the interfacial region[30] and two-dimensional preorganisation within the monolayer lead to enhancement of stability constants for host-guest interaction by factors as great as 10^6. With O-alkylated calix[4]arene derivatives, it becomes possible to detect nanomolar concentrations of basic and acidic proteins in aqueous solution.

3.1 Calix[4]arenetetraphosphonates - Receptor Molecules designed for Polar Solvents and Lipid Monolayers

It was envisaged that attachment of phosphonic acid groups to the wide rim of calix[4]arene would provide a receptor molecule which would be negatively charged at neutral pH and thus be suited to binding of cationic biological substrates, especially those incorporating sterically demanding guanidinium units. Synthesis of such calix[4]arene tetraphosphonates was accomplished by Ni-catalysed reaction of tetrabromocalix[4]arene with trialkylphosphites, the diester products being monodealkylated by subsequent reaction with LiBr to give species which are tetra-anionic at neutral pH. Alkylation at oxygen on the narrow rim with long aliphatic chains converts these materials to amphiphiles.[31]

Initial experiments with the receptor **15** in methanol solution showed the formation of 1:1 complexes and that **15** has a high affinity for N/C-protected arginine (K_{stab} ~10^4 M^{-1}) and lysine (K_{stab} ~10^3 M^{-1}) derivatives (Table 2), while interactions with other amino acids could not be detected.[31]

Interactions with the receptor presumably involve hydrogen bonding to the phosphonate centres and association of cationic and anionic centres, such presumptions being supported by the results of molecular modeling calculations (Fig. 12).

Significantly, ^{1}H NMR measurements indicate that the tetraphosphonate complexes must differ considerably from those of the calixarene sulfonates discussed previously, since no marked upfield shifts are seen for any of the guest resonances, indicating that the guest does not penetrate far into the calixarene cavity, unlike the case for the sulfonates.

Table 14-2. Association constants and free binding energies for the complexes between calix[4]arene tetraphosphonate **15** and basic amino acids. Arginine and lysine were examined in their free and N/C-protected form in methanol by NMR titrations at ambient temperature (20°C).

Entry	Guest	K_a [M^{-1}]	ΔG [kcal/mol]	Stoich.
1	Ts-Arg-OMe	$1{\bullet}10^4 \pm 12\%$	– 5.9	1:1
2	H-Arg-OH	$8{\bullet}10^2 \pm 28\%$	– 4.0	1:1
3	Ac-Lys-OMe	$7{\bullet}10^2 \pm 42\%$	– 3.9	1:1
4	H-Lys-OH	$3{\bullet}10^3 \pm 25\%$	– 4.8	1:1

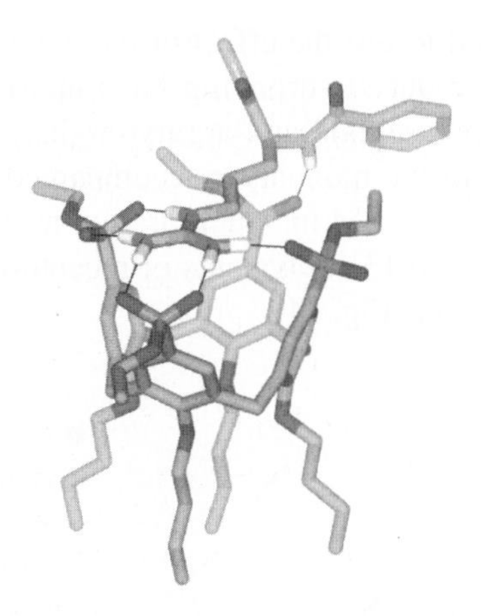

Figure 14-12. Optimised structure of the complex between *N/C*-protected arginine and calixarene tetraphosphonate **15**. After energy minimization, a Monte-Carlo simulation was carried out in water (MacroModel 7.0, 3000 steps).

Given the amphiphilic structure of tetraphosphonate **15**, it was expected that they should be surface active and they can, in fact, be readily incorporated into a stearic acid monolayer on water. Regular shifts in the pressure/area diagram for this monolayer on addition of increasing amounts of receptor are consistent with the expected expansion of the monolayer.

Subsequent additions of cationic analytes into the aqueous subphase produced moderate but distinct additional expansions of the monolayer which were not seen when the monolayer contained no calixarene. Thus, only the "doped" monolayer forms a complex with the aqueous analyte, presumably through multiple hydrogen-bonding interactions (Fig. 13).

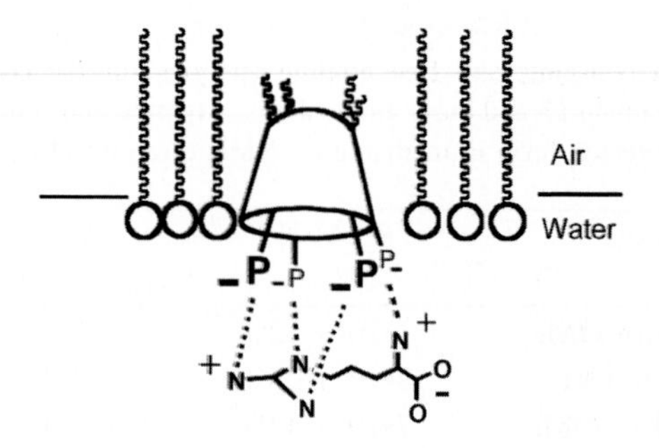

Figure 14-13. Schematic representation of the calixarene in the monolayer recognising an arginine guest in the subphase.

Experiments designed to test the effect of having a multifunctional guest in the aqueous phase produced surprising but important results.[32] Comparison of arginine, arginyl-arginine and arginyl-arginyl-arginine showed that remarkable expansions of the monolayer accompanied the change from one to the next. Thus, even at 10^{-7} M in water, arginyl-arginyl-arginine could be clearly detected with only 0.13 equivalents of receptor/stearic acid molecule embedded in the monolayer (Fig. 14).

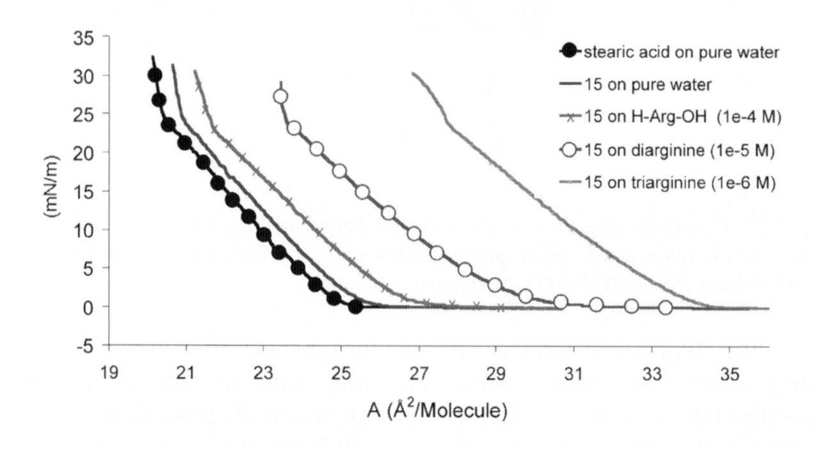

Figure 14-14. Pressure/area isotherms for arginine (10^{-4}M), diarginine (10^{-5}M) and triarginine (10^{-6}M) on the Langmuir film balance. 0.13 equivs. of receptor **15** were embedded in the stearic acid monolayer.

Such results indicate that the embedded receptors are not restricted in their motions throughout the monolayer and can thus assemble in groups over polytopic guests in the subphase.

A logical extension from this examination of small peptide guests was to proceed to large peptides and proteins with a high surface content of arginine or lysine residues. Thus, lysine-rich histone H1 and cytochrome-c both produce marked monolayer expansion at concentrations as low as 10^{-9} M. Normally, only the strongest of protein-protein interactions are conserved at such concentrations. Neither electrolytes nor buffer components seem to strongly influence the effects. While basic proteins (in their protonated forms) would be expected to interact with an anionic receptor, this could also be expected for neutral proteins (pI ~7) provided there is an elevated density of basic residues (*e.g.* arginine, lysine) on their surfaces. This is the case for BSA (pI 6.0), which has a barrel-like structure with abundant glutamate and aspartate residues on the "lids" but with mainly arginine and lysine residues on the "walls". Truly acidic proteins with anionic surfaces such as the ferritins should bind much more weakly to an anionic receptor, and the pA isotherms beautifully reflect these differences that may be discerned between proteins of these various types.

Table 14-3. Basic, neutral and acidic proteins on the Langmuir film balance.

Entry	Protein(10^{-8}M)	$c_{corr.}$ [M][a]	ΔA_{matrix} [A^2][b]	ΔA_{rec} [A^2][c]	pI	MW [kD]	Surface [kA^2]	Volume [kA^3][d]
1	histone H1	$4•10^{-9}$	0	5	10.4	7.7	43	10.2
2	cytochrome-c (horse)	$6•10^{-9}$	0	5	9.5	12.3	6.3	16.7
3	proteinase K	10^{-8}	1	6	8.1	38.4	11.0	46.0
4	chymotrypsin	10^{-8}	1	5	8.0	28.2	11.2	33.4
5	thrombin	10^{-9}	1	2	7.5	32.0	15.5	45.2
6	albumine (BSA)	$4•10^{-9}$	1	2	6.0	86.3	37.0	95.1
7	dodecamer	$7•10^{-9}$	1	1	5.9	190.0	75.0	
8	ferritin (24mer)	10^{-8}	1	2	5.5	455.3	175.5	555.1
9	acylcarrierprotein	10^{-8}	1	1	4.2	8.4	4.7	9.0

[a] $c_{corr.} = c_{exp.} \bullet surface_{prot.}/10kA^2$; [b] interaction with stearic acid alone; [c] additional interaction with the embedded receptor; [d] Connolly surface.

A useful tool for the analysis of charge effects on protein/protein and protein/receptor interactions is the electrostatic potential surface (EPS).[33] For the proteins examined to date, these surfaces are shown, for the proteins in their correct relative sizes, in Fig. 15.

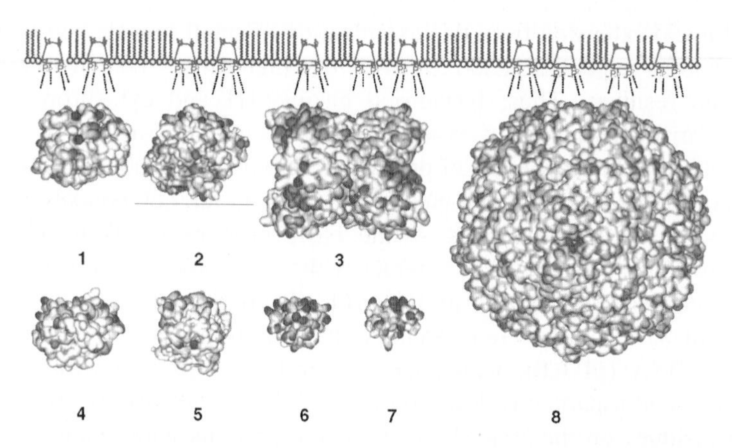

Figure 14-15. Proteins which are bound by receptor **10** in the stearic acid monolayer, depicted in their correct relative sizes. The Connolly surface is patterned with the electrostatic surface potential (ESP), showing basic and acidic domains on the protein surfaces. The basic proteins are shown in their proposed orientation relative to the monolayer, presenting their positive domains (blue) upwards. **1** Proteinase K, **2** thrombin, **3** ferritin, **4** chymotrypsin, **5** trypsin, **6** cytochrome C, **7** histone H1, **8** ferritin.

These "Connolly surface" representations[33] have positive charge coded as blue, negative charge as red, and obviously correlate well with the data reported in Table 3. That the observed effects cannot be ascribed to non-specific Coulombic interactions is indicated by the fact that incorporation of various simple mono- and di-phosphonates and sulfonates (Fig. 16) into the monolayer produces almost negligible changes in area.

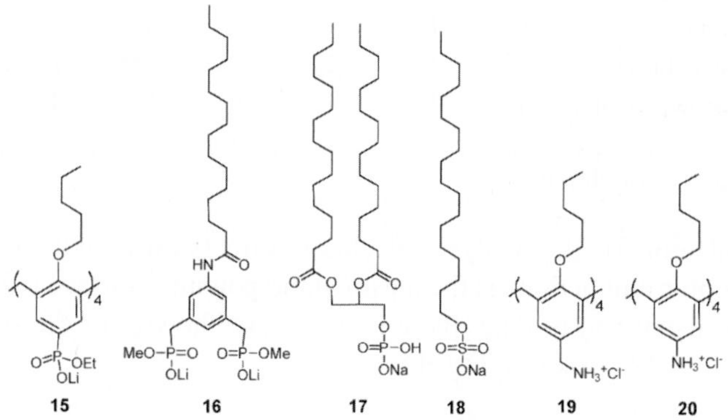

15 16 17 18 19 20

Figure 14-16. Host structures of all anionic and cationic amphiphiles embedded in a stearic acid monolayer.

3.1.1 Benzylammonium Calixarene 19 as a new Amphiphilic Receptor Molecule for the Recognition of Acidic Proteins at a Monolayer

An obvious means to enhance interactions of acidic proteins and nucleic acids with a receptor is to give that receptor a positive charge and this is simply done for a calix[4]arene scaffold by exchanging the phosphonate substituents of **15** for benzylammonium substituents in **19**. This molecule is relatively less polar and so concentrates more efficiently in the stearic acid monolayer than does the tetraphosphonate. As anticipated for a monolayer incorporating **19**, the presence of a basic protein such as histone H1 (pI 10.4) in the aqueous subphase has negligible effects on the monolayer area but the presence of an acidic protein such as the acylcarrier protein (ACP, pI 4.0) produces very large monolayer expansions. Data for these and other proteins illustrating the discriminatory power of the measurements are given in Fig. 17 and Table 4.

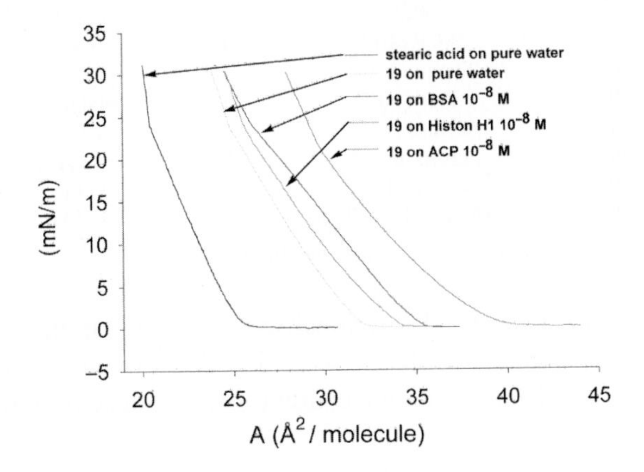

Figure 14-17. Langmuir isotherms obtained from 0.13 eq. of benzylammonium calixarene **19** embedded into a stearic acid monolayer at the air/water interface, over neutral (BSA), basic (Histone) and acidic proteins (ACP).

Table 14-4. Basic, neutral and acidic proteins recognized by tetrabenzylammonium calixarene **19** on the Langmuir film balance: dependence of the area increase ΔA on the pI values at concentrations around 10^{-8} M ACP = acyl carrier protein.

Entry	Host	Subphase [M]	pI	ΔA [A^2][a]	MW [kDa]
1	**19**	Water	—	0 (7.2)	—
2	**19**	Histone H1	10.4	0.8 (8.0)	7.7
3	**19**	Cytochr. C	9.5	1.4 (8.6)	12.3
4	**19**	Thrombin	7.5	2.0 (9.2)	32.0
5	**19**	Albumin	6.0	2.8 (10)	86.3
6	**19**	ACP	4.2	4 (11.2)	8.4

[a] Values in brackets refer to total area increase (host and protein)

3.1.2 Recognition of all Polar Proteins with the Anilinium Calixarene 20 in a Lipid Monolayer

The enhanced acidity of an anilinium ion relative to an ammonium ion means that around neutral pH appreciable concentrations of both the free base and its protonated form can be expected to be present, meaning that both H-bond acceptor and H-bond donor centres are present. The tetra-anilino species **20** is readily prepared from the tetranitro calix[4]arene precursor, and is readily incorporated into a stearic acid monolayer, where it leads to area changes in response to both acidic and basic proteins (Table 5).

Table 14-5. Basic, neutral and acidic proteins recognized by tetraanilinium calixarene **20** on the Langmuir film balance: dependence of the area increase ΔA on the pI values at concentrations around 10^{-8} M ACP = acyl carrier protein.

Entry	Host	Subphase [M]	pI	ΔA [A^2][a]	MW [kDa]
1	**20**	Water	—	0 (8.0)	—
2	**20**	ACP (10^{-8})	4.2	−1 (7)	8.4
3	**20**	Ferritin (10^{-8})	5.5	plateau	455.3
4	**20**	Dps (10^{-9})	5.9	plateau	190.0
5	**20**	Albumin (10^{-8})	6.0	1 (9)	86.3
6	**20**	Thrombin(10^{-9})	7.5	1 (9)	32.0
7	**20**	Cyt. c (10^{-8})	9.5	3 (11)	12.3
8	**20**	Histone H1(10^{-8})	10.4	5 (13)	7.7

[a] Values in brackets refer to total area increase (host and protein)

This dual sensitivity can be rationalised in terms of the presence of anilinium sites to interact with the anionic sites of acidic proteins and aniline-NH_2 sites to H-bond with the cationic sites of basic proteins (Fig. 18).

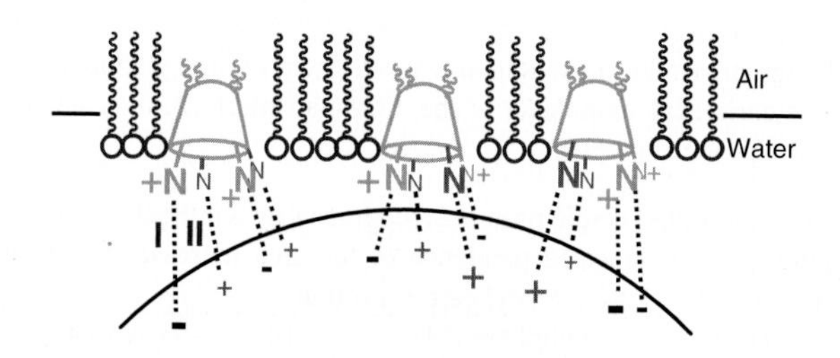

Figure 14-18. Proposed binding mode of the embedded anilinium calixarene halfsphere **20** with basic *and* acidic proteins.

3.1.3 Selective and Sensitive Protein Recognition with Monolayers Incorporating both Anionic and Cationic Calixarenes

While mixture of a cationic calixarene with an anionic calixarene might be expected simply to lead to their association to give an aggregate insensitive to the presence of other species, in fact it appears possible to incorporate the tetraphosphonate **15** in a monolayer along with either the tetra-ammonium species **19** or the tetra-anilinium species **20** and have both oriented such that the charged head groups contact the aqueous subphase and so respond to analytes therein. (Table 6). In the case of BSA, this means that it can be detected easily at subnanomolar concentrations, with a detection limit of ~10 pmol L^{-1}.

Table 14-6. Basic, neutral and acidic proteins recognized by a combination of 0.13 eq. calixarene tetraphosphonate **15** and 0.13 eq. **19** or 0.13 eq. **20** on the Langmuir film balance: dependence of the area increase *ΔA* on the pI values at concentrations around 10^{-8} M.

Entry	Receptors	Solute	pI	ΔA [A^2][a]	MW [kDa]
1	**15** and **19**	Water	—	0 (10)	—
2	**15** and **19**	Histone H1 (10^{-8})	10.4	8.7 (18.7)	7.7
3	**15** and **19**	Cytochr. c (10^{-8})	9.5	6 (16)	12.3
4	**15** and **19**	Thrombin (10^{-9})	7.5	2 (12)	32.0
5	**15** and **19**	Albumin (10^{-8})	6.0	5.2 (15.2)	86.3
6	**15** and **19**	ACP (10^{-8})	4.2	3 (13)	8.4
7	**15** and **19**	Water	—	0 (10)	—
8	**10** and **20**	Albumin (10^{-10})	6.0	4.2 (14.2)	86.3
9	**10** and **20**	Albumin (10^{-11})	6.0	0.5 (10.5)	86.3

[a] Values in brackets refer to total area increase (host and protein)

3.1.4 Direct Experimental Proof of Noncovalent Protein Binding to a Lipid Monolayer by Langmuir-Blodgett Techniques

Conventional dipping procedures using a quartz plate can be used to isolate the monolayers with their bound guests formed in experiments as described above. Measurement of the absorption spectra of these films (Fig. 19) shows the amount of guest present to be directly proportional to the amount of calixarene host added to the monolayer (except in the case of ACP, where the protein has an extremely weak absorption spectrum).

Measurements of the absorption as a function of the guest concentration for a given host concentration (Fig. 19) enabled the extraction of approximate values of the host-guest stability constants, which for BSA lie in the range 3-5 x 10^8 M^{-1}.[34]

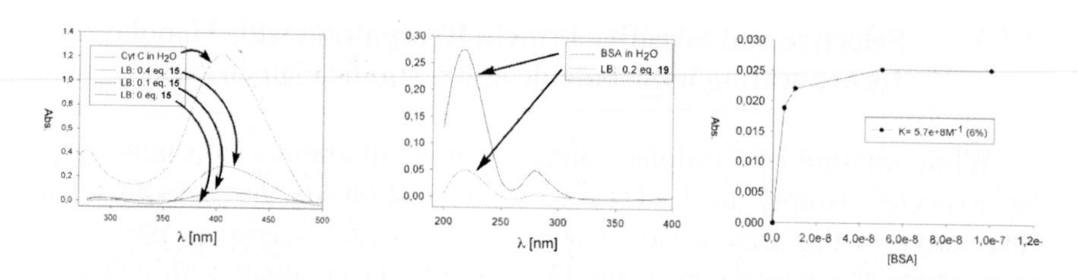

Figure 14-19. (left) UV/Vis spectra of cytochrome-c in water and in LB films drawn from the film balance experiments without, with 0.1 and with 0.4 eq. of embedded calixarene tetraphosphonate **15**; (centre) UV/Vis spectra of BSA in water and in LB films drawn from the film balance experiments with embedded benzylammonium calixarene **19**; (right) titration curve, corresponding to a stoichiometry of roughly 3-4 calixarenes per BSA protein unit,[37] for experiments with varied concentrations of **19**.

3.2 Pattern Recognition of Proteins

Protein sensing in a variety of environments, not necessarily simply biochemical, is of increasing importance,[35] but one example being the detection of signaling proteins in proteomics.[36] The method described herein based on readily accessible calixarene receptors and film balance area measurements is a simple and straightforward procedure useful for protein solutions in the lower nanomolar range (10^{-8} - 10^{-9} M). Since every protein reacts differently with the receptors **15**, **19** and **20** or mixtures thereof, the full variety of responses generates a recognition pattern (Fig. 20) which is specific for each protein.

Histone H1 and cytochrome-c, for example, are both basic proteins with a similar response to the tetraphosphonate receptor **15** but with very different responses to a monolayer containing both **15** and the anilinium calixarene **Z**, where now histone H1 produces much greater area changes. Similarly, thrombin and BSA are both neutral proteins with essentially identical responses to monolayers incoprorating either **15** or **20** but very diffferent responses to the mixture. Thus, each column in Fig. 20 constitutes a “fingerprint” for the given protein.

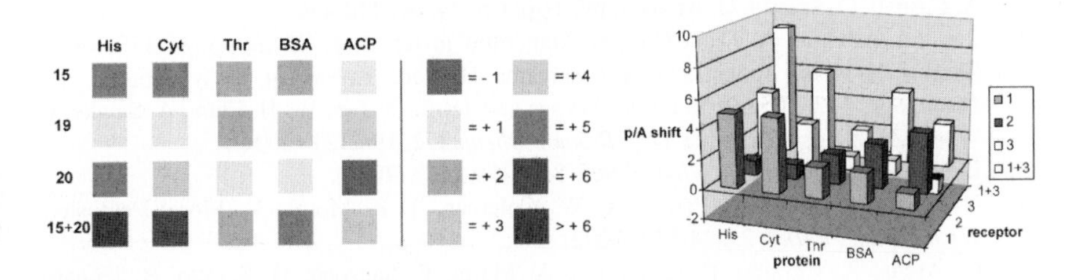

Figure 14-20. (Left): Recognition pattern evolving from film balance experiments with five different proteins (Histone H1, Cytochrome-c, Thrombine, BSA, ACP) and 4 different calixarene receptor cocktails: **15**, **19**, **20**, **15+20**. Color coding represents p/A shifts in A^2. (Right): Fingerprints of the five proteins based on the doped monolayer expansion assay. In all cases the protein concentration was 10^{-8} M.

4. CONCLUSIONS

The problem of generating huge receptors for huge molecules – "nanorecptors" for "nanoparticles" - appears to be one that is readily overcome by the use of monolayer embedding to allow relatively simple receptor units to aggregate appropriately in response to the presence of a guest at the interface. The exceptionally high stability of the resultant interfacial complexes means that not only can selectivity be controlled by the choice of receptor but also that the method is sensitive to extremely low protein concentrations

5. REFERENCES

1. (a) C. D. Gutsche, *Calixarenes,* Monographs in Supramolecular Chemistry No 1; J. F. Stoddart, Ed.; Royal Society of Chemistry: Cambridge, U. K., 1989 (Revised as *Calixarenes Revisited*, 1998); (b) *Calixarenes: A Versatile Class of Macrocyclic Compounds*; V. Böhmer, J. Vicens, Eds.; Kluwer Academic Publishers, Dordrecht, 1991. (c) R. Ungaro. A. Pochini in *Frontiers in Supramolecular Organic Chemistry and Photochemistry*; H.-J. Schneider, Ed.; VCH, Weinheim, Germany, 1991; pp. 57-81. (d). V. Böhmer, *Angew. Chem. Int. Ed. Engl.* **34**, 713-721 (1995) (e) L. Mandolini, R. Ungaro, Eds, *Calixarenes in Action*, Imperial College Press, London, 2000. (f) G. Lumetta, R. D. Rogers, A. Gopalan, Eds, *Calixarenes for Separation*, ACS Symposium Series 757, American Chemical Society, Washington, 2000. (g) *Calixarenes 2001*, Z. Asfari, V. Böhmer, J. M. Harrowfield, J. Vicens, Eds, Kluwer Academic Publishers, Dordrecht, 2001.
2. H.-J. Schneider in ref. 1(c), Ch. 2, pp. 29-56.

3. A. Casnati, D. Sciotto, G. Arena in ref. 1(g), Ch. 24, pp. 440-456.
4. See, for example, A. Dalla Cort, L. Mandolini in ref. 1(e), Ch. 3, pp. 85-110; (b) S. Shinkai, K. Araki, T. Matsuda, N. Nishiyama, H. Ikeda, L. Takasu, M. Iwamoto, *J. Am. Chem. Soc.* **112**, 9053-9058 (1990) but see also (c) S.-D. Tan, W.-H. Chen, A. Satake, B. Wang, Z.-L. Xu, Y. Kobuke, *Org. Biomol. Chem.* **2**, 2719-2721 (2004).
5. B. Nilius, G. Droogmans, *Physiol Rev.* **81**, 1415-1459 (2001).
6. N. Douteau-Guével, F. Perret, A. W. Coleman, J. P. Morel, N. Morel-Desrosiers, *J. Chem. Soc. Perkin 2*, 524-532 (2002).
7. G. Arena, A. Contino, F. G. Gulino, A. Margi, F. Sansone, D. Sciotto, R. Ungaro, *Tetrahedron Lett.* **40**, 1597-1600 (1999).
8. M. Selkti, A. W. Coleman, I. Nicolis, N. Douteau-Guével, F. Villain, A. Tomas, C, De Rango, *Chem. Commun.* 161-162 (2000).
9. N. Douteau-Guével, A. W. Coleman, J-P. Morel, N. Morel-Desrosiers, *J. Phys. Org. Chem.* **11**, 693-696 (1998).
10. F. Sansone, M. Segura, R, Ungaro in reference 1(g), Ch. 27, pp. 496-512.
11. F. Sansone, S. Barboso, A. Casnati, M. Fabbi, A. Pochini, F. Ugozzoli, R. Ungaro, *Eur. J. Org. Chem.* 897-905 (1998).
12. M. W. Peczuh, A. D. Hamilton, *Chem. Rev.* **100**, 2479-2494 (2000).
13. M. Maletic, H. Wennemers, D. Q. McDonald, R. Breslow, W. C. Still, *Angew. Chem. Int. Ed.* **35**, 1490-1492 (1996).
14. T. Mizutani, K. Wada, S. Kitagawa, *J. Am. Chem. Soc.* **121**, 11425-11431 (1999).
15. (a) M. A. Hossain, H.-J. Schneider, *J. Am. Chem. Soc.* **120**, 11208-11209 (1998); (b) M. Sirish, H.-J. Schneider, *Chem. Commun.* 907-908 (1999); (c) M. Sirish, V. A. Cherkov, H.-J. Schneider, *Chem. Eur. J.* **8**, 1181-1188 (2002).
16. R. Breslow, Z. Yang, R. Ching, G. Trojandt, F. Odobel, *J. Am. Chem. Soc.* **120**, 3536-3537 (1998).
17. R. A. Scott, A. G. Mauk, Eds, *Cytochrome-c: A Multidisciplinary Approach*, University Science Books, Sausalito (1996).
18. T. Aya, A. D. Hamilton, *Bioorg. Med. Chem. Lett.* **13**, 2651-2654 (2003).
19. A. C. Bach, C. J. Eyermann, J. D. Gross, M. J. Bower, R. L. Harlow, P. C. Weber, W. De Grado, *J. Am. Chem. Soc.* **116**, 3207-3219 (1994).
20. Q. Lin, H. S. Park, Y. Hamuro, C. S. Lee, A. D. Hamilton, *Biopolymers (Peptide Science)* **47**, 285-297 (1998).
21. (a) L. C. Petersen, R. P. Cox, *Biochem. J.* **192**, 687-693 (1980); (b) H. Pelletier, J. Kraut, *Science* **258**, 1748-1755 (1992).
22. (a) H. Zou, W. J. Henzel, X. Liu, A. Lutschg and X. Wang, *Cell.* **90**, 405 (1997); (b) P. Li, D. Nijhawan, I. Budihardjo, S. M. Srinivasula, M. Ahmad, E. S. Alnemri and X. Wang, *Cell.* **91**, 479 (1997); (c) C. Purring, H. Zou, X. Wang and G. McLendon, *J. Am. Chem. Soc.* **121**, 7435-7436 (1999).
23. T. Oshima, M. Goto, F. Shintaro, *Biomacromolecules* **3**, 438-444 (2002).
24. M. J. Pires, M. R. Aires-Barros, J. M. Cabral, *Biotechnol. Progr.* **12**, 290-301 (1996).
25. T. Peters, Jr. *All about Albumin*, Academic Press, san Diego (1996).
26. C. C. Annarelli, L. Reyes, J. Fornazero, J. Bert, R. Cohen, A. W. Coleman, *Cryst. Eng.* **3**, 173-177 (2000).
27. (a) L. Memmi, A. Lazar, A. Brioude, V. Ball, A. W. Coleman, *Chem. Commun.* 2474-2475 (2001).
28. E. Da Silva, P. Shahgaldian, A. W. Coleman, *Int. J. Pharmaceutics* **273**, 57-62 (2004).
29. H. S. Park, Q. Lin, A. D. Hamilton, *J. Am. Chem. Soc.* **121**, 8-13 (1999).
30. M. Sakurai, H. Tamagawa, Y. Inoue, K. Aiga, T. Kunitake, *J. Phys. Chem. B* **101**, 4810-4817 (1997).

31. R. Zadmard, M. Arendt, T. Schrader, *J. Am. Chem. Soc.* **126**, 7752-7753 (2004).
32. (a) T. Schrader, *Chem. Eur, J.* **3**, 1537-1541 (1997); (b) S. Rensing, A. Springer, T. Grawe, T. Schrader, *J. Org. Chem.* **66**, 5814-5821 (2001); (c) S. Rensing, T. Schrader, *Org. Lett.* 2161-2164 (2002).
33. M. L. Connolly, *Science* **221**, 709-712 (1983); *J. Appl. Crystallogr.* **16**, 548-550 (1983).
34. R. Zadmard, T. Schrader, *J. Am. Chem. Soc.* **127**, 904-915 (2005).
35. T. Kodadek, *Chem. Biol.* **8**, 105-115 (2001).
36. H. Zhu, M. Snyder, *Curr. Opin. Chem. Biol.* **5**, 40-45 (2001).
37. This corresponds well with ESI-MS measurements carried out by Coleman et al. on the related complex between BSA and tetrasulfonato-calixarenes: L. Memmi, A. Lazar, A. Brioude, V. Ball, A. W. Coleman, *Chem. Commun.*, 2474-2475 (2001).

Chapter 15

FLUORESCENT CHEMOSENSORS

The pathway to molecular electronics

Su Ho Kim,[a] Hyun Jung Kim,[a] JuyoungYoon,[b] and Jong Seung Kim[a]
[a]*Department of Chemistry, Dankook University, Seoul 140-714, Korea;* [b]*Department of Chemistry, Ewha Womans University, Seoul 120-750, Korea*

Abstract: The use of luminescence as a sensitive and selective response to molecular recognition is the focus of the present discussion. Calixarene-based ionophores bearing various fluorophores can be used as ion sensors due to binding responses including photo-induced electron transfer (PET), chelation-induced enhanced fluorescence (CHEF), fluorescence resonance energy transfer (FRET), photo-induced charge transfer (PCT) and excimer/exciplex formation.

Key words: Calixarenes, fluorescence, PET, FRET, PCT, excimer.

1. INTRODUCTION

The construction of sophisticated molecular sensors may be considered a basic exercise in nanochemistry since the fundamental requirements of high sensitivity and optimal selectivity can only be met by the combination of large molecular units, and calixarenes[1] are well-recognised as appropriate such units. Calix[4]arenes, in particular, have been prepared and characterised in all four possible conformations with an extraordinary variety of substituents, one of the major objectives of these syntheses being the use of the product ligands in many areas of both cation and anion coordination chemistry.[1]

Molecular sensors combine the properties of supramolecular receptors, selective for a given guest, with the ability to produce a measurable signal in response to binding.[2-4] Optical signals reflecting changes in absorption or fluorescence are the most frequently exploited because of the simplicity and low cost of the methods required. Absorption, often seen in the visible region as obvious colour changes, can commonly provide immediate

J. Vicens and J. Harrowfield (eds.), Calixarenes in the Nanoworld, 311–333.

identification of an analyte,[5] while fluorescence can provide ready detection even at nanomolar concentrations. Fluorescent chemosensors which combine such sensitivity with selectivity can, in the case of metal ions,[6-11] for which calixarenes already have established uses,[12-14] be particularly valuable in biological and environmental analysis.[15,16] Obvious applications are in the study of the toxicity of heavy metals such as Pb[17] and In.[18,19]

Since anions are also ubiquitous and play critical roles in many biological and chemical systems, there is an increasing interest in the design and development of receptors that selectively recognize specific anions.[20-22] Known synthetic anion sensors involve a chromophore or fluorophore linked to a neutral anion receptor containing urea,[23] thiourea,[24] amide,[25] phenol,[22(e),26] or pyrrole[27] subunits which can act as H-bond donor sites to bind simple anions such as F^-, $CH_3CO_2^-$, and $H_2PO_4^-$.

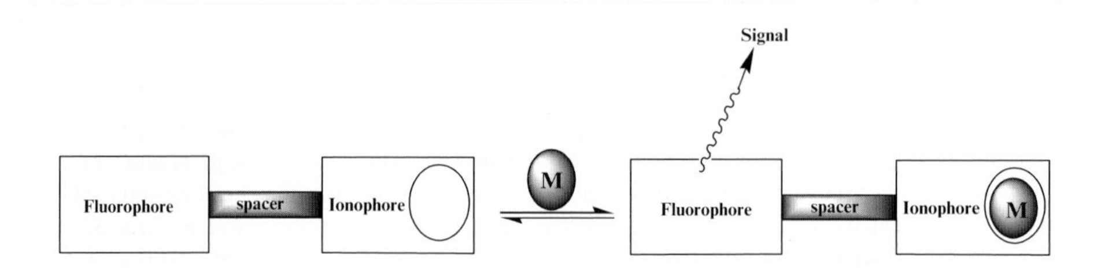

Figure 15-1. Fluorescence action upon metal ion complexation.

The essential structure of a fluorescent sensor is shown in Fig. 1.[15] A crucial aspect of the ionophore component is its binding selectivity but, once achieved, this binding can be detected through various responses such as PET (photo-induced electron transfer);[12,13,28,29] PCT (photo-induced charge transfer);[1] excimer/exciplex formation and extinction;[28,30] or FRET (fluorescence resonance energy transfer).[31] Since ion concentrations in the medium surrounding a sensor may be controlled in various ways, an ultimate application of fluorescent calixarene ionophore which goes beyond that of a simple sensor is in the area of “molecular electronics” as logic gates,[2] and there are prospects for such use of the molecules discussed herein.

2. PHOTO-INDUCED ELECTRON TRANSFER SENSORS

2.1 Principle

Photoexcitation generates strongly oxidising sites within molecules and, if an electron donor site, for example, a lone pair, is present, intramolecular electron transfer (PET) can occur, leading to quenching of the excited state luminescence. If substrate binding renders the donor less accessible, quenching does not occur and luminescence may be observed. Where the substrate is a metal which is chelated by the photosensitive molecule (Fig. 2), this is referred to as CHEF (chelation-enhanced fluorescence),[14] and the metal ion can be considered to act as a fluorescence "switch". Of course, since photoexcitation results in charge separation, it leads to the creation of reducing and well as oxidising sites and what is termed "reverse PET" may result if it is the reducing site which reacts. Although less common, there is evidence (see ahead) for its occurrence in calixarene systems.

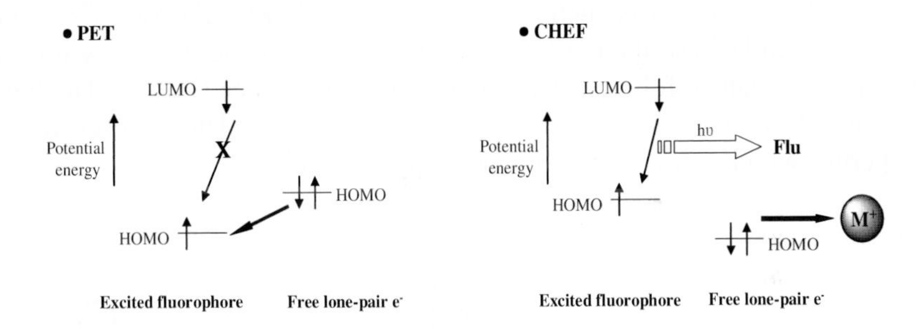

Figure 15-2. PET and CHEF systems.

2.2 Fluorophore-based PET Sensors

Bridging of a calix[4]arene by a polyether link can lead to selectivity in the binding of alkali metal ions and the attachment of 9-cyanoanthracene fluorophores to a 1,3-alternate calix[4]bis-*o*-benzocrown-6 unit as in **1**, for example, leads to a Cs^+-selective sensor.[13] In the free ligand, fluorescence is partly quenched by PET from the dialkoxybenzene unit of the bridge but the selective binding of Cs^+ (in preference to other alkali metal ions) to these oxygen atoms inhibits this electron transfer and allows emissive de-excitation (CHEF). Alkoxyaromatic centres are considerably less efficient than amino centres in PET, however, and greater CHEF effects can be obtained by appropriate modification of sensor structures.[3]

1

A simple means of introducing an amino centre is to include it within the metal-binding unit, as in the pyrene appended calixazacrown **2**. With two inequivalent metal-binding sites, this ligand shows different degrees of fluorescence enhancement depending on the site of metal ion coordination, so that metal ion exchanges such as of Ag^+–K^+, Cu^{2+}–K^+, and Ag^+–Cs^+, processes that can be described as "Molecular Taekwondo" because they involve site changes, can be followed by monitoring fluorescence (Fig. 3). The binding site selectivity is thus associated with differences in detection sensitivity, properties which can be put to practical application.[14a]

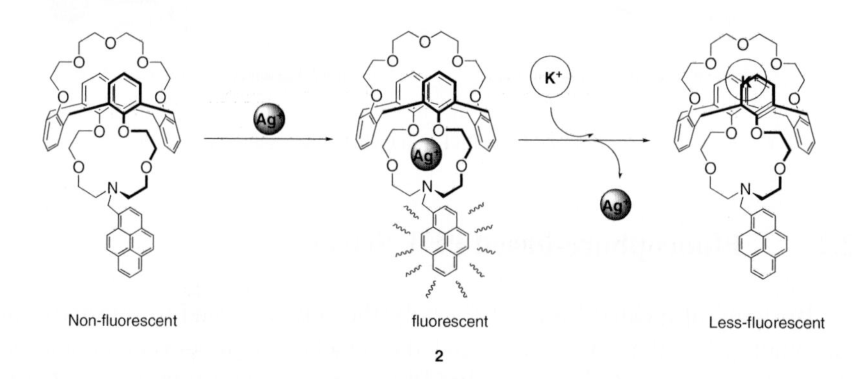

Figure 15-3. "Molecular Taekwondo 1" process of **2**.

Further effects of differences in metal ion coordination modes are seen with the pyrene-functionalised fluorogenic calix[4]arenes (**3**) and **4**, which have a pendent 2-aminoethyl group on the triazacrown ring. Their relatively weak luminescence compared to that of **5** indicates that PET quenching is primarily due to this pendent group. In the presence of Pb(II), both **3** and **4** show enhanced monomer emission, while excimer emission[32] is quenched. This CHEF effect can be attributed to a conformational change resulting from the metal ion binding as well as to coordination of the electron-donor *N*-centre. In contrast, alkali metal ion coordination causes both monomer and excimer emission to be enhanced, indicating that coordination is not associated with a significant change in the ligand conformation. Rather different consequences of metal ion binding are seen with the ligand **5**, which, when free, shows strong monomer (λ_{em} = 396 nm) and excimer (λ_{em} = 448 nm) emissions, consistent with the two pyrene units being in a face-to-face π-stacked arrangement, forming a dynamic excimer.[32] Binding of either Co(II) or Pb(II) quenches both emissions, consistent with the adoption of a new conformation where the two pyrene units are now rather remote from one another.[33]

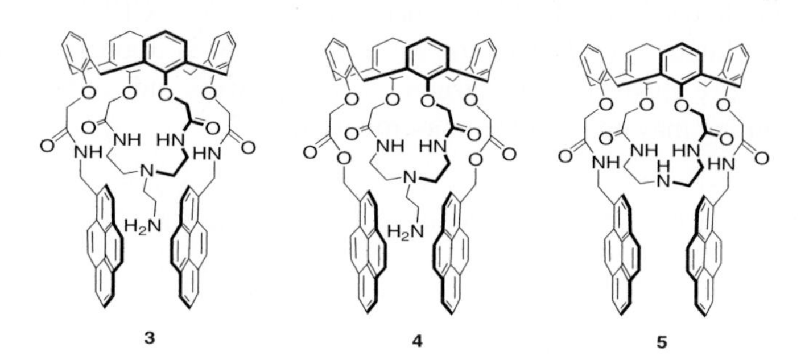

For the 9-cyanoanthracenyl-functionalised calix[4]crown-azacrown (**6**), particularly large CHEF effects were observed for Cs(I), Rb(I) and K(I). This can be explained as a consequence of the preferential location of these cations in the simple crown ring, to which the fluorophore is attached, rather than in the azacrown ring, the CHEF effect then being of the same origin as that described above for ligand **1**. "Taekwondo" effects opposite to those described for **2** can be observed for Cs(I)-Cu(II) and Cs(I)-Ag(I) exchanges.[14b]

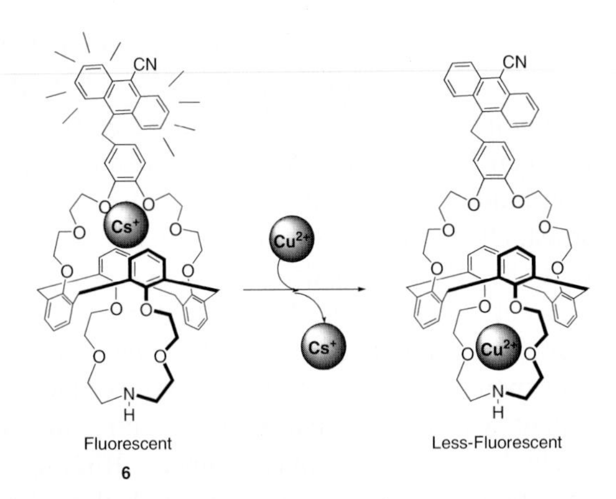

Figure 15-4. "Molecular Taekwondo 2" process of **6**.

The crystallographically characterised 1,3-alternate-calix[4]arene-based fluoroionophores **7** and **8** showed particularly marked CHEF effects with Pb(II) and In(III). For In(III), the *bis*-azacrown, **8**, is approximately x 20 more sensitive than the *mono*-azacrown **7**,[34] perhaps indicating that some In(III) binding may occur at the non-crown site in **7**.

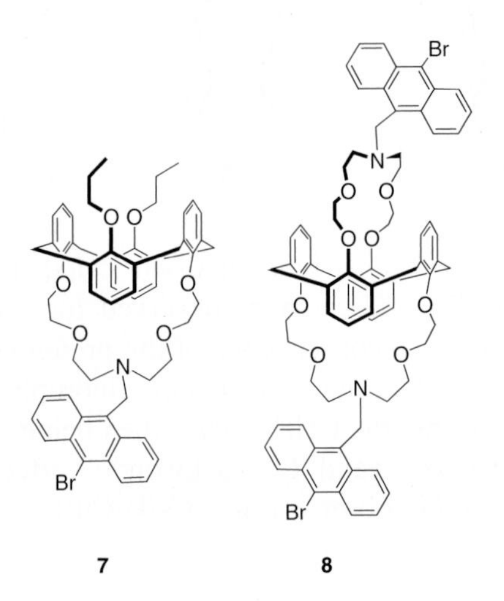

PET between the pyrene and nitrophenyl substituents of **9** is rendered less efficient by the selective binding of Na(I), which cause the two groups to become spatially more distant. The intensity of emission from **9** increases uniformly with increasing [Na(I)].[12]

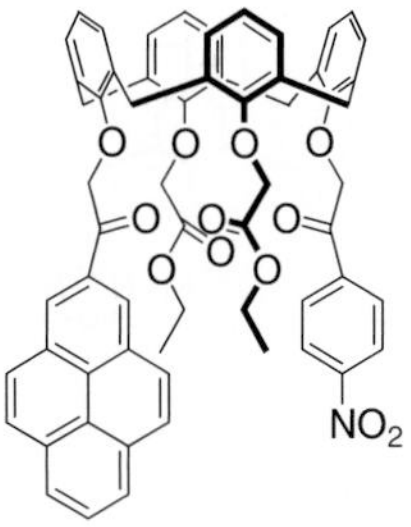

9

Polycyclic aromatic units are generally useful as fluorophores and association of imidazole coordination sites with a naphthyl substituent on a calix[4]-arene scaffold generates a sensor **14** with selectivity towards Cu(II) and Zn(II). The differing redox properties of these two metals explain the observations of fluorescence quenching with Cu(II) and fluorescence enhancement with Zn(II). These effects of the metals are eliminated by protonation of the imidazole sites and also by the addition of excess base, presumably as a consequence of hydroxo-complex formation, so that the pH dependence of the sensor fluorescence is complicated.[39] The naphthyl entity in the calix[4]crown sensors **15** and **16** also gives rise to metal sensitive luminescence, binding of K(I) in both sites of **15**, for example, causing fluorescence enhancement through CHEF. Pb(II) binding, however, causes fluorescence quenching, indicating that it is bound selectively to the simple crown-5 site.[40] Introduction of the very widely used substituted naphthyl group, "dansyl" (1-dimethylaminonaphthalene-8-sulfonate) as a fluorophore on a calix[4]crown as in **17** provides a sensor exhibiting high sensitivity and selectivity for Hg(II) in aqueous acetonitrile.[41] In the anthracenylcalix[4]arene **10**, the presence of a dioxotetraaza unit favours the chelation of transition metal cations such as Zn(II) and Ni(II), though again the consequences for emission differ, Zn(II) causing enhancement and Ni(II) quenching,[35] possibly through PET for the latter. The heterocyclic fluorophore "BODIPY" (boradipyrromethene) has recently become extremely commonly encountered because of its exceptional photophysical characteristics,[42] and its incorporation into the calix[4]crown ionophore **18** provides a sensor showing useful selectivity for Ca(II) over other physiologically important metal ions, its sensitivity reflecting a strong CHEF effect.[43]

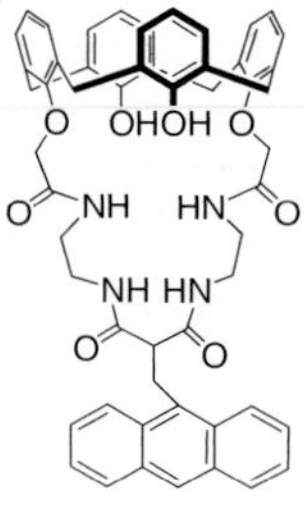

10 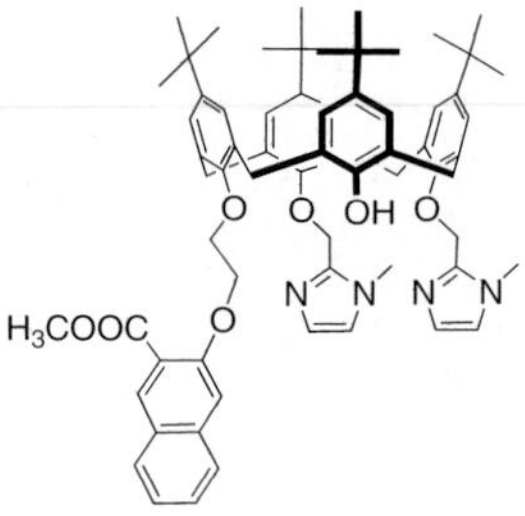 14 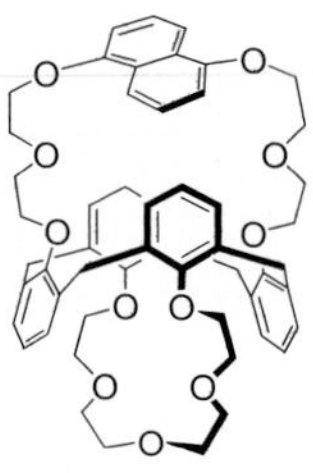 15

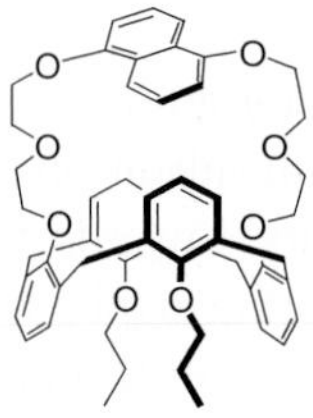

16 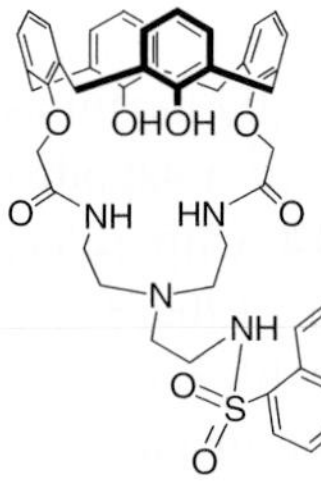 17 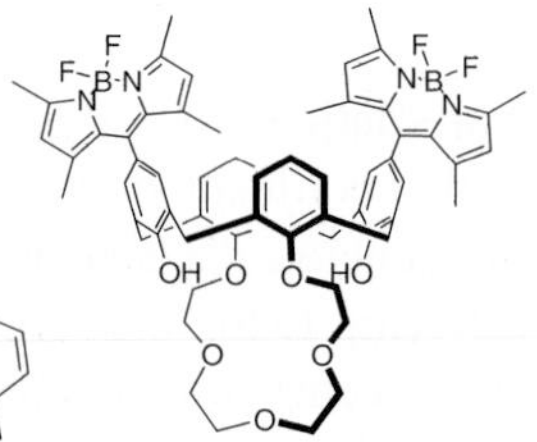 18

A fluorophore itself may be also a ligand and the bis(alkoxynaphthoic acid) derivative of calix[4]arene **13** shows a complicated pH dependence of metal ion induced luminescence quenching which can be rationalised in terms of carboxyl and phenolic group deprotonation reactions and different modes of complexation, by Fe(III) and Cu(II) in particular, to the changing array of oxyanionic sites.[38]

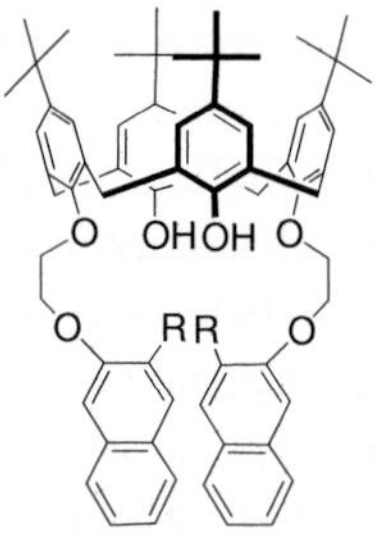

R= COOH

13

In analogy to metal ions, the proton may bind to an electron-donor site such as *N* and thus the emission from **11** is much stronger in acidic than in neutral or basic solutions due to the inhibition of PET (Fig.5). However, complexation of K(I) in acidic solutions leads to a diminished basicity at *N*, hence to a reduced degree of protonation and thus weaker luminescence.[36]

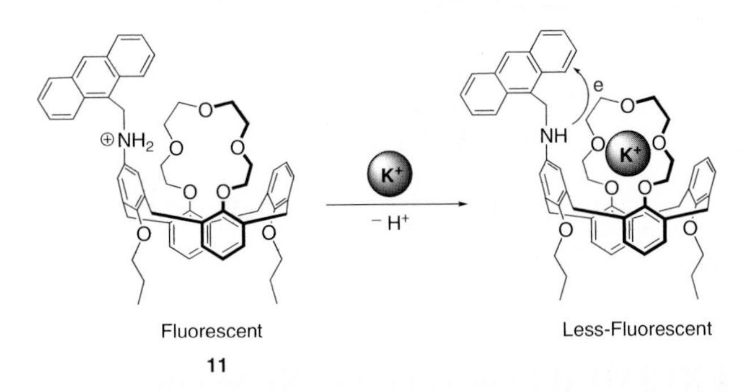

Figure 15-5. pH dependent PET process of **11**.

For the recognition of simple anions, interactions with bound hydrogen, especially amide NH, are useful and the tetramide-bridged derivative of calix[4]arene **12** exhibits a selective interaction with $CH_3CO_2^-$ (in comparison to F^-, Cl^-, Br^-, $H_2PO_4^-$, NO_3^-, I^-, and HSO_4^-). Here, acetate binding results in a quenching of fluorescence, ^{1}H NMR spectroscopy indicating that oxygen atoms of the anion interact with the anthracene-9H as well as the amide NH centres.[37] In analogy to metal ion complexation at the fluorophore, fluoride anion can bind to a phenylboron centre (at B), thus influencing emission from the phenyl group. Thus, the first known boron-appended calix[4]arene, **19**, is a sensitive and selective fluorescent fluoride sensor. The phenyl groups attached to the boron provide sufficient fluorescence intensity to allow binding events to be followed by fluorescence spectroscopy (λ_{ex} = 320 nm, λ_{em} = 395 nm), titration of **19** with Bu_4NF in $CHCl_3$ solution giving a fluorescence decrease.[44]

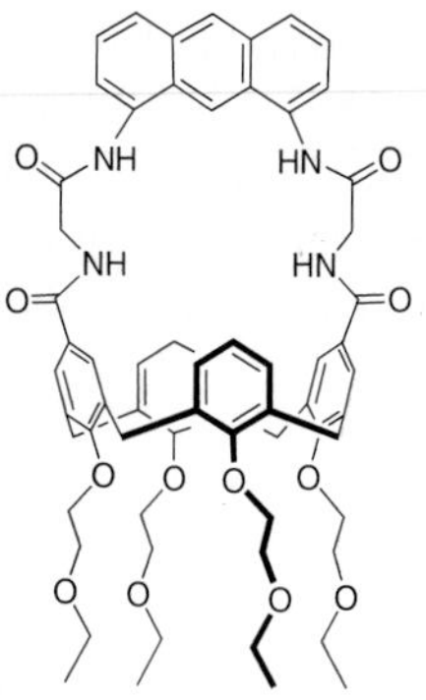

12 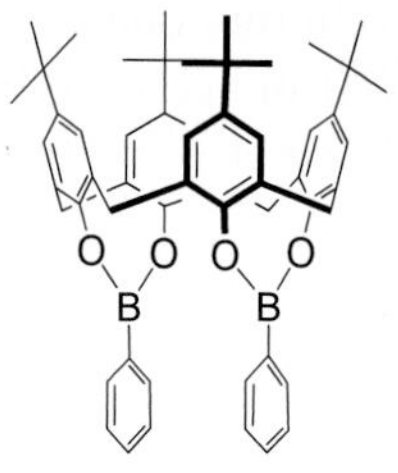 **19**

3. EXCIMER FORMATION SENSORS

3.1 Principle

The observation of excimer emission from molecules containing pyrene substituents has been noted above and indeed this type of emission is commonly found with large aromatic groups. An "excimer" or excited state dimer can be defined as a complex formed by the interaction of an excited fluorophore with the same fluorophore in its ground state and its formation can of course be facilitated if two fluorophore substituents are held in close proximity on a given molecular scaffold. Where excimer formation is competitive with decay of the excited state, emission may be seen from the "monomer" excited state and from the excimer, the excimer emission usually being broad and structureless.[33,45,46]

3.2 Fluorophore-based Excimer Formation Sensors

Pyrene displays not only a well-defined monomer emission at 370-430 nm but also an efficient excimer emission at around 480 nm.[45,47] The intensity ratio of excimer to monomer emission (I_E/I_M) is sensitive to conformational changes of the pyrene-appended receptors, so that variation in I_E/I_M values upon substrate binding to the receptor can be structurally informative.[3(b),32,45,48]

An example of what may be considered proton-controlled orientation of pyrene pairs, which may have practical application as a new type of imaging system, is provided by the behaviour of the polymer embedded calixarene derivative **30** containing two pyrene groups connected to a phenol Boc-protected calixarene. The pyrene groups in **30** are oriented face-to-face and, they give rise to strong excimer fluorescence as a result. Photo-induced removal of the Boc groups from **30** in the presence of acid generates **29**, in which intramolecular hydrogen bonding interactions stabilise a conformation in which the pyrene entities are remote and give rise to monomer emission only.[57]

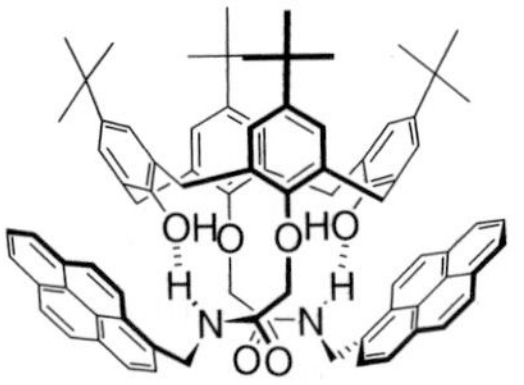

29 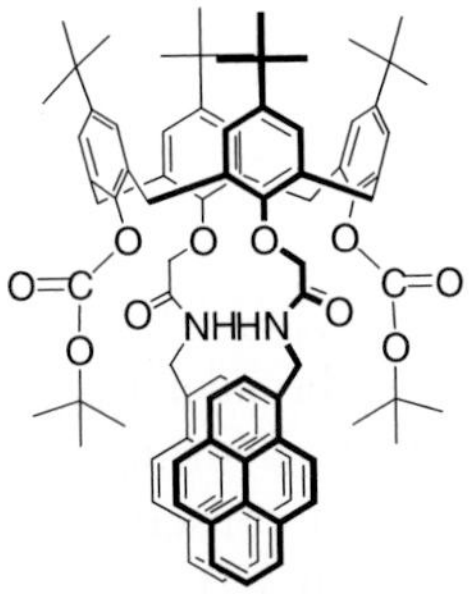 **30**

Concerning metal ion sensing, the calix[4]arene **20** with two pyrene substituents attached *via* amide links to distal rings,[49] for example, when surface-immobilised shows an excimer emission which is enhanced by Na(I) binding, a result interpreted as indicating that Na(I) coordination induces a conformational change which brings the pyrene units into closer contact. For a similar calixarene simply involving ester links instead of amide, the opposite effect of Na(I) coordination is observed, illustrating the subtlety of effects detectable by I_E/I_M measurements.[12(b)] Other systems sensitive to the alkali metals are provided by the calix[4]arenes **31** and **32**. In the absence of metal cations, **31** shows a strong excimer emission but the addition of Li(I), for example, leads to a marked decrease in this emission and a concomitant increase in the monomer emission. Na(I) causes similar effects with **32**, with a moderate selectivity for Na(I) over K(I).[28]

R= O–⟨⟩–NO$_2$

20 **31** **32**

In investigating a broader range of metal ions, Kim *et al.* reported that compound **21** displays a strong excimer emission in CH_3CN solution. This was extinguished as a consequence of coordination of Pb(II) but could be regained by the use of K(I) to displace the Pb(II), leading to an *ON-OFF* switch based on the metal exchange (Fig. 6). Computational studies of the complexes indicate that these effects can be explained in terms of HOMO-LUMO interactions in the excimer pyrene pair which are significant in the K(I) complex but not that of Pb(II).[46]

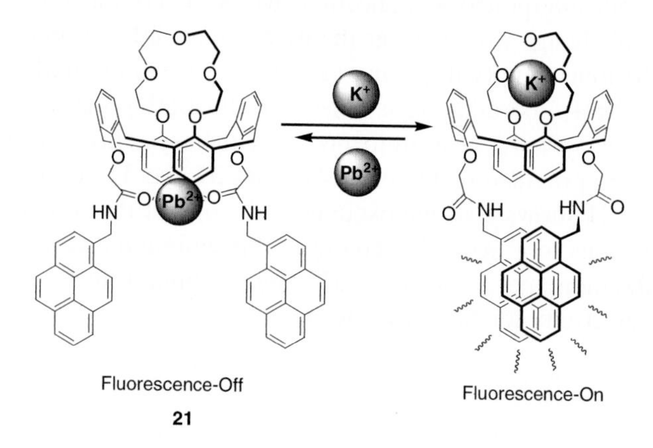

Figure 15-6. On-off excimer formation process of **21**.

Although the *N*-Methylpyreneamide-appended calix[4]crown-5 (**24**) shows minimal fluorescence responses to metal ion binding, the related crown-6 derivative **25** provides another system showing a strong fluorescence response to Pb(II). This has been rationalised in terms of a reverse-PET process involving the amide-O donor atoms bound to Pb(II) and the fluorescence properties of **25** as influenced by Pb(II), acid and base have been exploited for the operation of **25** as a "molecular logic gate" (see ahead).[54] Another pyrene-functionalised calixarene which is responsive to metal metal ions is the diazacrown derivative **26**. This has been found to exhibit selective *ON–OFF* type sensing behavior toward Hg^{2+} ion over other heavy metal ions.[55] The exceptional metal ion binding ability of thiacalixarenes[52(a),(b)] is an obvious basis for their use as a component of sensors of the type under discussion here, especially as the presence of S-donor centres should favour heavy-metal binding. Indeed, **22** does provide a Pb(II)-sensitive system, where again it appears the amide oxygen atoms not only participate in the binding with Pb(II) ion but also behave as a PET acceptor from the pyrene units.[52(c)]

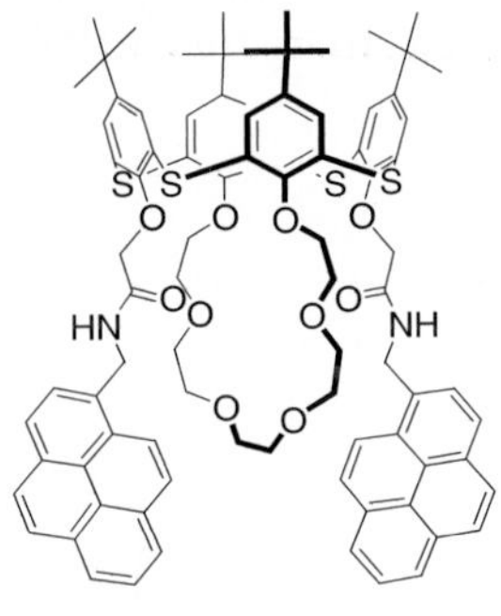

22

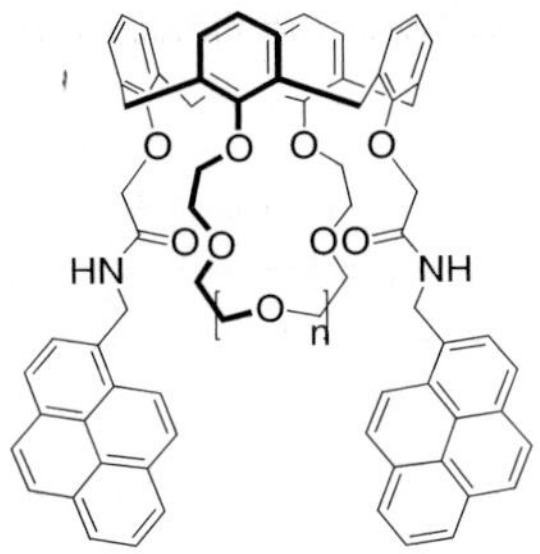

	n
24	1
25	2

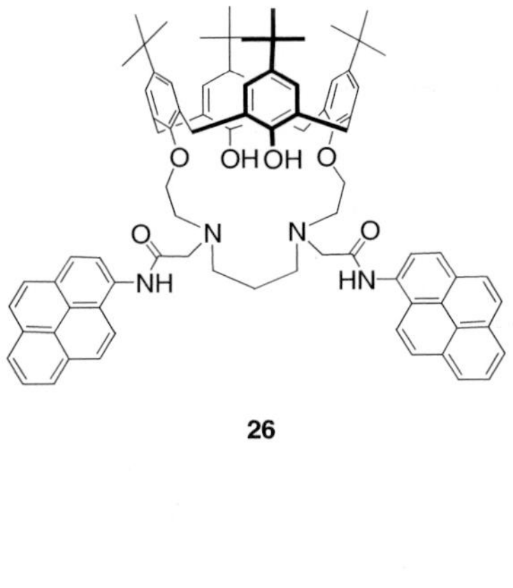

26

Incorporation of hydroxamate chelation sites within the bispyrenyl calix[4]arene-based receptor **23** provided a ligand suitable for binding of transition metal ions in organic/aqueous media. Both Ni(II) and Cu(II) are efficient quenchers of the monomer and excimer emissions, though the need for ligand deprotonation prior to binding and differences in the stability constants for the two metal complexes mean that this response is obtained in different ranges of pH.[53] Sensitivity to both transition and main-group metal ions is found with the *bis*- and *tris*-pyrenamide calix[4]arene derivatives **27** and **28**, where, interestingly, the ratio of monomer to excimer emission intensities was found to be strongly dependent on the different number of pyrene substituents.[56] Other systems showing responses to a range of metal ion types are found with **33** and **34**. The pyrene units are remote from each other in **33** and face-to-face π-stacked in **34**, and, in the excited state, the two pyrene units of **34** form a strong intramolecular excimer displaying an emission at 472 nm with a relatively weak monomer emission at 395 nm. In contrast, **33** exhibits only a monomer emission at 398 nm because intramolecular H-bonding between the phenolic OH oxygen atoms and the amide hydrogen atoms prevents π-stacking of the two pyrene groups. Fluorescence changes on addition of various metal ions to a solution of **33** showed that it has a remarkably high selectivity for In(III) over the other metal ions tested.[58]

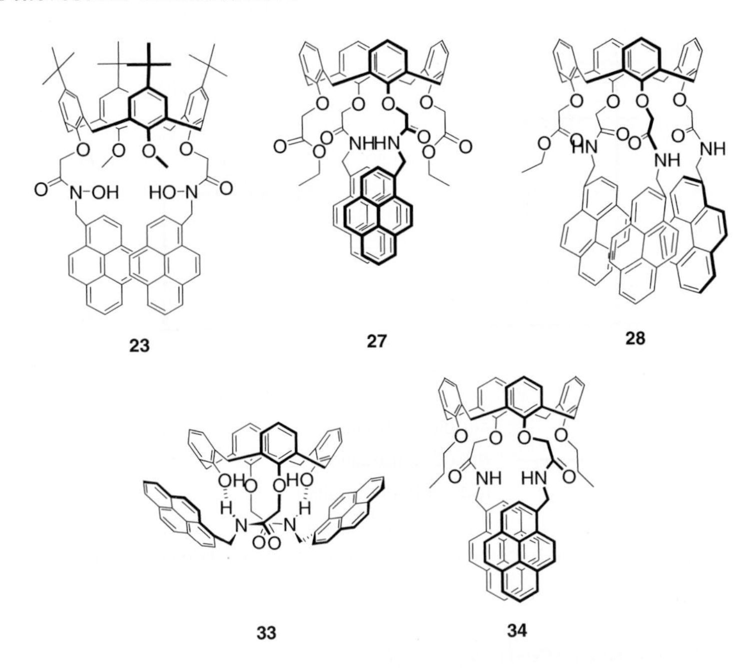

4. FRET (FLUORESCENCE RESONANCE ENERGY TRANSFER) SENSORS

4.1 Principle

FRET arises as a result of interaction between the excited states of two fluorophores when the emission spectrum of one fluorophore (the FRET donor) overlaps the absorption spectrum of another fluorophore (the FRET acceptor), by which excitation energy of the donor is transferred to the acceptor without emission of photon (Fig. 7).[59] FRET is a distance-dependent interaction, varying as the inverse sixth power of the intermolecular separation,[60a] making it useful over distances comparable with the dimensions of biological macromolecules. Thus, FRET is commonly considered as an important technique for investigating a variety of biological phenomena that produce changes in molecular proximity.[60b]

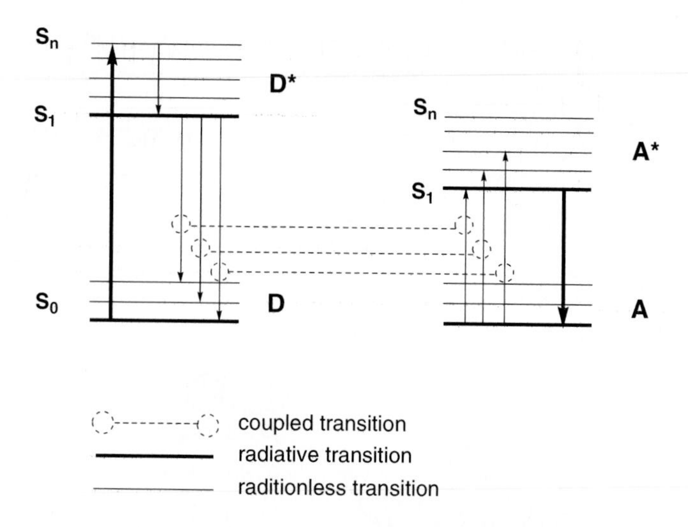

Figure 15-7. FRET (Fluorescence Resonance Energy Transfer) system.

4.2 Fluorophore-based FRET sensors

FRET requires a bichromophoric system and, as appropriate molecules, Kim *et al.*[58] developed calixarene derivatives with a diazo-group giving a visual color change as well as with a pyrenyl group providing a fluorescence change. In **35**, the pyrene units and *p*-nitrophenylazo groups can act as FRET donors and acceptors, respectively. Accordingly, FRET between them takes place to quench the pyrene monomer emission.[58]

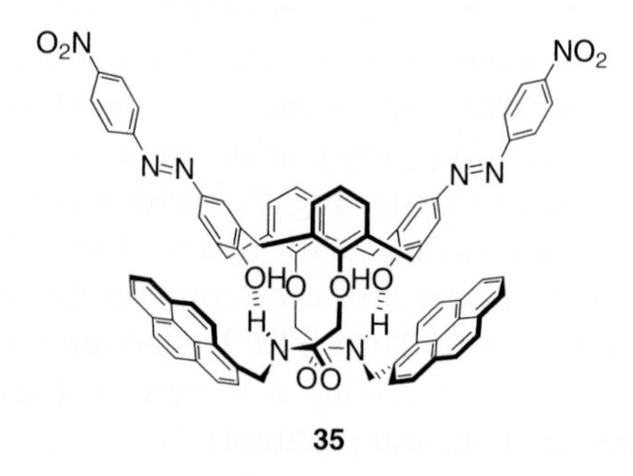

Compound **36**, with dual sensing probes, is another species showing sensitivity to Pb(II) over other metal ions. Complexation of **36** by Pb(II) causes significant enhancement of the pyrene emission, an effect attributed to poorer overlap of the chromophore manifolds in the complex, the p-nitrophenylazo absorptions in particular appearing to shift to shorter wavelengths in the presence of the metal (Fig. 8).[61]

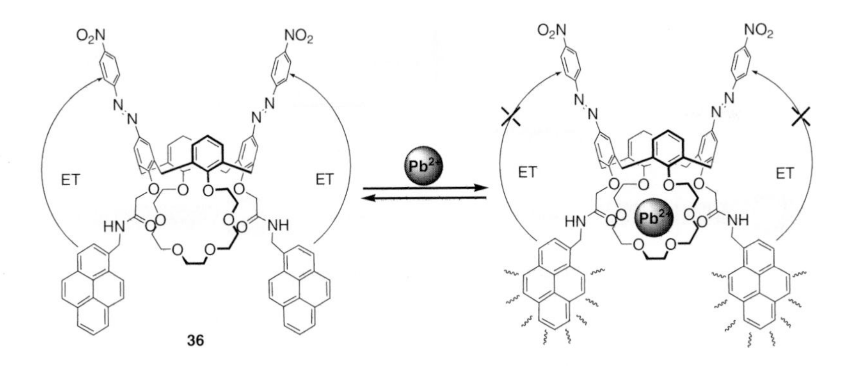

Figure 15-8. Metal ion induced FRET diminished in **36**. (ET = energy transfer).

5. PCT (PHOTO-INDUCED CHARGE TRANSFER) SENSORS

5.1 Principle

A fluorophore which contains an electron-donating group conjugated to an electron-withdrawing group performs intramolecular charge transfer from the donor to the acceptor upon excitation by light. The consequent change in dipole moment causes a Stokes shift that depends on the micro-environment of the fluorophore. It can thus be expected that cations in close interaction with the donor or the acceptor moiety will change the photophysical properties of the fluorophore since the complexed cation affects the efficiency of the intramolecular charge transfer.[15, 62-64]

When an electron donor group within the fluorophore interacts with a cation, the latter reduces the electron-donating character of donor group and due to the resulting reduction of conjugation, a blue-shifted absorption spectrum is expected together with a decrease of the extinction coefficient. In contrast, metal ion binding to the acceptor group enhances its electron-withdrawing character and the absorption spectrum should thus be red-shifted with an increase in the molar absorption coefficient (see Fig. 9). The fluorescence spectra are in principle shifted in the same direction as the absorption spectra. In addition to these shifts, changes in quantum yields and lifetimes may occur and all these photophysical effects should be dependent on the charge and the size of the cation.

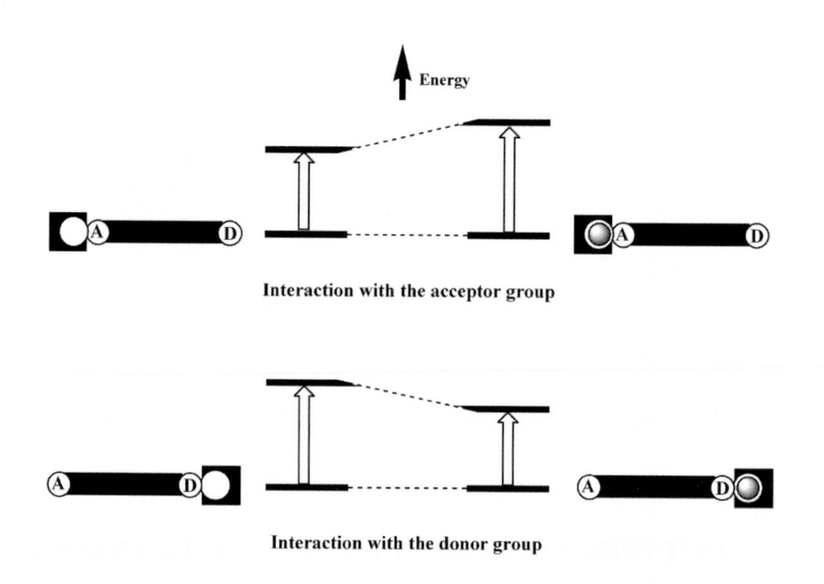

Figure 15-9. PCT (Photoinduced Charge Transfer) system.

5.2 Fluorophore-based PCT Sensors

Strong H-bonding within the pyrene-functionalised, anion-sensing calix[4]arene **37** causes a red-shift of the pyrene absorption. Under 346 nm excitation, the facing pyrene units form a dynamic excimer emitting at 482 nm, the monomer emitting at 385 nm. Complexation of F^- causes the absorption band to red-shift by 54 nm and excimer emission to blue-shift by 12 nm with enhanced intensity. The blue-shifted excimer emission is attributed to a *static excimer* as illustrated in Fig. 10.[65]

Cation sensing by PCT is observed with **38** and **39**. The fluorophore, 6-acyl-2-methoxy-naphthalene, contains an electron-donating substituent (methoxy group) conjugated to an electron-withdrawing substituent (carbonyl group) and this allows PCT to occur upon excitation. An increase in the molar absorption coefficients, together with red shifts of the absorption

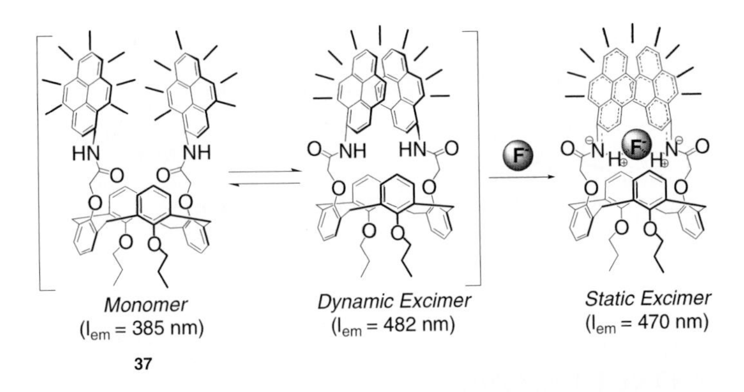

Figure 15-10. F^- induced dynamic and static excimer formation of **37**.

and emission spectra, are observed upon cation binding, as expected from the cation-induced enhancement of the electron-withdrawing character of the carbonyl group. The higher the charge density of the cation, the larger the observed effect. A large enhancement of the fluorescence quantum yield was observed under cation binding which can be explained in terms of the relative location of the singlet π-π* and n-π* states.[29]

OMe MeO MeO OMe OMe

38 **39**

Other cation sensing calixarenes responding by PCT have been based on the dansyl fluorophore, which is well known for its sensitivity to polarity of the medium owing to intramolecuar PCT. The water soluble, di- and tri-dansyl thiacalix[4]arene derivatives **40** and **41**, were prepared to investigate their metal sensing abilities in aqueous solution. Both are highly sensitive to Cd(II) though, possibly for steric reasons, **40** is the more sensitive.[50]

40: X_1=X_3=OH, X_2=X_4=

41: X_1=OH, X_2=X_3=X_4=

6. CONCLUSIONS

The selective and sensitive detection of both molecular and ionic species remains one of the major challenges in contemporary chemistry. The present review illustrates the utility of calixarene-derived sensors for cation and anion analysis in particular. The unique topology of calixarenes offers the potential for cavity design through selective functionalization as well as for manipulation of fluorescence characteristics arising from PET, CHEF, PCT and FRET effects. Study of the compounds described herein has provided greater understanding of the complex interactions involved in binding, thus offering the prospect of a number of industrial and environmental applications as extractants and sensors. The elaborate structures necessary to achieve the appropriate combination of characteristics required for sensing and numerous other possible functions[2] are the reason such studies fall within the domain of nanochemistry.

7. REFERENCES

1. (a) C. D. Gutsche, *Calixarenes,* Monographs in Supramolecular Chemistry No 1; J. F. Stoddart, Ed.; Royal Society of Chemistry: Cambridge, U. K., 1989 (Revised as *Calixarenes Revisited*, 1998); (b) *Calixarenes*: *A Versatile Class of Macrocyclic Compounds*; V. Böhmer, J. Vicens, Eds.; Kluwer Academic Publishers, Dordrecht, 1991. (c) R. Ungaro. A. Pochini in *Frontiers in Supramolecular Organic Chemistry and Photochemistry*; H.-J. Schneider, Ed.; VCH, Weinheim, Germany, 1991; pp. 57-81. (d). V. Böhmer, *Angew. Chem. Int. Ed. Engl.* **34**, 713-721 (1995) (e) L. Mandolini, R. Ungaro, *Calixarenes in Action*, Imperial College Press, London, 2000. (f) G. Lumetta, R. D. Rogers, A. Gopalan, *Calixarenes for Separation*, ACS Symposium Series 757, American Chemical Society, Washington, 2000. (g) *Calixarenes 2001*, Z. Asfari, V. Böhmer, J. M. Harrowfield, J. Vicens, Eds, Kluwer Academic Publishers, Dordrecht, 2001.
2. J. F. Callan, A. Prasanna de Silva, D. C. Magri, *Tetrahedron* **61**, 8551-8588 (2005).
3. (a) *Chemosensors of Ion and Molecule Recognition;* J. P. Desvergne, A. W. Czarnik, Eds., NATO ASI series, Kluwer Academic Publishers, Dordrecht, 1997; (b) A. Prasanna

de Silva, H. Q. Gunaratne, N. T. A. Gunnlaugsson, T. M. Huxley, C. P. McCoy, J. T. Rademacher, T. E. Rice, *Chem. Rev.* **97**, 1515 (1997).
4. (a) J. S. Kim, O. J. Shon, J. W. Ko, M. H. Cho, I. Y. Yu, J. Vicens, *J. Org. Chem.* **65**, 2386 (2000). (b) J. S. Kim, W. K. Lee, K. No, Z. Asfari, J. Vicens, *Tetrahedron Lett.* **40**, 3345 (2000); (c) J. S. Kim, O. J. Shon, W. Sim, S. K. Kim, M. H. Cho, J. G. Kim, I. H. Suh, D. W. Kim, *J. Chem. Soc., Perkin Trans. 1* 31 (2001). (d) J. S. Kim, P. Thuery, M. Nierlich, J. A. Rim, S. H. Yang, J. K. Lee, J. Vicens, *Bull. Kor. Chem. Soc.* **22**, 321 (2001).
5. (a) J. P. Dix, F. Vögtle: *Chem. Ber.* **113**, 457 (1980); (b) J. P. Dix, F. Vögtle, *Chem. Ber.* **114**, 638 (1981).
6. X. S. Xie, *Acc. Chem. Res.* **29**, 598 (1996).
7. P. M. Goodwin, W. P. Ambrose, R. A. Keller, *Acc. Chem. Res.* **29**, 607 (1996).
8. M. Orrit, J. Bernard, *Phys. Rev. Lett.* **65**, 2716 (1990).
9. U. Mets, R. Rigler, *J. Fluoresc.* **4**, 259 (1994).
10. (a) W. E. Moerner, T. Basche, *Angew. Chem., Int. Ed. Engl.* **32**, 457 (1993); (b) W. E. Moerner, *Acc. Chem. Res.* **29**, 563 (1996).
11. E. S. Yeung, *Acc. Chem. Res.* **27**, 209 (1994).
12. L. Aoki, T. Sakaki, S. Shinkal, *J. Chem. Soc., Chem. Commun.* 730 (1992).
13. (a) H. F. Ji, G. M. Brown, R. Dabestani, *Chem. Commun.* 609 (1999); (b) H. F. Ji, R. Dabestani, G. M. Brown, R. A. Sachleben, *Chem. Commun.* 833 (2000); (c) H. F. Ji, R. Dabestani, G. M. Brown, R. L. Hettich, *J. Chem. Soc., Perkin Trans. 2* 585 (2001).
14. (a) J. S. Kim, O. J. Shon, J. A. Rim, S. K. Kim, J. Yoon, *J. Org. Chem.* **67**, 2348 (2002); (b) J. S. Kim, K. H. Noh, S. H. Lee, S. K. Kim, S. K. Kim, J. Yoon, *J. Org. Chem.* **68**, 597 (2003).
15. B. Valeur, I. Leray, *Coord. Chem. Rev.* **205**, 3 (2000).
16. L. Fabbrizzi, A. Poggi, *Chem. Soc. Rev.* **24**, 197 (1995).
17. N. Rifai, G. Cohen, M. Wolf, L. Cohen, C. Faser, J. Savory, L. DePalma, *Ther. Drug Monit.* **15**, 71 (1993) and reference therein.
18. A. A. Moshtaghie, M. A. Ghaffari, *Iran Biomed. J.* **7,** 73 (2003).
19. R. E. Chapin, M. W. Harris, H. E. Sidney Hunter, B. J. Davis, B. J. Collins, A. C. Lockhart, *Fundam. Appl. Toxicol.* **27**, 140 (1995).
20. (a) *Supramolecular Chemistry of Anions*; A. Bianchi, K. Bowman-James, E. García-España, Eds., Wiley-VCH: New York, 1997. (b) P. D. Beer, P. A. Gale, *Angew. Chem., Int. Ed.* **40**, 486 (2001).
21. (a) M. Scherer, J. L. Sessler, A. Gebauer, V. Lynch, *Chem. Commun.* 85 (1998); (b) C. Dusemund, K. R. A. S. Sandanayake, S. Shinkai, *J. Chem. Soc., Chem. Commun.* 333 (1995); (c) P. Anzenbacher, Jr. K. Jursíková, V. M. Lynch, P. A. Gale, J. L. Sessler, *J. Am. Chem. Soc.* **121**, 11020 (1999); (d) S. Yun, H. Ihm, H. G. Kim, C. W. Lee, B. Indrajit, K. S. Oh, Y. J. Gong, J. W. Lee, J. Yoon, H. C. Lee, K. S. Kim, *J. Org. Chem.* **68**, 2467 (2003); (e) S. K. Kim, J. Yoon, *Chem. Commun.* 770 (2002).
22. (a) Y. Kubo, M. Yamamoto, M. Ikeda, M. Takeuchi, S. Shinkai, S. Yamaguchi, K. Tamao, *Angew. Chem., Int. Ed.* **42**, 2036 (2003); (b) K. Choi, A. D. Hamilton: *Angew. Chem., Int. Ed.* **40**, 3912 (2001); (c) D. H. Lee, S. U. Son, J.-I. Hong, *Angew. Chem., Int. Ed.* **43**, 4777 (2004); (d) J.-H. Liao, C.-T. Chen, J.-M. Fang, *Org. Lett.* **4**, 561 (2002); (e) D. H. Lee, J. H. Im, S. U. Son, Y. K. Chung, J.-I. Hong, *J. Am. Chem. Soc.* **125**, 7752 (2003); (f) M. Vázquez, L. Fabbrizzi, A. Taglietti, R. M. Pedrido, A. M. González-Noya, M. R. Bermejo, *Angew. Chem., Int. Ed.* **116**, 1996 (2004).
23. (a) E. J. Cho, J. W. Moon, S. W. Ko, J. Y. Lee, S. K. Kim, J. Yoon, K. C. Nam, *J. Am. Chem. Soc.* **125**, 12376 (2003); (b) H. Xie, S. Yi, X. Yang, S. Wu, *New J. Chem.* **23**, 1105 (1999).

24. (a) V. Thiagarajan, P. Ramamurthy, D. Thirumalai, V. T. Ramakrishnan, *Org. Lett.* **7**, 657 (2005); (b) D. A. Jose, D. K. Kumar, B. Ganguly, A. Das, *Org. Lett.* **6**, 3445 (2004); (c) B. P. Hay, T. K. Firman, B. A. Moyer, *J. Am. Chem. Soc.* **127**, 1810 (2005); (d) L. Nie, Z. Li, J. Han, X. Zhang, R. Yang, W. Liu, F. Wu, J. Xie, Y. Zhao, Y. Jiang, *J. Org. Chem.* **69**, 6449 (2004).
25. (a) P. D. Beer, F. Szemes: *J. Chem. Soc., Chem. Commun.* 2245 (1995); (b) P. D. Beer, F. Szemes, V. Balzani, C. M. Sala, M. G. B. Drew, S. W. Dent, M. Maestri, *J. Am. Chem. Soc.* **119**, 11864 (1997); (c) L. Kuo, J. Liao, C. Chen, C. Huang, C. Chen, J. Fang, *Org. Lett.* **5**, 1821 (2003).
26. (a) D. H. Lee, K. H. Lee, J. I. Hong, *Org. Lett.* **3**, 5 (2001); (b) D. H. Lee, H. Y. Lee, K. H. Lee, J. I. Hong, *Chem, Commun.* 1188 (2001).
27. (a) P. A. Gale, J. L. Sessler, V. Král, V. Lynch, *J. Am. Chem. Soc.* **118**, 5140 (1996); (b) J. L. Sessler, *Coord. Chem. Rev.* **222**, 57 (2001); (c) C. B. Black, B. Andrioletti, A. C. Try, C. Ruiperez, J. L. Sessler, *J. Am. Chem. Soc.* **121**, 10438 (1999).
28. T. Jin, K. Ichikawa, T. J. Koyama, *J. Chem. Soc., Chem. Commun.* 499 (1992).
29. I. Leray, F. O'Reilly, J. L. Habib Jiwan, J. Ph. Soumillion, B. Valeur, *Chem. Commun.* 795 (1999).
30. S. Nishizawa, H. Kaneda, T. Uchida, N. Teramae, *J. Chem. Soc. Perkin Trans. 2* 2325 (1998).
31. (a) M. A. Hossain, H. Mihara, A. Ueno, *J. Am. Chem. Soc.* **125**, 11178 (2003); (b) H. Takakusa, K. KiKuchi, Y. Urano, T. Higuchi, T. Nagano, *Anal. Chem.* **73**, 939 (2001); (c) M. Arduini, F. Felluga, F. Mancio, P. Rossi, P. Tecilla, U. Tonellato, N. Valentinuzzi, *Chem. Commun.* 1606 (2003).
32. C. J. Broan, *Chem. Commun.* 699 (1996).
33. (a) J. Y. Lee, S. K. Kim, J. H. Jung, J. S. Kim, *J. Org. Chem.* **70**, 1463 (2005); (b) S. H. Lee, S. H. Kim, S. K. Kim, J. H. Jung, J. S. Kim, *J. Org. Chem.* **70**, 9288 (2005).
34. S. H. Yang, O. J. Shon, K. M. Park, S. S. Lee, H. J. Park, M. J. Kim, J. H. Lee, J. S. Kim, *Bull. Korean Chem. Soc.* **23**, 1585 (2002).
35. F. Unob, Z. Asfari, J. Vicens, *Tetrahedron Lett.* **39**, 2951 (1998).
36. J. Bu, Q. Zheng, C. Chen, Z. Huang, *Org. Lett.* **6**, 3301 (2004).
37. R. Miao, Q. Zheng, C. Chen, Z. Huang, *Tetrahedron Lett.* **46**, 2155 (2005).
38. J. Liu, Q. Zheng, J. Yang, C. Chen, Z. Huang, *Tetrahedron Lett.* **43**, 9209 (2002).
39. Y. Cao, Q. Zheng, C. Chen, Z. Huang, *Tetrahedron Lett.* **44**, 4751 (2003).
40. J. H. Bok, H. J. Kim, J. W. Lee, S. K. Kim, J. K. Choi, J. Vicens, J. S. Kim, *Tetrahedron Lett.* **47** (2006) 1237.
41. Q. Chen, C. Chen, *Tetrahedron Lett.* **46**, 165 (2005).
42. (a) G. Beer, K. Rurack, J. Daub, *Chem. Commun.* 1138 (2001); (b) T. Gareis, C. Huber, O. S. Wolfbeis, J. Daub, *Chem. Commun.* 1717 (1997); (c) K. Rurack, M. Kollmannsberger, J. Daub, *New J. Chem.* **25**, 289 (2001); (d) N. Dicesare, J. R. Lakowicz, *Tetrahedron Lett.* **42**, 9105 (2001); (e) C. N. Baki, E. U. Akkaya, *J. Org. Chem.* **66**, 1512 (2001).
43. N. R. Cha, S. Y. Moon, S. K. Chang, *Tetrahedron Lett.* **44**, 8268 (2003).
44. S. Arimori, M. G. Davidson, T. M. Fyles, T. G. Hibbert, T. D. James, G. I. Kociok-Köhn, *Chem. Commun.* 1640 (2004).
45. F. M. Winnik, *Chem. Rev.* **93**, 587 (1993).
46. S. K. Kim, S. H. Lee, J. Y. Lee, J. Y. Lee, A. Bartsch, J. S. Kim, *J. Am. Chem. Soc.* **126**, 16499 (2004).
47. J. B. Birks, *Photophysics of Aromatic Molecules*; Wiley-Interscience: London, 1970.
48. (a) F. D. Lewis, Y. Zhang, R. L. Letsinger, *J. Am. Chem. Soc.* **119**, 5451 (1997). (b) J. Lou, T. A. Hatton, P. E. Laibinis, *Anal. Chem.* **69**, 1262 (1997). (c) A. T. Reis e Sousa,

E. M. S. Castanheira, A. Fedorov, J. M. G. Martinho, *J. Phys. Chem. A* **102**, 6406 (1998). (d) Y. Suzuki, T. Morozumi, H. Nakamura, M. Shimomura, T. Hayashita, R. A. Bartsch, *J. Phys. Chem. B* **102**, 7910 (1998).

49. N. J. van der Veen, S. Flink, M. A. Deij, R. J. M. Egberink, F. J. M. van Veggel, F. D. N. Reinhoudt, *J. Am. Chem. Soc.* **122**, 6112 (2000).
50. M. Narita, Y. Higuchi, F. Hamada, H. Kumagai, *Tetrahedron Lett.* **39**, 8687 (1998).
51. (a) J. K. Lee, W. Sim, S. K. Kim, J. H. Bok, M. S. Lim, S. W. Lee, N. S. Cho, J. S. Kim, *Bull. Korean Chem. Soc.* **23**, 314 (2004). (b) J. K. Lee, S. K. Kim, S. H. Lee, P. Thuery, J. Vicens, J. S. Kim, *Bull. Korean Chem. Soc.* **24**, 524 (2003). (c) J. M. Lehn, *Supramolecular Chemistry*; VCH: Weinheim, 1995.
52. S. K. Kim, J. K. Lee, J. M. Lim, J. W. Kim, J. S. Kim, *Bull. Korean Chem. Soc.* **25**, 1247 (2004).
53. B. Bodenant, T. Weil, M. Businelli-Pourcel, F. Fages, B. Barbe, I. Pianet, M. Laguerre, *J. Org. Chem.* **64**, 7034 (1999).
54. S. H. Lee, J. Y. Kim, S. K. Kim, J. H. Lee, J. S. Kim, *Tetrahedron.* **60**, 5171 (2004).
55. J. H. Kim, A. R. Hwang, S. K. Chang, *Tetrahedron Lett.* **45**, 7557 (2004).
56. H. J. Kim, J. H. Bok, J. Vicens, I. H. Suh, J. Ko, J. S. Kim, *Tetrahedron Lett.* **46**, 8765 (2005).
57. J. M. Kim, S. J. Min, S. W. Lee, J. H. Bok, J. S. Kim, *Chem. Commun.* 3427 (2005).
58. S. K. Kim, S. H. Kim, H. J. Kim, S. H. Lee, S. W. Lee, J. Ko, R. A. Bartsch, J. S. Kim, *Inorg. Chem.* **43**, 2906 (2004).
59. J. R. Lakowicz *Principles of Fluorescence Spectroscopy*, 2nd Ed.; Plenum: New York, 1999.
60. (a) L. Stryer, R. P. Haugland, *Proc. Natl. Acad. Sci. USA.* **58**, 719 (1967). (b) C. Berney, G. Danuser, *Biophys. J.* **84**, 3992 (2003).
61. S. H. Lee, S. K. Kim, J. H. Boc, S. H. Lee, J. Yoon, K. Lee, J. S. Kim, *Tetrahedron Lett.* **46**, 8163 (2005).
62. B. Valeur, J. Bourson, J. Pouget, A. W. Czarnik: *Fluorescent Chemosensors for Ion and Molecule Recognition*, ACS Symposium Series 538, American Chemical Society, Washington DC, 1993, 25.
63. W. Rettig, R. Lapouyade: *Probe design and chemical sensing*, J. R. Lakowicz, Topics in Fluorescence Spectroscopy, Vol. 4, Plenum, New York, 1994, 109.
64. B. Valeur, F. Badaoui, E. Bardez, J. Bourson, P. Boutin, A. Chatelain, I. Devol, B. Larrey, J. P. Lefévre, A. Soulet, J. P. Desvergne, A. W. Czarnik (Eds.). *Chemosensors of Ion and Molecule Recognition*, NATO ASI Series, Kluwer, Dordrecht, 195 (1997).
65. S. K. Kim, J. H. Bok, R. A. Bartsch, J. Y. Lee, J. S. Kim, *Org. Lett.* **7**, 4839 (2005).

Chapter 16

NANOPOROUS ARCHITECTURES

Calixarene assembly in the solid state

Carmine Gaeta, Consiglia Tedesco, and Placido Neri

Dipartimento di Chimica, Università di Salerno, Via Ponte don Melillo, I-84084 Fisciano (Salerno), Italy, e-mail:neri@unisa.it

Abstract: Porous materials with molecule-sized channels or cavities have been extensively investigated because of their scientific and technological interest. Selected examples of solid-state, porous, supramolecular architectures obtained from calixarene derivatives are here reviewed as a broad illustration of their properties. Both purely organic and hybrid metal-organic materials are considered.

Key words: Calixarenes, supramolecular architectures, crystal engineering, nanoporous materials, channels, nanotubes, solid-state.

1. INTRODUCTION

The fundamental aim of supramolecular chemistry[1] is the control of noncovalent intermolecular interactions to organize two or more species into stable multicomponent assemblies with well-defined architectures and functions. These interactions can be confined in the space to give discrete supramolecular entities (host-guest complexes[2] or self-assembling species[3]) or can be periodically distributed to originate solid-state supramolecular architectures (crystals).[4]

In the latter instance, in addition to the scientific challenge behind the idea of *making crystals by design* (crystal engineering),[4,5] the main purpose is the preparation of new solid-state materials with specific functions or properties.[6] These materials can have potential applications in nano-technology,[7] molecular sieving, catalysis,[8] drug delivery, gas-storage[9] or – purification,[10] or as sensors,[11] and switches, discussed also elsewhere in this book. For most of these applications, stable supramolecular architectures

J. Vicens and J. Harrowfield (eds.), Calixarenes in the Nanoworld, 335–354.

endowed with nanodimensioned pores or cavities are required, implying the capability to surmount Nature's abhorrence for vacuum. Two further questions associated to this problem concern the tendency to fill the space by network interpenetration[12] or by guest species with important structural supporting role and whose removal causes the architecture to collapse.[13]

Nanoporous supramolecular architectures can be conveniently prepared by using three fundamental approaches, namely inorganic,[14] hybrid metal-organic,[9d,15] and purely organic.[16] In the two latter instances, the use of building blocks (or "tectons")[17] containing a preformed cavity, such as calixarenes,[18] can be particularly convenient to overcome the aforementioned problems.

Examples of calixarene-based nanoporous supramolecular architectures are still relatively limited, probably due to the difficulty in handling three-dimensional objects with low symmetry. However, the work reviewed in this chapter shows that very interesting calixarene-based materials have already been obtained which often show surprising properties.

A complete survey of all porous architectures obtained from calixarenes and resorcarenes is beyond the scope of this chapter. In particular, we will focus our attention on selected examples of the former class of compounds, which may serve as a broad illustration of the possible general approaches.

2. PURELY ORGANIC ARCHITECTURES

2.1 Unsubstituted Calixarenes

The solid-state features of unsubstituted calixarenes have been extensively investigated.[18] Such calixarenes usually give nonporous assemblies containing no channels but the capacity of the solids to include and exchange a wide range of guests is well known.[19] For the purposes of a comparison with true porous materials, a brief description of their properties is included in this chapter.

The sublimation *in vacuo* at 300°C of pure *p*-H-calix[4]arene (**1**) gives rise to crystals built by means of weak dispersive forces but which show unusual thermal stability.[9c] In the solid state three molecules of **1** adopt a cyclic, mutually included arrangement of a pseudo-spherical shape (Fig. 1). These trimers form a hexagonal close-packed (hcp) lattice involving simple van der Waals interactions without intermolecular H-bonds. The deviation of the trimers from the ideal spherical shape gives rise to a nonporous architecture with non-contiguous voids of ~153 $Å^3$. Interestingly, the same hcp lattice of pure **1** is obtained in its inclusion compounds with dichloromethane, chloroform, or acetonitrile, where in all cases the solvent is situated in the interstitial voids.

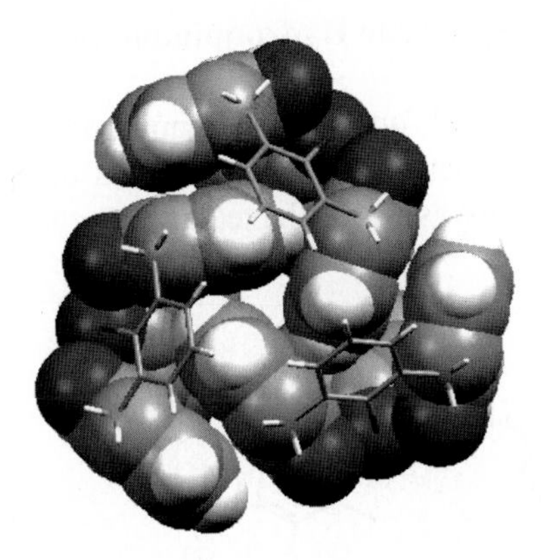

Figure 16-1. Mutually included cyclic trimer of **1** as viewed along the *c* axis (mixed space-filling and stick representation).

When the guest is CCl_4, CF_3Br, C_2F_6, or CF_4 a remarkable thermal stability of the host-guest system is observed by thermogravimetry. In fact, these guests are released only at temperatures well above their normal boiling point (higher by 120-370°C). Very interestingly, with CH_4 as guest, heating at 150°C (320°C higher than its b.p.) is required to empty the interstitial voids, thus making this material well suited for gas storage.[9c]

Very different is the solid-state structure of *p-tert*-butylcalix[4]arene (**2**), obtained by sublimation under conditions similar to those of **1**.[19a] It forms a bilayer-type structure, with facing calixarene molecules, providing large voids of ~235 Å^3. This material is able to absorb gases such as N_2, O_2, air, and CO_2.[10b] In particular CO_2 is absorbed more rapidly than N_2 or O_2, while H_2 is not absorbed. The low affinity for H_2 can be exploited in the production process for its purification by removal of the CO_2 main impurity.[10b] For further discussion of this chemistry, see Chapter 8.

2.2 Calixarenes Bearing Quinone/Hydroquinone Subunits

Quinone/hydroquinone moieties are particularly interesting for designing nano-architectures because they may give rise to hydrogen bond networks or to charge transfer interactions.[20] In addition, the resulting material should display electrochemical and photochemical properties exploitable in nano-technology, catalysis, and molecular sensing.

2.2.1 Partial Calix[4]arene Hydroquinone Derivatives

Recently several partial quinone/hydroquinone derivatives have been prepared and, amongst them, both proximal and distal dimethoxycalix[4]-dihydroquinone derivatives **3-5** have been characterized by X-ray crystallography.[21]

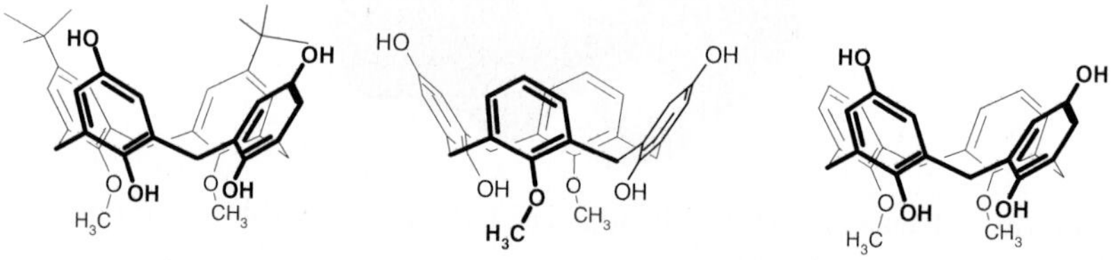

HO OH OH OH OH H_3C CH_3 **3** HO OH OH O OH H_3C CH_3 **4** HO OH O OH O OH H_3C CH_3 **5**

Proximal *p-tert*-butylcalix[4]dihydroquinone (**3**) forms cubic crystals by means of a particular interplay of H-bond and van der Waals-like interactions which results in the unprecedented simultaneous existence of water channels and very large unoccupied hydrophobic cavities (900 Å^3).[21b] The unit cell contains 48 calix[4]arene molecules in the *cone* conformation, stabilized by intramolecular H-bonds, and 155 water molecules mostly inside nanodimensioned channels. The nanoporous architecture is generated by a fundamental [6+2] supramolecular unit constituted by six calixarene and two water molecules all linked by means of H-bonds (Fig. 2a).

Two calix[4]arenes of different supramolecular units interact by mutually including one *tert*-butyl group into the cavity of the other through CH-π and van der Waals interactions (Fig. 2b). In this way the [6+2] supramolecular unit propagates in space by interaction with six octahedrally disposed others (Fig. 2c), leaving empty spaces among them (Fig. 2d). The crystalline framework is preserved after removal of channel water molecules under vacuum at 50°C. These features may foreshadow potential applications of this solid-state assembly in the development of new functional materials of nanotechnological interest.

p-H-1,3-calix[4]dihydroquinone (**4**) forms two polymorphs, **4-A** and **4-B**, both displaying calix[4]arene molecules in the cone conformation.[21a,c] The former is built up from mutually-included calixarene trimers, very similar to those described for **1** (Fig. 1), and connected each other by water molecules. The latter is formed by alternate layers of crystallographically independent calixarene molecules.

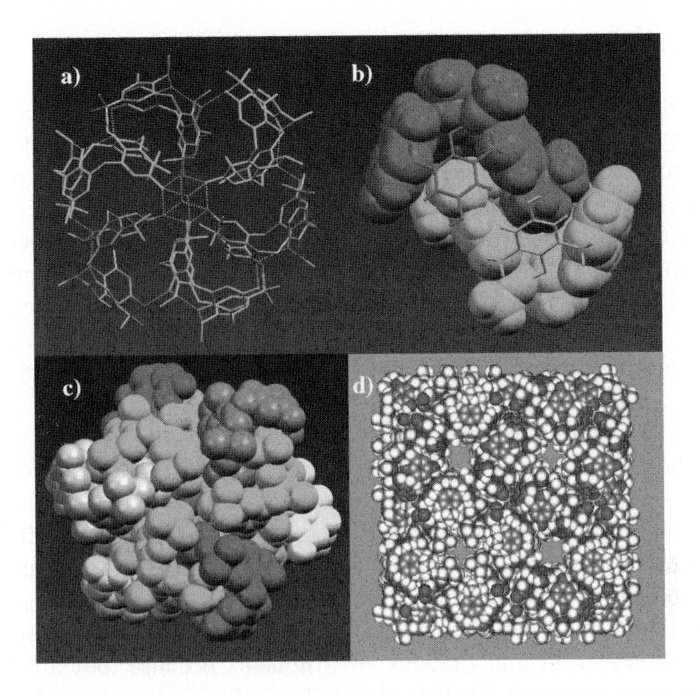

Figure 16-2. Solid state assembly of proximal calix[4]dihydroquinone **3**: a) the fundamental [6 + 2] supramolecular unit; b) locking interaction by mutual inclusion of two calixarenes belonging to two different supramolecular units; c) octahedral coordination of a central (green) supramolecular unit with six other ones (different colors); d) space-filling model of the unit cell showing the polar channels (upper left and lower right) and the apolar cavities (upper right and lower left).

In **4-A** each trimer is connected to the next one, along the *c* axis, through two water molecules located exactly on the crystallographic ternary axis. Each chain is surrounded by six other chains shifted alternatively by 1/3 of the chain repeat distance.

In **4-B** each independent calixarene molecule interacts with three other ones; no interaction exists in the same layer between crystallographically equivalent molecules. One methanol molecule is connected through hydrogen bonds to the calixarene upper rim and to another methanol molecule. The chloroform molecule exhibits only van der Waals interactions.

Proximal *p*-H-calix[4]dihydroquinone (**5**) gives rise to a structure composed of bilayers of facing calixarene molecules interacting by means of H-bonds.[21c] Within each bilayer, solvent molecules (ethyl acetate) are included and are also H-bonded to the upper-rim OH groups.

2.2.2 Calix[4]arene Tetrahydroquinone Derivative

A beautiful example of an organic nanotube is provided by calix[4]arene tetrahydroquinone (**6**).[7,22] In fact, X-ray crystallography of a crystal of **6**, grown from a water-acetone (1:1) solution, shows the presence of nanotubes formed by a cyclic repeating unit composed of eight molecules of **6** stabilized by H-bond-bridges between free OH groups at the upper rim and water molecules. The tubular octamer forms a linear tubular polymer, where the stability is further enhanced by the formation of H-bonded bridges between repeating units and well-ordered intertubular π-π stacking interactions. Each nanotube has a 17x17 Å cross section with a 6x6 Å square pore.

Calix[4]hydroquinone nanotubes are organized into bundles to form a novel chessboard-like rectangular structure (Fig. 3). These nanotubes are able to capture silver ions from aqueous solution by means cation-π interaction with their aromatic walls.[7] Under photochemical irradiation, the hydroquinone moieties of **6** reduce silver ions to silver neutral atoms. The silver atoms join to form nanowires of 0.4 nm diameter inside the pores of self-assembled calix[4]hydroquinone nanotubes. The silver nanowires are very stable under air and aqueous environments because they are protected by the organic structure. They exist as coherently oriented three-dimensional arrays of ultrahigh density and thus could be used as model systems for investigating anisotropic conductivity phenomena and as nanoconnectors for designing nanoelectronic devices.

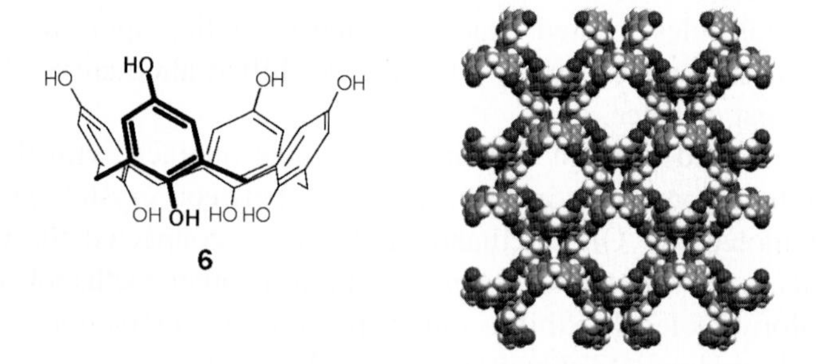

Figure 16-3. Chessboard like supramolecular architecture of **6** showing a cross-section of the square pores.

2.3 A Calix[4]arene Phosphonate

A tubular structure has been obtained by reacting the calix[4]arene dihydroxyphosphonic acid derivative **7** with propane diamine in ethanol.[23] The crystal structure is based on dimeric units of **7**, in which two calix[4]arenes are mutually included head to head by means of π-π interactions (Fig. 4). Six dimeric units form a hexagonal array of 15 Å radius and 16 Å depth having but threefold symmetry as a result of alternation of the dimer orientations (Fig. 4). Adjacent dimers interact by means of H-bond-bridges between phosphonate groups of the two units and diammonium cations. Ethanol molecules are also bonded to phosphonate groups via H-bonds. Bridging of the hexameric units via spanning propane diammonium cations and ethanol molecules leads to channels of 40 Å length which contain all the water molecules. The channel is amphiphilic, showing alternately a hydrophobic zone, the region of the aromatic rings, a hydrophilic zone of the phosphate/ammonium interactions, and another hydrophobic zone of the alkyl chains of the propane diammonium cations, along its 40 Å length. This channel can be compared to the water channel structure of aquaporin 1,[24] which is formed by 6 completely spanning α-helices, and two shorter, linked helices. The geometrical features and the combination of the variable polarity of the cavity surface of both systems show clear analogies. Thus, the multimolecular assembly of **7** can be considered a mimetic of biological membrane transport channels.

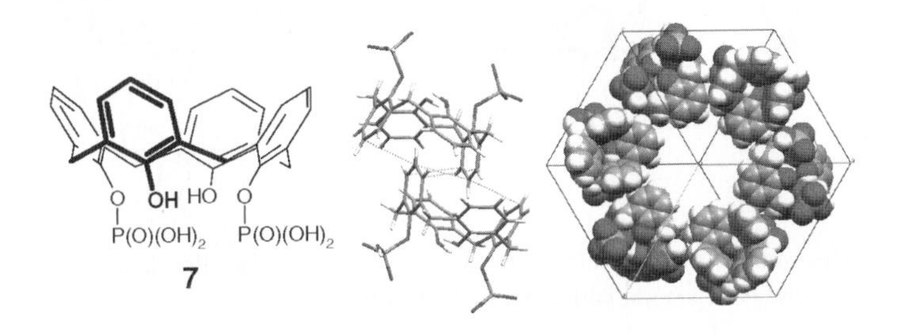

Figure 16-4. Short van der Waals contacts in the mutually included dimer of **7** (centre) and space-filling model of the crystal packing as viewed along the [111] direction (right).

2.4 Peptidocalixarenes

Calixarene derivatives conjugated with amino acids or peptides (peptidocalixarenes) have found interesting applications in molecular and anion recognition,[25] self-assembly to form dimeric capsules[26] and protein surface recognition (see Chapter 12).[27]

The *N*-linked peptidocalix[4]arenes (see Fig. 5), in which the amino acids are linked to the calixarene upper rim through nitrogen atom (*e.g.* **9**), generally show different properties from the *C*-linked analogues (*e.g.* **8**) both in solution[25] and in the solid state.[28] In fact, *C*-linked peptidocalix[4]arene (**8**) crystallizes from a mixture of ethyl acetate/acetone (1:1, v/v) to form an organic nanotube through a two-dimensional network of hydrogen bonds between the amide chains of adjacent conformers. Very different is the X-ray crystal structure of the *N*-linked peptidocalix[4]arene (**9**), where the molecules are assembled through van der Waals interactions.

Two conformers, ABA′B′ and EFE′F′, of the *C*-linked peptidocalix-[4]arene (**8**) are present in the crystal lattice, their difference being in the orientation of the amino acid units. In the ABA′B′ conformer the Cbz-alanine moieties are oriented upward with respect to the calixarene cavity; in the EFE′F′ unit the same moieties are oriented downward. Nanotubes are generated by means H-bonding interactions between alanine-moieties of identical conformers: ABA′B′ with ABA′B′ and EFE′F′ with EFE′F′. In this way, the packing of the two conformers causes the calixarenes cavities to be stacked one over the other to form aromatic nanotubes.

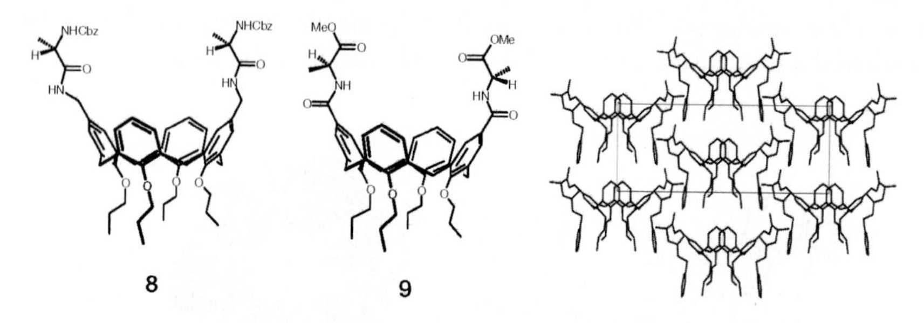

Figure 16-5. The simple peptidocalixarenes **8** and **9** and (right) the layer of EFE'F' conformers of **8** found in its crystal lattice, viewed along the *c* axis.

2.5 Calixcrowns

An additional example of a solid-state array of nanocylinders has been recently provided by a calix[4]arene-crown ether derivative (calixcrown[29]). In fact, molecular nanotube **10**, obtained by linking three 1,3-alternate calix[4]arene moieties with crown-3 bridges, is able to pack in a head-to-tail fashion to give straight, infinitely long nanocylinders (see Chapters 7 and 8).[30] Along the cylinder, each molecule of **10** is rotated by 90° with respect to the next one.

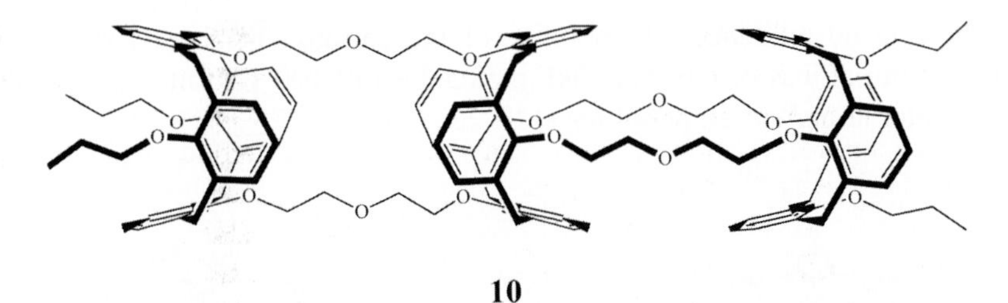

10

The nanocylinders are packed parallel in a square arrangement maximizing the intermolecular van der Waals interactions and leaving empty spaces among them, which are filled with chloroform solvent molecules. Considering the capability of calixarenes to interact with NO_x gases,[31] it can be anticipated that solid-state bundles of **10** can be used to transport and detect NO^+ ions by means of conductivity or color changes.

3. ARCHITECTURE OF ALKALI METAL COMPLEXES OF PHENOXO-CALIXARENES

The chemical and structural properties of complexes formed by simple deprotonation of phenolic calixarenes with appropriate alkalis have been widely investigated.[32] In some instances, interesting porous architectures have been described.

3.1 Common Calixarenes

Reaction of *p-tert*-butylcalix[8]arene (**11**) with K_2CO_3 in a two-phase H_2O/THF system led to the corresponding monopotassium salt, which crystallized at the interface.[33] The monodeprotonated calix[8]arene ligand adopts the pleated-loop conformation previously observed for neutral *p-tert*-butylcalix[8]arene.[34] The macrocycles stack one over the other to give a one-dimensional system with a "coaxial" channel in the middle (Fig. 6). In between two consecutive calix[8]arenes there are two K^+ ions involved in interactions with the phenolic oxygens and with bridging water molecules. Sandwiched between the next pair of calixarenes is a 10-molecule water cluster with a distorted cubane core. The structure of this $(H_2O)_{10}$ cluster was unprecedented in the literature, but was predicted as possible by *ab initio* studies.[35] The interior of the channel system is polar because it is lined by calixarene OH groups, whereas the exterior is apolar, being composed of calixarene *tert*-butyl groups and THF molecules, both acting to shield the H_2O molecules and K^+ cations which fill the channel. NMR experiments indicate that water in the channels undergoes exchange with D_2O in solution but there is no evidence for mobility of the cations. TGA measurements show the initial loss of the four THF molecules (40-100°C) followed by that of the water at higher temperatures (110-220°C).

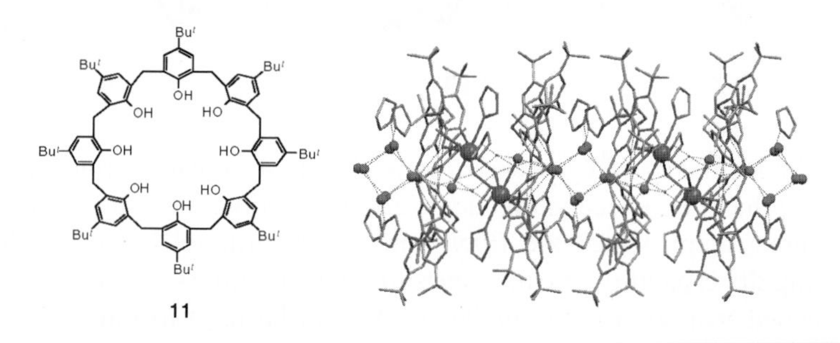

Figure 16-6. Channel structure of the monopotassium salt of **11** along the *b* axis. The $(H_2O)_{10}$ cluster and K^+ ions are highlighted as spheres (right).

3.2 Thiacalixarenes

Thiacalixarenes are a relatively new class of calixarenes in which the CH_2 bridges are replaced by S atoms that provide additional coordination sites and modify the cavity dimensions.[36] The presence of these new coordination sites gives rise to additional interactions with metal atoms which can generate huge cavities in the crystalline state.

The tetra-anionic salt of the *p-tert*-butylthiacalix[6]arene (**12**), obtained by reaction with KH in THF, has been crystallized from methanol.[37] The X-ray structure of the tetra-anionic host shows that it is bound to four potassium cations, with three methanol molecules incorporated in a cavity constituted by three phenol units, while four methanol molecules and one water molecule are outside this cavity. Neighbouring host molecules are bound by extensive S···K···(O,S) bridging coordination to form a ladder-like, polymeric 1D chain. Interchain hydrophobic interactions between lateral *tert*-butyl groups complete a crystal lattice with a huge cavity (19 Å widest span) filled by the guests. In air, the methanol is easily lost and the material looses its crystallinity to give the solvent-free apohost with a collapsed cavity. The apohost can recover the guest by exposure to the appropriate vapours and thus restore the original crystal structure.

Crystallization of the tetra-potassium salt of *p-tert*-butylthiacalix[8]arene[38] (**13**) from methanol leads to a very similar crystal lattice (Fig. 7).[39]

The tetra-anionic host binds four potassium cations; two of them are in the middle of the host loop surrounded with four oxygen and three sulfur atoms and the other two are located either at the upper or lower rim. Four molecules of methanol are incorporated in a groove constituted by four phenol units, while another four are outside this cavity. (O,S)···K···(O,S) interactions between adjacent hosts and interchain hydrophobic interactions between lateral *tert*-butyl groups give rise to a crystal lattice in which the channels are narrower with respect to the corresponding *p*-*tert*-butylthiacalix[6]arene salt.[37] Because of the zeolitic-like structure, the cavities easily lose methanol to give a guest-free apohost with a different microcrystalline structure. Again, the apohost can recover the guest and the original crystal structure after exposure to methanol or benzene vapours. The behavior of the apohost is reversible.

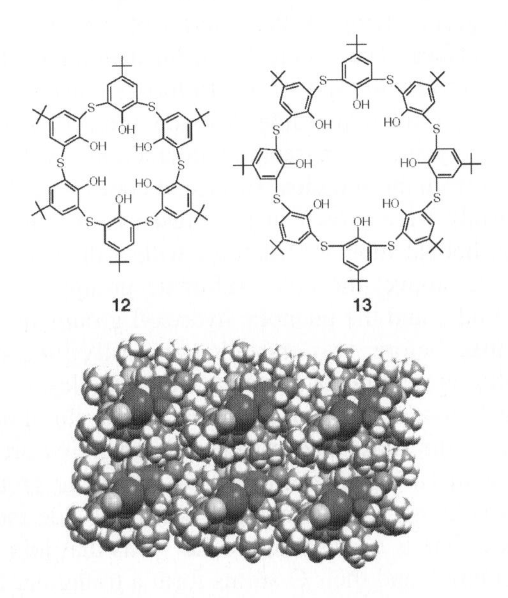

Figure 16-7. The hexa- and octa-thiacalixarenes **12** and **13**, and the crystal packing of the tetrapotassium salt of **13** as viewed along the *b* axis (right).

4. HYBRID ORGANIC-INORGANIC ARCHITECTURES

4.1 *p*-Sulfonatocalixarenes

Water soluble *p*-sulfonatocalixarenes[40] are amphiphilic molecules containing a hydrophobic cavity able to host organic guests both in solution[40,41] and in the solid state,[42] and polar sulfonate-groups at the upper rim available for metal cation coordination. These characteristics make them interestings "molecular building blocks" to fabricate nanoporous supramolecular architectures.

4.1.1 *p*-Sulfonatocalixarenes and Lanthanides

p-Sulfonatocalix[4]arene (**14**) forms spherical and helical tubular structures in the presence of pyridine *N*-oxide and lanthanide ions.[43] When crystals were grown from an aqueous solution of Na_5**14**, pyridine *N*-oxide and $La(NO_3)_3{\cdot}6H_2O$, 12 *p*-sulfonatocalix[4]arene molecules were assembled in a spherical fashion at the vertices of an icosahedron, with 48 negatively charged sulfonate groups lying on the exterior of the sphere and defining a polar outer shell surface. The sphere has a inner volume of ~1700 Å^3 and contains ~30 water molecules and 12 Na^+ to form a cluster that stabilizes the spherical assembly. Twelve pyridine *N*-oxide molecules are held in the calixarene cavities by means π-stacking interactions, with their O atoms extending outward from the spherical surface to coordinate the La^{3+} ions.

When the crystals were grown in the presence of a greater amount of pyridine *N*-oxide, helical tubular structures with a diameter of about 15 Å were obtained.[43] As above, the polar sulfonate groups lie on the external surface of the cylinder and the phenolic hydroxyl groups lie in the inside of the structure. Unlike before, two crystallographically independent pyridine *N*-oxide molecules are intercalated between aromatics rings of adjacent calixarenes through π-stacking interactions, to form a chiral helical assembly along the length of the tube. The cylindrical structures are arranged in a hexagonal array with Na^+ ions coordinated to sulfonate groups of adjacent turns of the helix (Fig. 8). Two types of pyridine *N*-oxide molecules fill the calixarene cavities. One is coordinated to La^{3+} ions that join adjacent tubes. The other is disordered, and their O atoms form a triangular tunnel bounded by three adjoining tubes. Each channel contains La^{3+} ions in addition to the hydrated Na^+ ions observed in the spherical core.

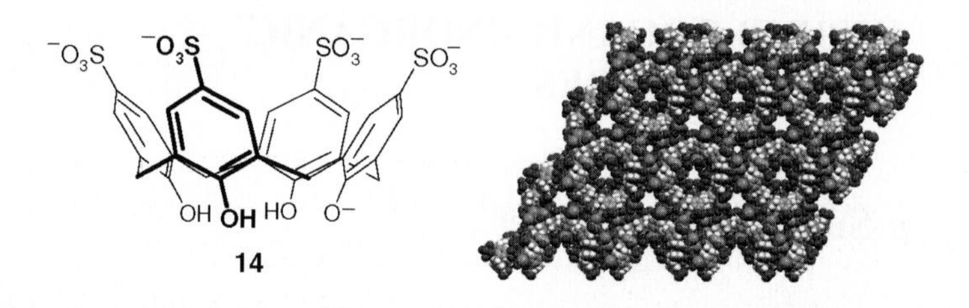

Figure 16-8. Top view of the hexagonal array of helical nanotubes obtained from aqueous solution of Na_5**14**, pyridine N-oxide and $La(NO_3)_3{\bullet}6H_2O$.

When pyridine *N*-oxide was replaced by 18-crown-6, a different assembly was obtained.[44] Thus, 12 calixarene molecules are arranged at the vertex of a cuboctahedron. This spherical arrangement contains a internal volume of about 1258 Å^3 in which are present six aquated lanthanide ions. Each sphere contains six pores in which disordered water molecules are present. The pores have a van der Waals diameter of 4.18 Å.

It is worthy of mention that the X-ray crystal structure of pure Na_5**14**[45] shows a bilayer arrangement consisting of alternating aqueous and organic layers. The former is composed by the polar surface of the sulfonate groups, counterions, and water molecules. The organic layer consists of a π-stacked, two dimensional bilayer grid composed of calixarene molecules arranged in an alternating "up-down" antiparallel fashion.

4.1.2 *p*-Sulfonatocalixarenes and Porphyrins

Nanotubes result from solid-state self-assembly when the *p*-sulfonato-calix[4]arene-tetraacid (**15**) reacts with the cationic porphyrin **16.**[46] The crystal structure of material obtained from pH 2 solution showed the presence of trimeric units of porphyrin **16**, with the central molecule sandwiched by the other two, which are rotated by 45° with respect to it. Each methylpyridinium moiety of the central porphyrin is inserted into the calixarene cavity (Fig. 9) to form a supramolecular complex composed of three porphyrin units and four calixarene molecules. The Na^+ ions are hepta-coordinated by oxygen atoms of the hydroxycarbonylmethoxy residues. Solid-state self-assembly of these neutral supramolecules creates nanopores of about 19 x 32 Å (Fig. 9) occupied by disordered solvent molecules.[46]

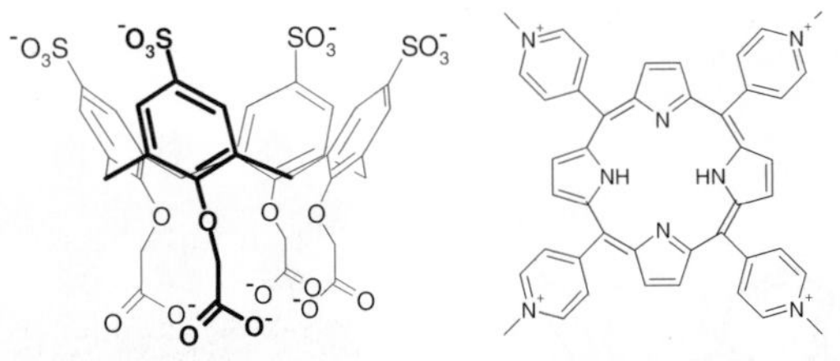

15 **16**

When the crystals were obtained from a solution at pH = 6, the asymmetric units contained two hepta-anionic calixarene molecules, three porphyrin systems and two Na^+ ions. Here, the same trimeric porphyrin "core" is stacked with two additional porphyrins, one above and one below, to form a 4:5 calixarene/porphyrin supramolecule, which is associated with a sixth isolated porphyrin. Further, the packing gives rise to two series of interconnected channels in which the solvent molecules are disordered.[46]

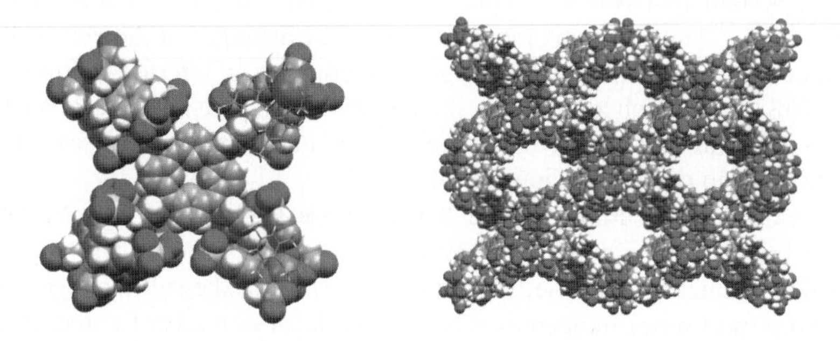

Figure 16-9. Interaction of the centrosymmetric porphyrin ring with the four calixarene units in the 4:3 supramolecular complex of **15/16** (left) and crystal packing as viewed along the *c* axis (right).

4.2 Phosphonate Derivatives

A crystal lattice with large pores has been obtained for a calixarene-diphosphonic acid derivative **17** coordinated to Nd(III).[47] Crystals of limited stability were obtained from a methanolic solution of **17** and $Nd(NO_3)_3$, acidified with 1M HNO_3. Each calixarene-diphosphonic acid molecule in the cone conformation serves as a bridging ligand in the μ^4 manner, connecting four Nd(III) atoms. Four of these Nd–calixarene–Nd units are connected to form a supramolecular ring. These rings are connected in a two-dimensional net and through the O–P–O bridges in the third dimension forming a porous three-dimensional polymeric structure (Fig. 10). The pores mainly have a hydrophobic nature, with a diameter of approximately 17 Å, and do not contain solvent molecules.[47]

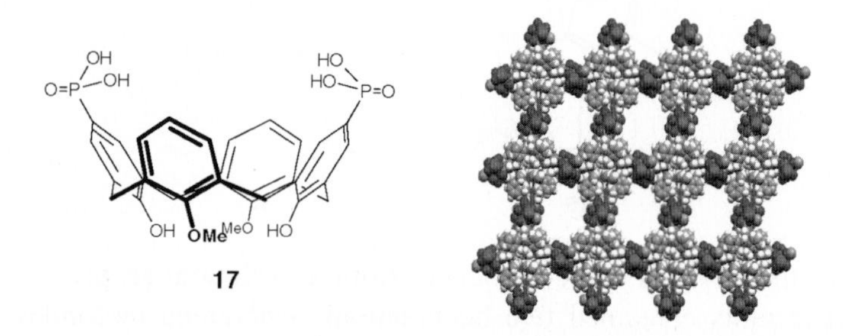

Figure 16-10. Crystal packing of the Nd(III) complex of **17** as viewed along the *a* axis (right).

4.3 Calixquinones

Crystallization of 1,3-calix[4]diquinone bis(acid) (**18**), from a ethanol/dichloromethane solution containing excess $Pb(ClO_4)_2 \cdot xH_2O$ gives rise to a three-dimensional coordination network containing channels.[48] The X-ray structure shows a trimeric unit [Pb_9(**18**-2H)$_3(ClO_4)_6(OH)_6$], with crystallographic *3/m* symmetry, containing three unique Pb(II) atoms. One Pb(II) atom is bound to four oxygen atoms to the lower rim of **18**, two acid oxygen atoms, two perchlorate anions and a water molecule. The other two Pb(II) atoms are bonded to quinone oxygen atoms, water molecules and hydroxide ions. These trimeric units are interconnected *via* upper rim quinone oxygen-Pb(II) interactions to form a three-dimensional network containing channels of *ca.* 14 Å in diameter probably filled with disordered solvent molecules (Fig. 11). Because of the presence of the quinone subunits, these channels should be able to display redox activity.

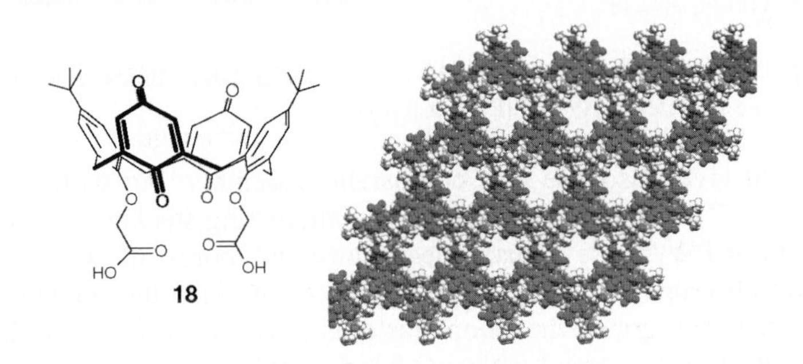

Figure 16-11. Channel structure of the Pb(II) complex of **18** as viewed along the *c* axis (right).

4.4 Cationic Calixarene Complexes/Polyoxometalate Anions

An interesting example of porous organic-inorganic assembly was constructed by coupling a cationic calixarene-Na^+ complex with an inorganic polyoxotungstate anionic cluster.[49] Single crystals were obtained by mixing a solution of tetraester **19** in $CHCl_3$ and $Na_3PW_{12}O_{40}$ in MeOH. The crystal lattice contains head-to-tail columnar arrays of **19-Na$^+$** complexes which propagate in three directions along the *a*, *b*, and *c* axes. They are stacked alternately and orthogonally in a three-dimensional framework where α-$[PW_{12}O_{40}]^{3-}$ anions are inserted. Six **19-Na$^+$** units surround the anion cluster, without bonding interaction between **Na$^+$** ions and the cluster oxygen atoms. In this way an empty cavity is created between four **19-Na**$^+$ complexes, generating nanopores with size of about 6 × 9 Å along the three axes (Fig. 12). They intersect each other to form a three-dimensional nanoporous network, occupied by solvent molecules.

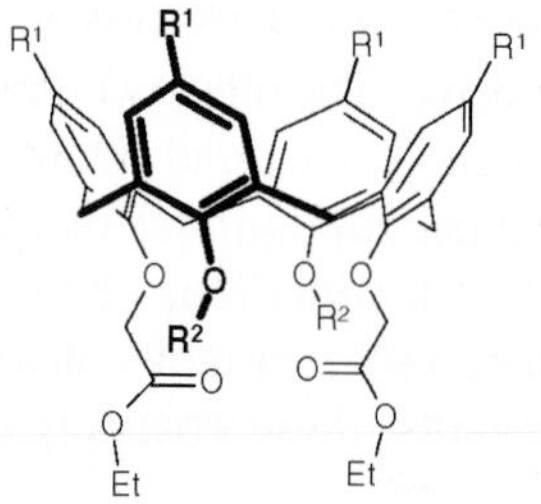

19 $R^1 = t$-Bu, $R^2 = CH_2COOEt$
20 $R^1 = H$, $R^2 = H$

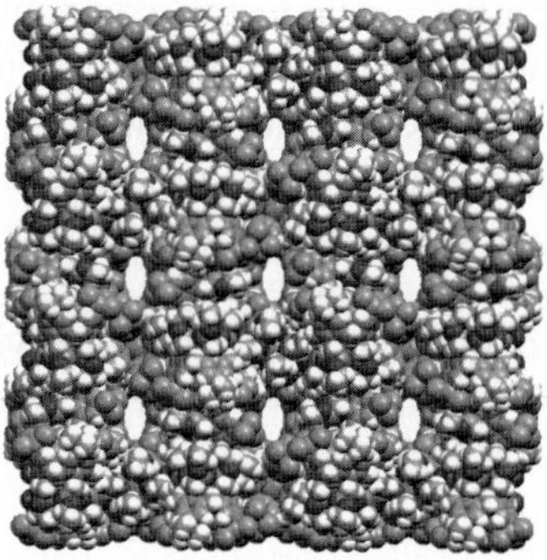

Figure 16-12. The sodium ion complexants **19** and **20** and the porous structure of the hybrid organic-inorganic assembly of [**19**-Na]$_3$[$PW_{12}O_{40}$] (right).

Diester **20** gives also rise to a comparable assembly, but with different characteristics.[49] In this instance, there is an alternating stacking of layers of **20-Na**$^+$ and α-$PW_{12}O_{40}^{3-}$. Thus, one-dimensional pores of 4 × 8 Å are formed, which propagate along the [101] direction. Also the orientation of cationic complexes around the anion cluster is different. In fact, two **20-Na**$^+$ units are linked to the anion by Na—O bonds, while four units are linked to the anion by weak intermolecular interactions. After treatment of the crystal of [**20**-Na]$_3$[$PW_{12}O_{40}$] at 100°C for 2 h, it loses the solvent molecules while the nanoporous structure is maintained. The material thus obtained is able to sorb reversibly small alcohol molecules into the one-dimensional pores with higher binding activities than molecules of water or higher alcohols such as 1-butanol.[49] The porous structure of the sorbent is retained during and after several sorption/desorption cycles.

5. CONCLUSIONS

The work reviewed in this chapter clearly shows that calixarenes are excellent "tectons" to produce solid-state supramolecular architectures characterized by the presence of large pores or cavities. Their preformed three-dimensional molecular cavity coupled to its easy chemical modification make them more attractive than the comparable planar building blocks. However, these particularities also make their handling as three-dimensional bricks more difficult. Therefore, the design or "engineering" of calixarene-based supramolecular solids is more empirical and certainly less predictable than in other areas of crystal engineering.

Notwithstanding that numerous examples of beautiful supramolecular architectures with very interesting properties have already been obtained, it seems that the research efforts in this field are still limited and somewhat scattered. It is conceivable that more systematic studies may lead to a higher level of predictability and transferability in the "engineering" of calixarene-based supramolecular solids, thus further expanding the great potentialities of these materials.

6. REFERENCES

1. a) J.-M. Lehn, *Supramolecular Chemistry: Concepts and Perspectives* (VCH, Weinheim, 1995); b) J. L. Atwood, J. E. D. Davies, D. D. MacNicol, F. Vögtle, J.-M. Lehn, Eds. *Comprehensive Supramolecular Chemistry* (Pergamon, New York, 1996), vols. 1–11; c) J. W. Steed, J. L. Atwood, *Supramolecular Chemistry* (John Wiley & Sons, Chichester, UK, 2000).
2. D. J. Cram, *Angew. Chem., Int. Ed. Engl.* **27**, 1009-1020 (1988).
3. a) G. M. Whitesides, J. P. Mathias, C. T. Seto, *Science* **254**, 1312-1319 (1991); b) D. S. Lawrence, T. Jiang, M. Levett, *Chem. Rev.* **95**, 2229-2260 (1995); c) D. Philip, J. F. Stoddart, *Angew. Chem., Int. Ed. Engl.* **35**, 1154-1196 (1996); d) R. H. Vreekamp, J. P. M. van Duynhoven, M. Hubert, W. Verboom, D. N. Reinhoudt, *Angew. Chem., Int. Ed. Engl.* **35**, 1215-1218 (1996); e) M. M. Conn, J. Rebek, Jr., *Chem. Rev.* **97**, 1647-1668 (1997); f) K. A. Jolliffe, M. Crego-Calama, R. Fokkens, N. M. M. Nibbering, P. Timmermann, D. N. Reinhoudt, *Angew. Chem., Int. Ed.* **37**, 1247-1250 (1998); g) L. J. Prins, J. Huskens, F. de Jong, P. Timmerman, D. N. Reinhoudt, *Nature* **398**, 498-502 (1999); h) J. Rebek, Jr., *Acc. Chem. Res.* **32**, 278-286 (1999); h) F. Hof, L. C. Stephen, C. Nuckolls, J. Rebek, Jr., *Angew. Chem., Int. Ed.* **41**, 1488-1508 (2002).
4. a) G. R. Desiraju, *Crystal Engineering. The Design of Organic Solids* (Elsevier, New York, 1989); b) G. R. Desiraju, *Angew. Chem., Int. Ed. Engl.* **34**, 2311-2327 (1995); c) G. R. Desiraju, *The Crystal As a Supramolecular Entity* (Wiley, New York, 1996); d) *Molecular Architectures,* Special issue, *Acc. Chem. Res.* **38**, 215-378 (2005); e) *Funtional Nanostructures*, Thematic issue, *Chem. Rev.* **105**, 1023-1562 (2005).
5. a) D. Braga, *Chem. Commun.* 2751-2754 (2003); b) J. D. Dunitz, *Chem. Commun.* 545-548 (2003).
6. K. R. Seddon, M. Zaworotko, *Crystal Engineering: The Design and Application of Functional Solids* (Kluwer Academic, Dordrecht, 1999).
7. B. H. Hong, S. C. Bae, C.-W. Lee, S. Jeong, K. S. Kim, *Science* **294**, 348-351 (2001).
8. K. Endo, T. Koike, T. Sawaki, O. Hayashida, H. Masuda, Y. Aoyama, *J. Am. Chem. Soc.* **119**, 4117-4122 (1997).
9. a) D. V. Soldatov, E. V. Grachev, J. A. Ripmeester, *Cryst. Growth Des.* **2**, 401-408 (2002); b) M. Eddaoudi, J. Kim, N. Rosi, D. Vodak, J. Wachter, M. O'Keeffe, O. M. Yaghi, *Science* **295**, 469-472 (2002); c) J. L. Atwood, L. J. Barbour, A. Jerga, *Science*

296, 2367-2369 (2002); d) J. L. C. Rowsell, O. M. Yaghi, *Angew. Chem., Int. Ed.* **44**, 4670-4679 (2005).

10. a) G. Férey, *Science* **292**, 994-995 (2001); b) J. L. Atwood, L. J. Barbour, A. Jerga, *Angew. Chem., Int. Ed.* **43**, 2948-2950 (2004).
11. M. Albrecht, M. Lutz, L. A. Spek, G. van Koten, *Nature* **406**, 970-974 (2000).
12. S. R. Batten, R. Robson, *Angew. Chem., Int. Ed.* **37**, 1461-1494 (1998).
13. J. A. Swift, V. A. Russell, M. D.Ward, *Adv. Mater.* **9**, 1183-1186 (1997).
14. a) A. K. Cheetham, G. Ferey, T. Loiseau, *Angew. Chem., Int. Ed.* **38**, 3268-3292 (1999); b) D. Bradshaw, J. B. Claridge, E. J. Cussen, T. J. Prior, M. J. Rosseinsky, *Acc. Chem. Res.* **38**, 273-282 (2005).
15. a) S. Kitagawa, R. Kitaura, S. Noro, *Angew. Chem., Int. Ed.* **43**, 2334-2375 (2004); b) C. N. Gianneschi, S. M. Masar, C. A. Mirkin, *Acc. Chem. Res.* **38**, 825-837 (2005).
16. a) P. J. Langley, J. Hulliger, *Chem. Soc. Rev.* **28**, 279-291 (1999); b) T. Hertzsch, S. Kluge, E. Weber, F. Budde, J. Hulliger, *Adv. Mater.* **13**, 1864-1867 (2001); c) A. V. Nossov, D. V. Soldatov, J. A. Ripmeester, *J. Am. Chem. Soc.* **123**, 3563-3568 (2001); d) D. A. Plattner, A. K. Beck, *Helv. Chim. Acta* **85**, 4000-4011 (2002); e) D. V. Soldatov, E. V. Grachev, J. A. Ripmeester, *Cryst. Growth Des.* **2**, 401-408 (2002); f) T. Hertzsch, F. Budde, E. Weber, J. Hulliger, *Angew. Chem., Int. Ed.* **41**, 2282-2284 (2002).
17. a) M. Simard, D. Su, J. D. Wuest, *J. Am. Chem. Soc.* **113**, 4696-4698 (1991); b) S. Mann, *Nature* **365**, 499-505 (1993); c) M. W. Hosseini, A. De Cian, *Chem. Commun.* 727-733 (1998); d) G. R. Desiraju, *Nat. Mater.* **1**, 77-79 (2002); e) M. W. Hosseini, *Cryst. Eng. Comm.* **6**, 318-322 (2004); f) M. W. Hosseini, *Acc. Chem. Res.* **38**, 313-323 (2005).
18. For general reviews on calixarenes, see: a) V. Böhmer, *Angew. Chem., Int. Ed. Engl.* **34**, 713-745 (1995); b) C. D. Gutsche, *Calixarenes Revisited*, (Royal Society of Chemistry, Cambridge, 1998); c) Z. Asfari, V. Böhmer, J. Harrowfield, J. Vicens, Eds. *Calixarenes 2001* (Kluwer, Dordrecht, 2001).
19. a) J. L. Atwood, L. J. Barbour, A. Jerga, B. L. Schottel, *Science* **298**, 1000-1002 (2002). b) P. K. Thallapally, G. O. Lloyd, J. L. Atwood, L. J. Barbour, *Angew. Chem., Int. Ed.* **44**, 3848-3851 (2005).
20. a) F. D'Souza, G. R. Deviprasad, *J. Org. Chem.* **66**, 4601–4609 (2001); b) S. Meddeb-Limem, B. Malezieux, P. Herson, S. Besbes-Hentati, H. Said, J.–C. Blais, M. Bouvet, *J. Phys. Org. Chem.* **18**, 1176-1182 (2005).
21. a) For a very preliminary communication, see: Tedesco, L. Gregoli, R. Pisacane, E. Gavuzzo, P. Neri, "*Crystal design based on calix[4]dihydroquinone units*" CrystEngComm Discussion 1: Innovation in Crystal Engineering, University of Bristol, United Kingdom (June 29 - July 1, 2002). b) C. Tedesco, I. Immediata, L. Gregoli, L. Vitagliano, A. Immirzi, P. Neri, *CrystEngComm* **7**, 449-453 (2005); c) C. Tedesco, L. Gregoli, I. Immediata, E. Gavuzzo, P. Neri, manuscript in preparation.
22. a) B. H. Hong, J. Y. Lee, C.-W. Lee, J. C. Kim, S. C. Bae, K. S. Kim, *J. Am. Chem. Soc.* **123**, 10748-10749 (2001); b) S. B. Suh, J. C. Kim, Y. C. Choi, S. Yun, K. S. Kim, *J. Am. Chem. Soc.* **126**, 2186-2193 (2004).
23. A. W. Coleman, E. Da Silva, F. Nouar, M. Nierlich, A. Navaza, *Chem. Commun.* 826-827 (2003).
24. a) K. Murata, K. Mitsuoka, T. Hirai, T. Walz, P. Agre, J. B. Heymann, A. Engel, Y. Fujiyoshi, *Nature* **407**, 599-607 (2000); b) T. Walz, T. Hirai, K. Murata, J. B. Heymann, K. Mitsuoka, Y. Fujiyoshi, L. B. Smith, P. Agre, A. Engel, *Nature* **387**, 627-630 (1997).
25. A. Casnati, F. Sansone, R. Ungaro, *Acc. Chem. Res.* **36**, 246-254 (2003).
26. a) R. E. Brewster, S. B. Shuker, *J. Am. Chem. Soc.* **124**, 7902-7903 (2002); b) F. Sansone, L. Baldini, A. Casnati, E. Chierici, G. Faimani, F. Ugozzoli, R. Ungaro, *J. Am. Chem. Soc.* **126**, 6204-6205 (2004).

27. a) M. W. Peczuh, A. D. Hamilton, *Chem. Rev.* **100**, 2479-2494 (2000); b) R. K. Jain, A. D. Hamilton, *Org. Lett.* **2**, 1721-1723 (2000); c) H. S. Park, Q. Lin, A. D. Hamilton, *Proc. Natl. Acad. Sci. U.S.A.* **99**, 5105-5109 (2002); d) S. N. Gradl, J. P. Felix, E. Y. Isaco, M. L. Garcia, D. Trauner, *J. Am. Chem. Soc.* **125**, 12668-12669 (2003); e) T. Mecca, G. M. L. Consoli, C. Geraci, F. Cunsolo, *Bioorg. Med. Chem.* **12**, 5057-5062 (2004); f) S. Francese, A. Cozzolino, I. Caputo, C. Esposito, M. Martino, C. Gaeta, F. Troisi, P. Neri, *Tetrahedron Lett.* **46**, 1611-1615 (2005).
28. L. Baldini, F. Sansone, A. Casnati, F. Ugozzoli, R. Ungaro *J. Supramol. Chem.* **2**, 219-226 (2002).
29. A. Casnati, R. Ungaro, Z. Asfari, J. Vicens, in: *Calixarenes 2001*, edited by Z. Asfari, V. Böhmer, J. Harrowfield and J. Vicens (Kluwer, Dordrecht, 2001), chapter 20, pp. 365-384.
30. V. G. Organo, A. V. Leontiev, V. Sgarlata, H. V. R. Dias, D. M. Rudkevich, *Angew. Chem., Int. Ed.* **44**, 3043-3047 (2005).
31. a) G. V. Zyryanov, Y. Kang, D. M. Rudkevich, *J. Am. Chem. Soc.* **125**, 2997-3007 (2003). Y. Kang, G. V. Zyryanov, D. M. Rudkevich, *Chem. Eur. J.* **11**, 1924-1932 (2005).
32. a) G. M. L. Consoli, F. Cunsolo, C. Geraci, P. Neri, *Org. Lett.* **3**, 1605-1608 (2001); b) G. M. L. Consoli, F. Cunsolo, C. Geraci, E. Gavuzzo, P. Neri, *Tetrahedron Lett.* **43**, 1209-1211 (2002); c) T. A. Hanna, L. Liu, L. N. Zakharov, A. L. Rheingold, W. H. Watson, C. D. Gutsche, *Tetrahedron* **58**, 9751-9757 (2002); d) T. A. Hanna, L. Liu, A. M. Angeles-Boza, X. Kou, C. D. Gutsche, K. Ejsmont, W. H. Watson, L. N. Zakharov, C. D. Incarvito, A. L. Rheingold, *J. Am. Chem. Soc.*, **125**, 6228-6238 (2003).
33. R. D. Bergougnant, A. Y. Robin, K. M. Fromm, *Cryst. Growth Des.* **5**, 1691-1694 (2005).
34. C. D. Gutsche, A. E. Gutsche, A. I. Karaulov, *J. Incl. Phenom.* **3**, 447-451 (1985).
35. a) S. Maheshwary, N. Patel, N. Sathayamurthy, A. D. Kulkarny, S. R. Gadre, *J. Phys. Chem. A* **105**, 10525-10537 (2001); b) U. Buck, F. Huisken, *Chem. Rev.* **100**, 3863-3890 (2000).
36. a) M. W. Hosseini in: *Calixarenes 2001*, edited by Z. Asfari, V. Böhmer, J. Harrowfield and J. Vicens (Kluwer, Dordrecht, 2001), chapter 6, pp. 110-129; b) P. Lhotak, *Eur. J. Org. Chem.* 1675-1692 (2004).
37. K. Endo, Y. Kondo, Y. Aoyama, F. Hamada, *Tetrahedron Lett.* **44**, 1355-1358 (2003).
38. Y. Kondo, K. Endo, N. Iki, S. Miyano, F. Hamada, *J. Incl. Phenom. Macrocyclic Chem.* **52**, 45-49 (2005).
39. Y. Kondo, K. Endo, F. Hamada, *Chem. Commun.* 711-712 (2005).
40. a) A Casnati, D. Sciotto, G. Arena, in: *Calixarenes 2001*, edited by Z. Asfari, V. Böhmer, J. Harrowfield and J. Vicens (Kluwer, Dordrecht, 2001), chapter 24, pp. 440-456 and refs. therein. b) W. Abraham, *J. Inclusion Phenom. Macrocyclic Chem.* **43**, 159-174 (2002); c) N. Iki, T. Suzuki, K. Koyama, C. Kabuto, S. Miyano, *Org. Lett.* **4**, 509-512 (2002); d) N. Kon, N. Iki, S. Miyano, *Org. Biomol. Chem.* **1**, 751- 755 (2003); e) F. Perret, J.-P. Morel, N. Morel-Desrosiers, *J. Supramol. Chem.* **15**, 199-206 (2003).
41. a) K. Ito, M. Kida, A. Noike, Y. Ohba, *J. Org. Chem.* **67**, 7519-7522 (2002); b) A. Mendes, C. Bonal, N. Morel-Desrosiers, J.-P. Morel, P. Malfreyt, *J. Phys. Chem. B* **106**, 4516-4524 (2002); c) N. Douteau-Guével, F. Perret, A. W. Coleman, J.-P. Morel, N. Morel-Desrosiers, *J. Chem. Soc. Perkin Trans. 2* 524-532 (2002); d) A. Specht, P. Bernard, M. Goeldner, L. Peng, *Angew. Chem., Int. Ed.* **41**, 4706-4710 (2002) ; e) Y. Liu, E.-C. Yang, Y. Chen, D.-S. Guo, F. Ding, *Eur. J. Org. Chem.* 4581-4588 (2005); f) H. Bakirci, A. L. Koner, W. M. Nau, *J. Org. Chem.* **70**, 9960-9966 (2005); g) A. Specht, F. Ziarelli, P. Bernard, M. Goeldner, L. Peng, *Helv. Chim. Acta* **88**, 2641-2653 (2005).

42. a) M. Selkti, A. W. Coleman, I. Nicolis, N. Douteau-Guevel, F. Villain, A. Tomas, C. de Rango, *Chem. Commun.* 161-162 (2000); b) J. L. Atwood, T. Ness, P. J. Nichols, C. L. Raston, *Cryst. Growth Des.* **2**, 171-176 (2002); c) P. J. Nichols, C. L. Raston, *Dalton Trans.* **14**, 2923-2927 (2003); d) A. Lazar, E. Da Silva, A. Navaza, C. Barbey, A.W. Coleman, *Chem. Commun.* 2162-2163 (2004); e) J. L. Atwood, S. J. Dalgarno, M. J. Hardie, C. L. Raston, *Chem. Commun.* 337-338 (2005).
43. G.W. Orr, L. J. Barbour, J. L. Atwood, *Science* **285**, 1049-1052 (1999).
44. J. L. Atwood, L. J. Barbour, S. J. Delgarno, M. J. Hardie, C. L. Raston, H. R. Webb, *J. Am. Chem. Soc.* **126**, 13170-13171 (2004).
45. J. L. Atwood, A. W. Coleman, H. Zhang, S. G, Bott, *J. Inclusion Phenom.* **5**, 203-211 (1989).
46. L. Di Costanzo, S. Geremia, L. Randaccio, R. Purrello, R. Laceri, D. Sciotto, F. G. Gulino, V. Pavone, *Angew. Chem., Int. Ed.* **40**, 4245-4247 (2001).
47. J. Plutnar, J. Rohovec, J. Kotek, Z. Žák, I. Lukeš, *Inorg. Chim. Acta* **335**, 27-35 (2002).
48. P. D. Beer, M. G. B. Drew, P. A. Gale, M. I. Ogden, H. R. Powell, *Cryst. Eng. Comm.* **2**, 1-5 (2000).
49. Y. Ishii, Y. Takenaka, K. Konishi, *Angew. Chem., Int. Ed.* **43**, 2702-2705 (2004).

AUTHOR INDEX

D

E

G

R

T

SUBJECT INDEX

Q

U

It is terribly amusing how many different climates of feeling one can go through in one day.

—ANNE MORROW LINDBERGH, b. 1906
American writer & aviator

 Printed in the United States of America. International representatives: Worldwide Media Services, Inc., 115 East Twenty-third Street, New York, New York 10010. Canadian representatives: General Publishing Co., Ltd., 30 Lesmill Road, Don Mills, Ontario M3B 2T6. ISBN 0-89471-514-3 (paper). Cover illustration by K.B. Hwang. Interior illustrations by Ann Bedrick. This book may be ordered by mail from the publisher. Please add $2.50 for postage and handling for each copy. *But try your bookstore first!* Running Press Book Publishers, 125 South Twenty-second Street, Philadelphia, Pennsylvania 19103.

REFLECTIONS

Memory is more indelible than ink.

—ANITA LOOS (1893–1981)
American writer

You can't go around hoping that most people have sterling moral characters. The most you can hope for is that people will pretend that they do.

—FRAN LEBOWITZ, b. 1951
American writer

The way I see it, if you want the rainbow,
you gotta put up with the rain.

—DOLLY PARTON, b. 1946
American songwriter & entertainer

Millions long for immortality who do not know what to do with themselves on a rainy Sunday afternoon.

—SUSAN ERTZ (1894–1985)
American writer

Show me a person who has never made a mistake and I'll show you somebody who has never achieved much.

—JOAN COLLINS, b. 1933
English actress

One can never speak enough of the virtues, the dangers, the power of shared laughter.

—FRANÇOISE SAGAN, b. 1935
French writer

There is not one big cosmic meaning for all, there is only the meaning we each give to our life. . . . To give as much meaning to one's life as possible is right to me.

—ANAÏS NIN (1903-1977)
French-born American writer

The only interesting answers are those which destroy the questions.

—SUSAN SONTAG, b. 1933
American writer

If you have material things, that's fine. God has blessed you. But things can be taken away from you, so you'd best hold on to the things that mean something, like nature, or just having each other.

—JUNE CARTER CASH, b. 1929
American singer

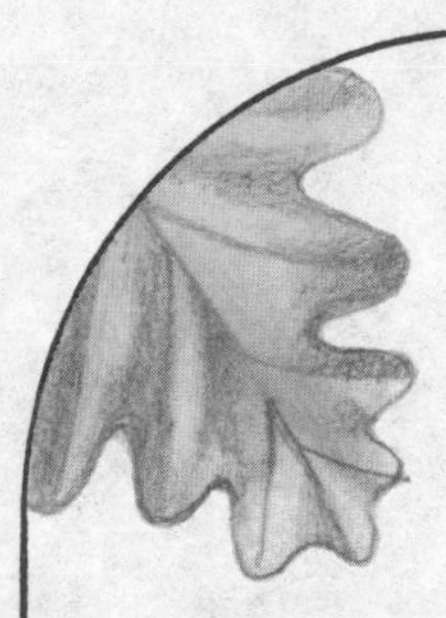

Never economize on luxuries.

—ANGELA THIRKELL (1890-1961)
English writer

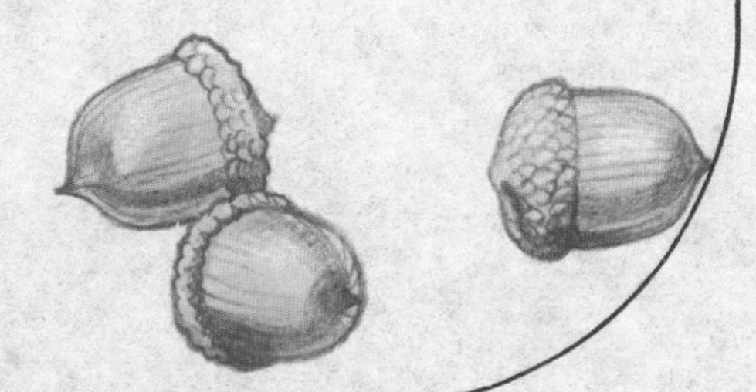

One is not born a woman, one becomes one.

—SIMONE de BEAUVOIR (1908-1986)
French writer

It's all right for a woman to be, above all, human.
I am a woman first of all.

—ANAÏS NIN (1903-1977)
French-born American writer

Each friend represents a world in us, a world possibly not born until they arrive, and it is only by this meeting that a new world is born.

—ANAÏS NIN (1903-1977)
French-born American writer

Trouble is a sieve through which we sift our acquaintances. Those too big to pass through are our friends.

—ARLENE FRANCIS, b. 1908
American actress

I don't need an overpowering, powerful, rich man to feel secure. I'd much rather have a man who is there for me, who really loves me, who is growing, who is real.

—BIANCA JAGGER, b. 1945 (?)
Nicaraguan-born American actress

A woman has got to love a bad man once or twice in her life, to be thankful for a good one.

—MARJORIE KINNAN RAWLINGS (1896-1953)
American writer

I know I'm in a time of transition, being somewhere between vaguely and acutely disrupted; I know the old ways I've lived are no longer adequate and the new stage of life has not yet emerged. It's a painful yet challenging place to be.

—MARSHA, age 49

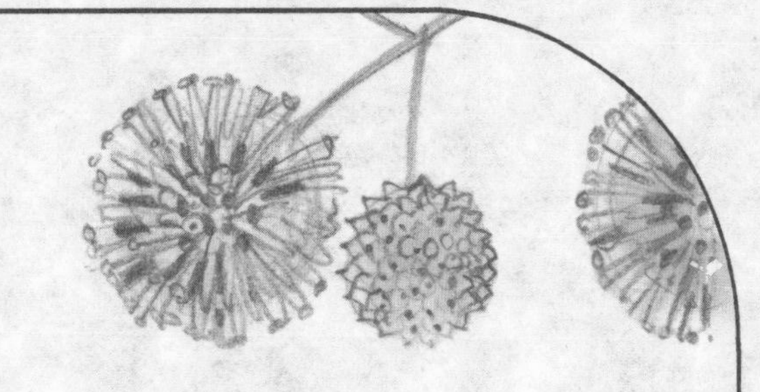

In the effort to give good and comforting answers to the young questioners whom we love, we very often arrive at good and comforting answers for ourselves.

—RUTH GOODE, b. 1905
American writer

Once, power was considered a masculine attribute.
In fact, power has no sex.
—KATHERINE GRAHAM, b. 1917
American journalist

True strength is delicate.

—LOUISE NEVELSON (1900–1988)
American sculptor

God made men stronger but not necessarily more intelligent. He gave women intuition and femininity. And used properly, that combination easily jumbles the brain of any man I've ever met.

—FARRAH FAWCETT, b. 1947
American actress

The average man is more interested in a woman who is interested in him than he is in a woman with beautiful legs.

—MARLENE DIETRICH, b. 1904
German actress

A woman is like a teabag. You don't know her strength until she is in hot water.

—NANCY REAGAN, b. 1923
First Lady

If I have to, I can do anything.

—HELEN REDDY, b. 1941
Australian-born American singer

I don't think of myself as a poor deprived ghetto girl who made good. I think of myself as somebody who from an early age knew I was responsible for myself, and I had *to make good.*

—OPRAH WINFREY, b. 1954
American entertainer

You have to . . . learn the rules of the game.
And then you have to play it better than anyone else.

—DIANNE FEINSTEIN, b. 1933
American politician

Perhaps one has to be very old before one learns to be amused rather than shocked.

—PEARL S. BUCK (1892-1973)
American writer

I have no romantic feelings about age. Either you are interesting at any age or you are not. There is nothing particularly interesting about being old—or being young, for that matter.

—KATHARINE HEPBURN, b. 1909
American actress

It had long since come to my attention that people of accomplishment rarely sat back and let things happen to them. They went out and happened to things.

—ELINOR SMITH, b. 1908
Pioneer aviator

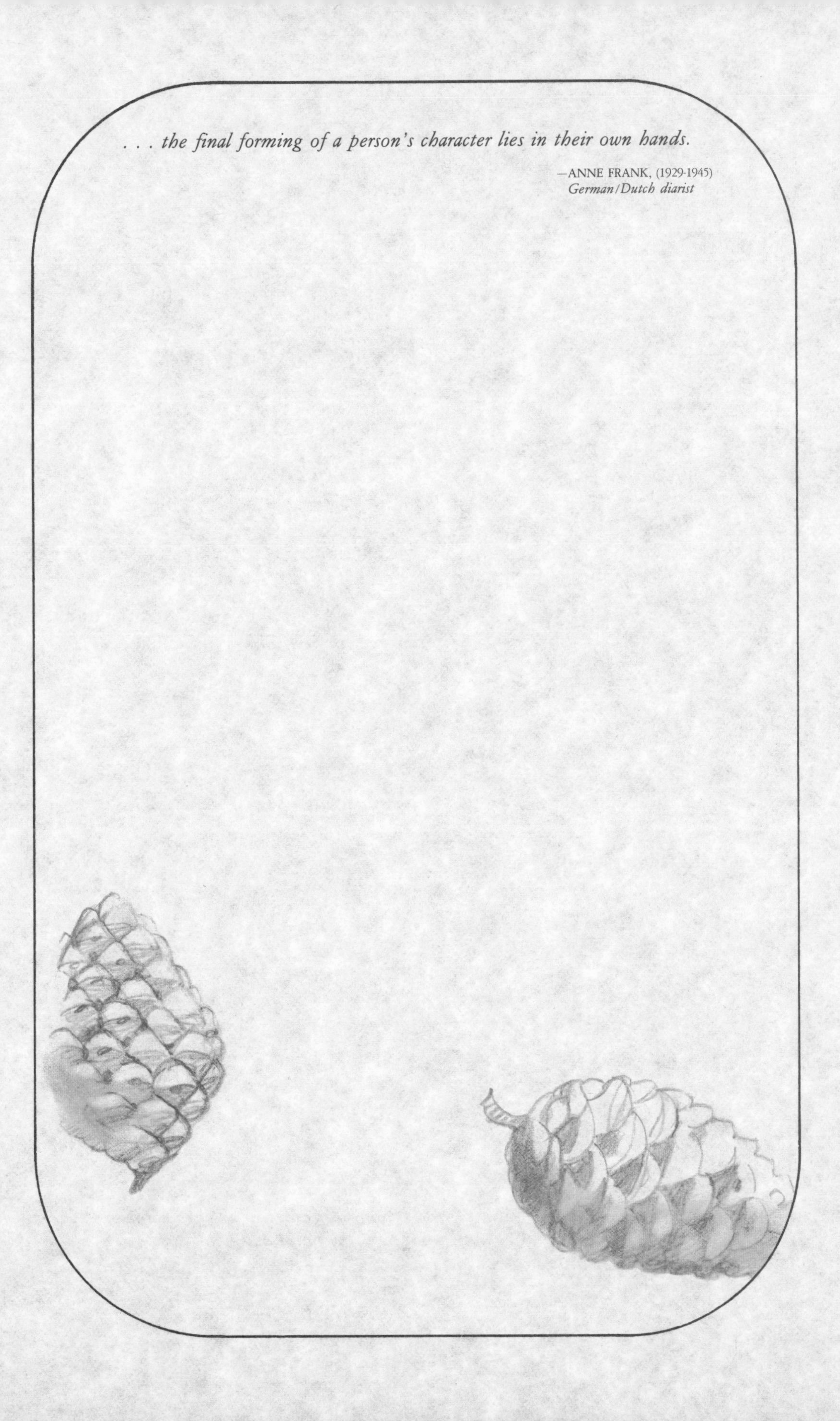

. . . the final forming of a person's character lies in their own hands.

—ANNE FRANK, (1929-1945)
German/Dutch diarist

For a certain type of woman who risks losing her identity in a man, there are all those questions . . . until you get to the point and know that you really are living a love story.

—ANOUK AIMEE, b. 1932
French actress

A woman without a man is like a fish without a bicycle.
—GLORIA STEINEM, b. 1934
American writer

Life is so constructed, that the event does not, cannot, will not, match the expectation.

—CHARLOTTE BRONTË (1816-1855)
English writer

. . . it's so much better to desire than to have. The moment of desire is the most extraordinary moment. The moment of desire, when you know *something is going to happen—that's the most exalting.*

—ANOUK AIMÉE, b. 1932
French actress

When she stopped conforming to the conventional picture of femininity she finally began to enjoy being a woman.

—BETTY FRIEDAN, b. 1921
American writer

No one should have to dance backward all their lives.

—JILL RUCKELSHAUS, b. 1937 (?)
American lecturer

The especial genius of women I believe to be electrical in movement, intuitive in function, spiritual in tendency.

—MARGARET FULLER (1810-1850)
American journalist

I often think that a slightly exposed shoulder emerging from a long satin nightgown packed more sex than two naked bodies in bed.

—BETTE DAVIS (1908–1989)
American actress

It is an ill wind that blows when you leave the hairdresser.

—PHYLLIS DILLER, b. 1917
American comedienne

One's life has value so long as one attributes value to the life of others, by means of love, friendship, indignation and compassion.

—SIMONE de BEAUVOIR (1908-1986)
French writer

Whoever is happy will make others happy too.

—ANNE FRANK (1929-1945)
German/Dutch diarist

I think anybody bright would realize that you've got to be charming and wonderful and fun and adorable and pleasant and terrific—some of the time.

—HELEN GURLEY BROWN, b. 1922
American writer, editor

. . . but surely for everything you love you have to pay some price.

—AGATHA CHRISTIE (1891-1975)
English writer

I don't think that . . . one gets a flash of happiness once, and never again; it is there within you, and it will come as certainly as death . . .

—ISAK DINESEN (1885-1962)
Danish writer

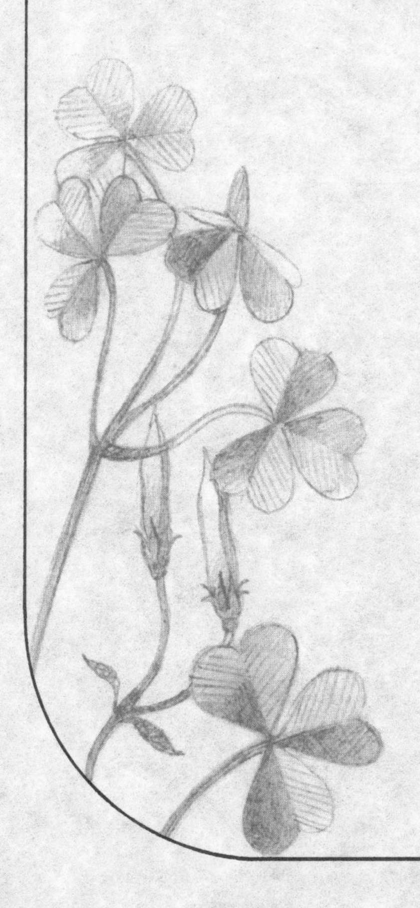

You fall in love with someone, and part of what you love about him are the differences between you; and then you get married and the differences start to drive you crazy.

—NORA EPHRON, b. 1941
American writer

How strange when an illusion dies.
It's as though you've lost a child . . .

—JUDY GARLAND (1922-1969)
American actress

For some ill-defined reason, lovers have a particular penchant for travelling, perhaps in the hope that by exchanging backdrops for that of the unknown, those fleeting dreams will be retained a little longer.

—CAROLE CHESTER, b. 1937
English writer

Relationship is a pervading and changing mystery . . . brutal or lovely, the mystery waits for people wherever they go, whatever extreme they run to.

—EUDORA WELTY, b. 1909
American writer

If only we could all accept that there is no difference between us where human values are concerned. Whatever sex. Whatever the life we have chosen to live.

—LIV ULLMAN, b. 1938
Norwegian actress

There is more difference within the sexes than between them.

—IVY COMPTON-BURNETT (1892-1969)
English satirist

One of my theories is that men love with their eyes; women love with their ears.

—ZSA ZSA GABOR, b. 1923
Hungarian-born American actress

A diamond is the only kind of ice that keeps a girl warm.

—ELIZABETH TAYLOR, b. 1932
English-born American actress

Marriage isn't a 50-50 proposition very often. It's more like 100-0 one moment and 0-100 the next.

—BILLIE JEAN KING, b. 1943
American tennis star

The only time a woman really succeeds in changing a man is when he's a baby.

—NATALIE WOOD (1938-1981)
American actress

When you love someone all your saved-up wishes start coming out.

—ELIZABETH BOWEN (1899-1973)
Anglo-Irish writer

The story of a love is not important—what is important is that one is capable of love. It is perhaps the only glimpse we are permitted of eternity.

—HELEN HAYES, b. 1900
American actress

Besides learning to see, there is another art to be learned—not to see what is not.

—MARIA MITCHELL, (1818-1889)
American astronomer

The trouble is all in the knob at the top of our bodies.
—MARGARET ATWOOD, b. 1939
Canadian writer

. . . it isn't until you come to a spiritual understanding of who you are—not necessarily a religious feeling, but deep down, the spirit within—that you can begin to take control.

—OPRAH WINFREY, b. 1954
American entertainer

Never confuse movement with action.

—BILLIE JEAN KING, b. 1943
American tennis star

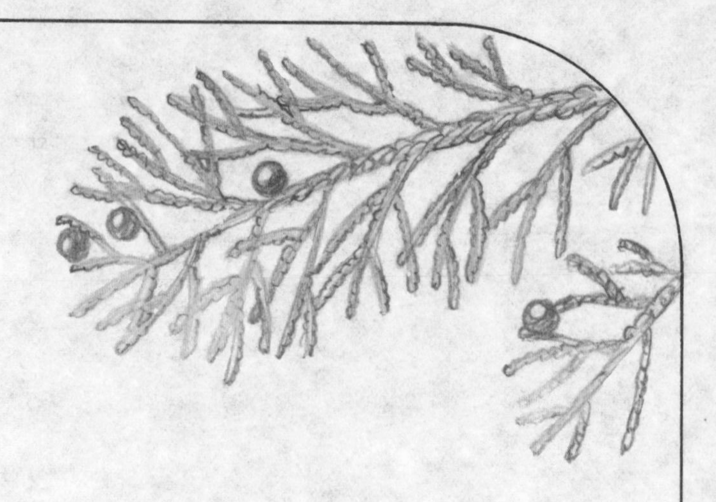

You mustn't force sex to do the work of love, or love to do the work of sex.

—MARY McCARTHY (1912–1989)
American writer

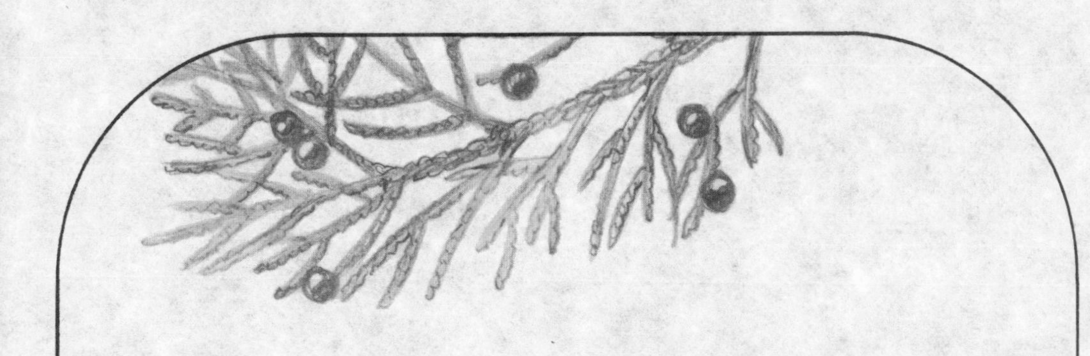

The mark of a true crush . . . is that you fall in love first and grope for reasons afterward.

—SHANA ALEXANDER, b. 1925
American writer

No partner in a love relationship . . . should feel that he has to give up an essential part of himself to make it viable.

—MAY SARTON, b. 1912
Belgian-born American writer

I was not looking for my dreams to interpret my life, but rather for my life to interpret my dreams.

—SUSAN SONTAG, b. 1933
American writer

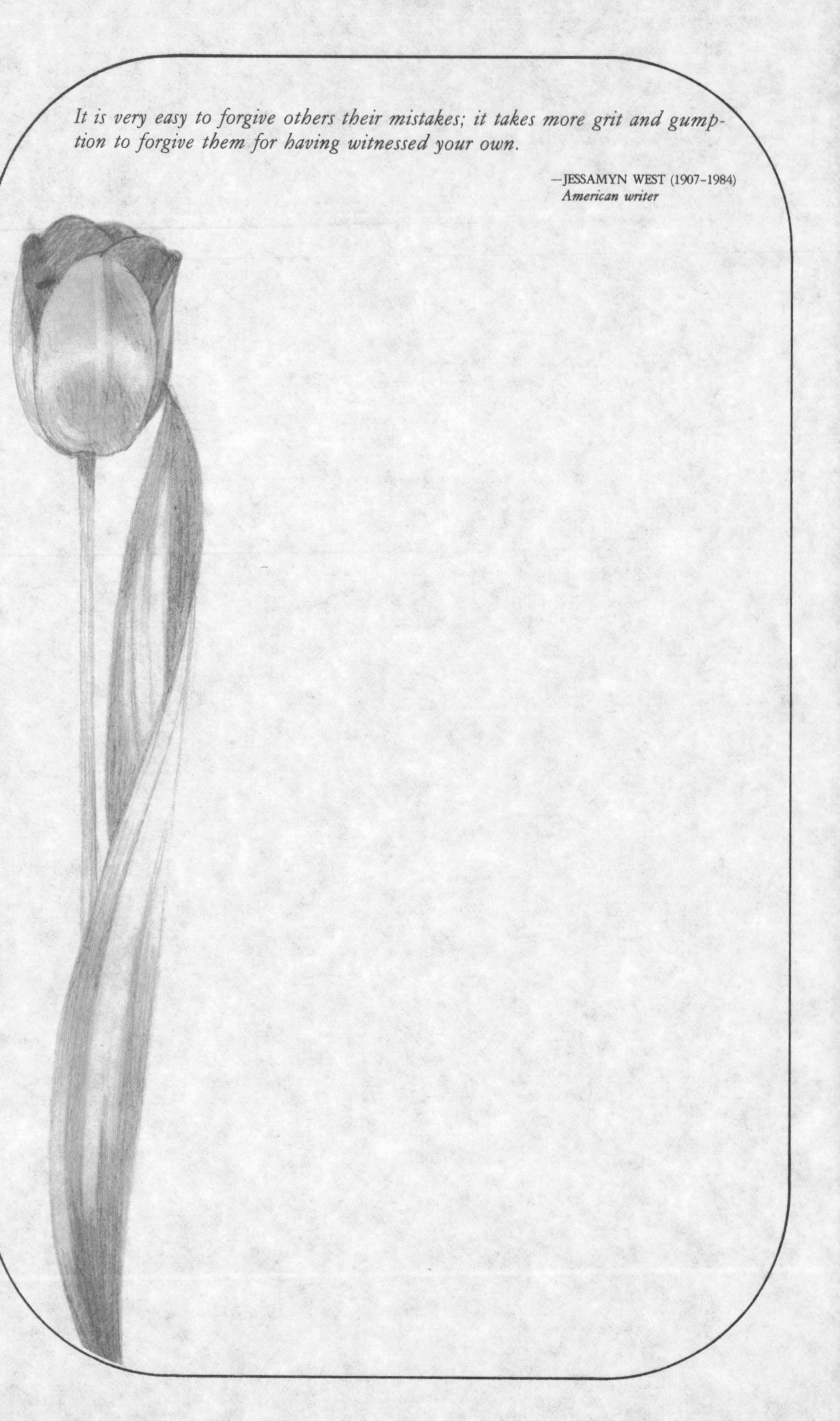

It is very easy to forgive others their mistakes; it takes more grit and gumption to forgive them for having witnessed your own.

—JESSAMYN WEST (1907–1984)
American writer

Comedy is tragedy plus time.

—CAROL BURNETT, b. 1936
American entertainer

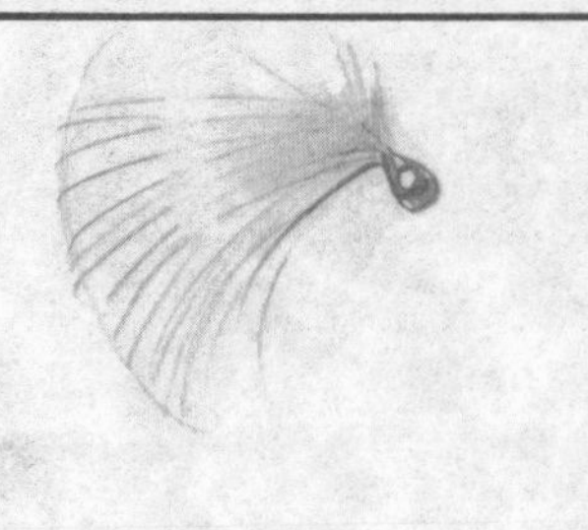

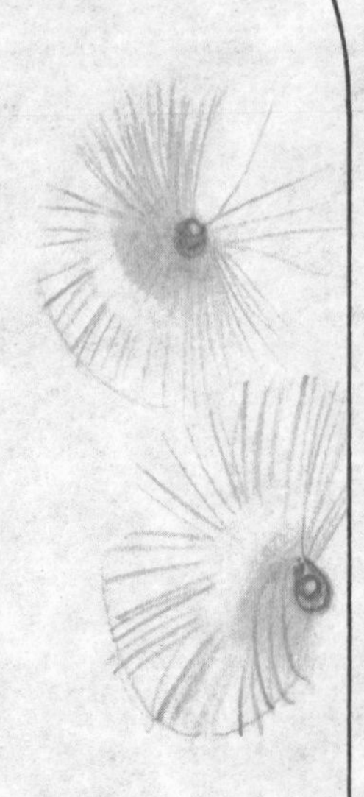

Some memories are realities, and are better than anything that can ever happen to one again.

—WILLA CATHER (1873-1947)
American writer

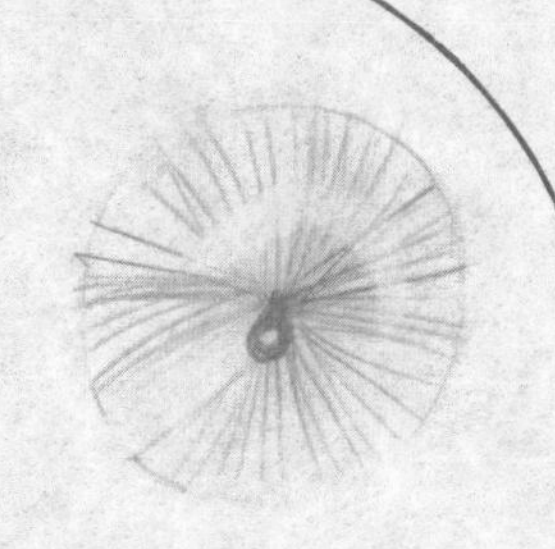

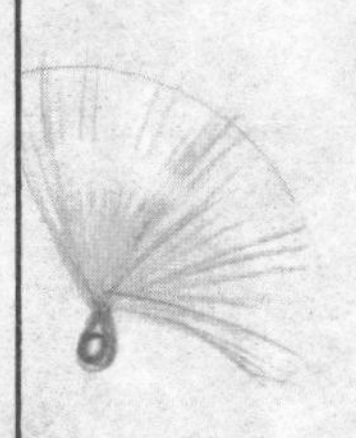

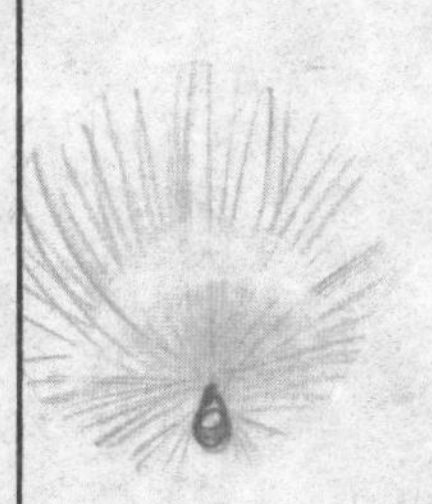

Never go back to a place where you have been happy. Until you do it remains alive for you.

—AGATHA CHRISTIE (1891-1975)
English writer

. . . friendships aren't perfect and yet they are very precious. For me, not expecting perfection all in one place was a great release.

—LETTY COTTIN POGREBIN, b. 1939
American writer

. . . I have learned that to have a good friend is the purest of all God's gifts, for it is a love that has no exchange of payment.

—FRANCES FARMER (1910-1970)
American actress & writer

I have never understood why "hard work" is supposed to be pitiable . . . You get tired, of course, often in despair, but the struggle, the challenge, the feeling of being extended as you never thought you could be, is fulfilling and deeply, deeply satisfying.

—RUMER GODDEN, b. 1907
British novelist

I don't regret anything I've ever done, so long as I enjoyed doing it at the time.

—KATHERINE HEPBURN, b. 1909
American actress

To have one's individuality completely ignored is like being pushed quite out of life. Like being blown out as one blows out a light.

—EVELYN SCOTT (1893-1963)
American writer

People are their own people. They've got their own personalities, and if you try to put strings on them, you run that person away. You've got to give; you've got to trust; there must be respect.

—TINA TURNER, b. 1940
American entertainer

Great loves too must be endured.

—COCO CHANEL (1883-1970)
French designer

Love is not enough. It must be the foundation, the cornerstone—but not the complete structure. It is much too pliable, too yielding.

—BETTE DAVIS (1908–1989)
American actress

If you do not tell the truth about yourself you cannot tell it about other people.

—VIRGINIA WOOLF (1882-1941)
English writer

The most exhausting thing in life is being insincere.

—ANNE MORROW LINDBERGH, b. 1906
American writer & aviator

"I think patience is what love is," he said, "because how could you love somebody without it?"

—JANE HOWARD, b. 1935
American writer

In real love you want the other person's good. In romantic love you want the other person.

—MARGARET ANDERSON (1893-1973)
American publisher

When one is a stranger to oneself then one is estranged from others too.

—ANNE MORROW LINDBERGH, b. 1906
American writer & aviator

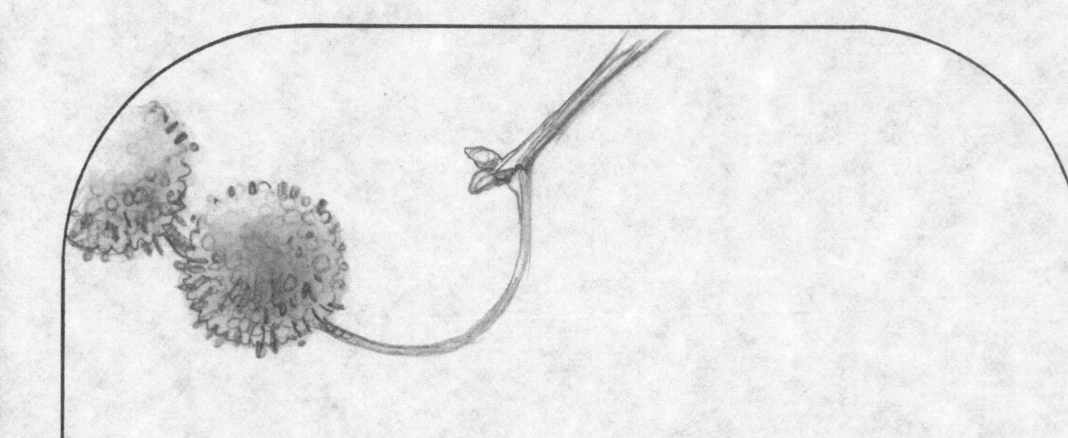

I didn't belong as a kid, and that always bothered me. If only I'd known that one day my differentness would be an asset, then my early life would have been much easier.

—BETTE MIDLER, b. 1945
American entertainer

The more the years go by, the less I know. But if you give explanations and understand everything, then nothing can happen. What helps me go forward is that I stay receptive, I feel that anything can happen.

—ANOUK AIMÉE, b. 1932
French actress

Who has words at the right moment?

—CHARLOTTE BRONTË (1816-1865)
English writer

Joy seems to me a step beyond happiness—happiness is a sort of atmosphere you can live in sometimes when you're lucky. Joy is a light that fills you with hope and faith and love.

—ADELA ROGERS ST. JOHNS (1894–1988)
American journalist

Because of their age-long training in human relations—for that is what feminine intuition really is—women have a special contribution to make to any group enterprise. . .

—MARGARET MEAD (1901-1978)
American anthropologist

My writing is full of things seen, not heard. I get more material staring out at the world, not overhearing things.

—ANN BEATTIE, b. 1948
American writer

Although the world is full of suffering,
it is also full of the overcoming of it.

—HELEN KELLER (1880-1968)
American writer

Happiness is not a matter of events; it depends on the tides of the mind.

—ALICE MEYNELL (1847-1922)
English poet

The excursion is the same when you go looking
for your sorrow as when you go looking for your joy.
—EUDORA WELTY, b. 1909
American writer

I think more *is the cry of the curious person.*

—BETTE MIDLER, b. 1945
American entertainer

I wish that every human life might be pure transparent freedom.

—SIMONE de BEAUVOIR (1908-1986)
French writer

In search of my lost innocence, I walked out a door. At the time I believed I was looking for a purpose, but I found instead the meaning of choice.

—LIV ULLMAN, b. 1938
Norwegian actress